Sources and Techniques in Boat Archaeology

Papers based on those presented to a Symposium at Greenwich in September 1976, together with edited discussion

edited by

Sean McGrail

National Maritime Museum, Greenwich, Archaeological Series No. 1

BAR Supplementary Series 29

1977

British Archaeological Reports

122, Banbury Road, Oxford OX2 7BP, England

Details of all issues of British Archaeological Reports will be sent free of charge and without any obligation to purchase, on request from the above address.

Cheques and postal orders should be made payable to "British Archaeological Reports" and sent to the above address.

For a list of other B.A.R. publications, please see the last page.

B.A.R. Supplementary Series 29, 1977: "Sources and Techniques in Boat Archaeology."

Price £6 (post free). Payments made in currency other than sterling must be calculated at the current rate of exchange and an extra 10% added to cover the cost of bank charges.

ISBN 0 904531 82 1

Printed in Great Britain

CONTENTS

LIST OF ILLUSTRATIONS

EDITOR'S INTRODUCTION

The suggestion that a Symposium on Boat Archaeology should be held came from E. V. Wright, excavator of the Bronze Age Ferriby boats, and a Trustee of the National Maritime Museum. The idea was warmly welcomed by Basil Greenhill, Director of the Museum, who nominated a small committee to determine the themes to be discussed and to invite the principle speakers. This organising committee aimed to bring together all those involved in the study of the form, structure, function, and operational performance of ancient boats. In this context the word boat was used as a convenient shorthand for all forms of water transport built by hand. It was hoped that this gathering would take stock of the "state of the art", and critically evaluate the methods and techniques currently in use.

The Symposium was held from 20th to 24th September 1976, at the National Maritime Museum Greenwich, and was attended by 85 people from 10 different countries, as far apart as Norway and Australia, Switzerland and North America. Chemists, Botanists, Naval Architects, Engineers, Computer and Information Scientists, Historians, Divers, Television Producers, Editors, Boatbuilders, and Ethnographers as well as Archaeologists, were there, reflecting not only the wide ranging appeal of this subject, but also the multi-disciplinary approach now required in Boat Archaeology.

The organising committee did not attempt to include in the programme all forms of evidence or all techniques used by boat archaeologists, but concentrated on those topics it believed to be especially important. Our non-nautical colleagues should note that these are not exclusively nautical topics beyond the ken of the land archaeologist, for of the 8 themes considered at the Symposium, conservation, ethnography, quantitative methods, recording, hypothetical reconstructions, and experimental work, are all of direct concern to other branches branches of archaeology; whilst the remaining themes - a glossary of terms, and ancient boatbuilding - must be of wide interest, for the problems of the one and the methodology of the other are common to the whole archaeological discipline.

Authors presented the main points of their papers to the Symposium in talks of 20 minutes duration. Each main theme, consisting of two papers, was then discussed in hour long sessions which were recorded. Edited versions of these discussions are published here, immediately after their associated papers.

The papers by C. W. Gregson and J. Roelfzema were not presented at the Symposium, but they are published now because they add to the theme on the conservation of waterlogged wood, which is, after all, the principle raw material of boat archaeology and thus of great importance. The papers by Carl Olof Cederlund, Timothy Severin, and David Sturdy were presented in a session for selected short papers, chaired by E. V. Wright; they are here published under their relevant theme headings.

It was planned that, after the formal conclusion of the Symposium, Hag Ahmed Youssef Mustafa the archaeologist in charge of the Cheops Ship, should give an illustrated lecture on this most important find. In the event, the Hag was unable to attend the Symposium, and at the last minute, his British collaborator, Richard Walter, deputised for him. Regretably, it has not proved possible to obtain permission from the Egyptian authorities for Walter's presentation to be published.

I have not standardised the systems of references used by this international band of authors, nor have I attempted to ensure uniform usage of technical terms. At the present stage of development of boat archaeology, this latter would be an impossible task, as the Glossary paper by J. E. G. McKee, and the subsequent discussion show. Terms such as rib, luting, carvel, can mean different things to different authors. Nevertheless, to help the non-nautical reader, extract from two of the most recent glossaries - by Basil Greenhill, and by E. V. Wright - are appended. These should resolve the meaning of most terms, except possibly the word carvel, which, if we follow Basil Greenhill's advice, may in future fall into disuse, or be restricted to vessels built of planks not joined together at their edges, but fastened to pre-erected frames which determine the shape of the finished vessel.

One of the most pleasing aspects of the Symposium was the proposal by Professor Sverre Marstrander, voicing the spontaneous suggestion of many delegates, that a permanent organisation of boat archaeologists should be formed. A provisional committee was elected by the Symposium to undertake this task, and it is envisaged that this organisation will not only arrange future Symposia, but also tackle such international problems as the standardisation of technical terms and the harmonisation of recording methods, and investigate the possibilities of an international data bank.

I wish to thank Mrs. Tricia Hitchcock who transcribed the tapes of the discussion periods and coped most effectively with this often difficult task, and Mrs. Giannella Nicol who typed the edited discussions and other supporting sections of this publication in defiance of my handwriting.

Sean McGrail
Editor

LIST OF CONTRIBUTORS

Carl Olof Cederlund fil.lic.
Curator, Statens Sjøhistoriska Museum, Stockholm

Arne Emil Christensen M.A.
Curator, University Collection of National Antiquities,
Viking Ship Museum, Oslo

J. F. Coates M.A., F.R.I.N.A., R.C.N.C.
Deputy Director of Naval Construction,
Ministry of Defence, Bath

J. M. Coles M.A., Ph.D., F.S.A.
Reader in European Prehistory, University of Cambridge

Ole Crumlin-Pedersen
Director, Viking Ship Museum,
Roskilde, Denmark

J. de Jong
IJsselmeerpolders Development Authority,
Lelystad, Netherlands

Angela Care Evans B.A., F.S.A.
Senior Research Assistant,
Dept. of Mediaeval & Later Antiquities, British Museum, London

Octavio Lixa Filgueiras
Professor of Architecture, Escola Superior de Belas Artes, Oporto,
and Director, Department of Naval Ethnology, Centre for Historical Studies,
University of Oporto

J. M. Graham, B.A., M.Sc.
Lecturer, Department of Computer Science,
University of Reading

Basil Greenhill C.M.G., B.A., F.S.A., F.R.Hist.S.
Director, National Maritime Museum, Greenwich

C. W. Gregson N.D.D., A.T.C., Dip. Cons.
Conservation Officer, Archaeological Research Centre,
National Maritime Museum, Greenwich

Olof Hasslöf Ph.D.
Formerly Curator at the Sjøhistoriska Museum, Stockholm, and
Professor at the University of Copenhagen

J. F. Levy D.Sc., A.R.C.S., F.I.Biol., F.I.W.Sc., F.I.A.W.S., F.L.S.
Reader in Wood Science, Dept. of Botany, Imperial College, London

Sean McGrail B.A., Master Mariner
Chief Archaeologist, Archaeological Research Centre,
National Maritime Museum, Greenwich.

J. E. G. McKee O.B.E.
Caird Research Fellow,
National Maritime Museum, Greenwich

Peter Marsden F.S.A.
Field Officer, Department of Urban Archaeology,
Museum of London

J. Roelfzema Nav. Arch.
Lecturer, Engineering College of Science and Technology (H.T.S.)
Amsterdam

Timothy Severin, M.A., B.Litt.
16 Thanet Street, London WC1

David Sturdy M.A., B.Litt.
25, Rona Road, London NW3

David M. Wilson Litt.D., F.S.A.
Director, The British Museum, London

E. V. Wright M.B.E., M.A., F.S.A.
Trustee, National Maritime Museum, Greenwich

1. INTRODUCTORY ADDRESS

David Wilson

On the first evening of this Symposium we were privileged to be present at the opening of an exhibition of archaeological research in the Zuider Zee; this was a splendid occasion, stimulating to all and a fitting prelude to our proceedings. Only after the war was serious archaeological work on nautical archaeology initiated in the Netherlands - since then more than 300 wrecks have been reported - mostly vessels used in the fishing industry, but some few being medieval and later merchant ships, which (with the famous Bremen Cog) enable us to understand in some measure the methods and construction and appearance of vessels otherwise only known from paintings and literary texts. The exhibition shows this, but also illuminates many areas of economic speculation - trade, fishing and so on - which are not perhaps the immediate concern of this Symposium.

Not every country has been either as fortunate or as far-seeing as Holland. But gradually we are beginning to appreciate much more of the maritime history of Europe through archaeological investigations. The work of the maritime museum in Gdansk and of the newly-built Museum at Roskilde, for example, have added enormously to our knowledge. But all of us could mention many finds of new boats and new evidence, from Mikulcice in Czechoslovakia to Bergen in Norway, which are now being studied by scholars.

In this country we have not perhaps been so fortunate nor so energetic until comparatively recently. Mr. Greenhill's recent book and the foundation of the new Archaeological Research Centre at Greenwich, under Mr. McGrail's leadership, augur well for the future. We hope these are harbingers of better times ahead, as I am sure are the highly professional approaches of underwater archaeologists like Commander Bax and his team who work out of Fort Bovisand, the newly-founded unit at St. Andrew's University under Colin Martin, and the Mary Rose team.

To people like myself who work in the field of early medieval archaeology in north-western Europe, the problems of sea and river transport are ever-present. We are fortunate in having a fairly broad-based knowledge of ships and ship-building from surviving remains and from literature. The Osberg ship, the Skuldeley finds, the Antwerp vessels and the Graveney boat have given us a fair idea of early medieval shipping in northern waters. When we look at our knowledge of smaller boats we are, however, in a very different situation. Small boats are little noticed, their performance is the subject of guess-work and their remains are rarely cherished. This surely is something we must look at.

It is I think interesting and hopeful that so many sessions of this conference should be devoted to discussions of definition and recording. Here surely are some of the major problems. Some answers are provided here. How does one define a langskip for example? A recently published book by a distinguished Icelandic specialist says that the Gokstad boat is a langskip - whereas many people would say that is was a karfi. In many ways I am sorry we have no Hjalmer Falk - a philologist - to help us in our study.

Basic measurement is also a problem. Wood shrinks, but the reconstructions produced by practically all scholars rarely take account of the way in which it does shrink. A dug-out boat, for example, which has lain for years in the basement of a centrally-heated museum will nowadays bear little relation to the boat which once ferried people across a lake or river.

I am pleased too, to see that experimental archaeology is so much to the fore in the discussions of the conference. I am deeply sorry that Paul Johnston's untimely death has robbed us of one of the most enthusiastic replica builders. Many will have seen, or will have the opportunity to see, the films he made of Irish curraghs and (more particularly) of the Bronze Age skin boats postulated by Professor Marstrander on the basis of the Norwegian rock engravings. But the work which he was so good at goes on, and Sean McGrail will be talking about the practical aspects of this very important subject and John Coles will talk about the principles involved in making replicas.

As an Englishman I particularly commend the papers presented by Professors Filgueiras and Hasslöf; for in this country one of the things most neglected in archaeology is the study of ethnographical parallels. Our methods here are sometimes muddled, as we have very little contemporary ethnographic materials in many of the fields found on the continent, but I am sure we can learn much here from our continental colleagues.

In welcoming you to what is a most useful and lively Symposium, I would like to express our thanks to Mr. Greenhill and the members of his staff for whetting our appetites in the production of this programme. Particularly, however, I think we must be most grateful to be meeting in this Museum, surely one of the most splendid sites of any nautical museum in the world.

2. SETTING THE SCENE

E.V. Wright

Summary

This paper offers an outline of the material available for the study of Boat Archaeology in Britain in chronological order from the Mesolithic to the beginning of precise historical records. Although sparse except for dugouts and until recently painfully ill-recorded, the range of types is wide. Dating evidence is also often inadequate. Conditions in the British Isles for survival of actual boat-remains are good and this is attested by the growing list of finds in the last quarter century.

The second part of the paper suggests in brief a systematic approach to control, investigation, recovery and research following a discovery.

I am privileged as an "involved trustee" of the National Maritime Museum to offer an introductory paper for this our fourth symposium on topics related to archaeology. The first on "Aspects of the History of Wooden Shipbuilding" took place in March 1969; the second on "Three Major Ancient Boat Finds in Britain" in 1971; and the third on the often critical techniques of "Waterlogged Wood Conservation" in 1973. The theme for this more ambitious gathering is "Archaeology of the Boat" itself and it provides us with an opportunity to establish or influence in this country a number of basic disciplines relevant to our specialised branch of the archaeological science or art.

Since the 1971 Symposium a number of important events have occurred in this country. Publications include the definitive report on the Sutton Hoo Ship in Volume 1 of "The Sutton Hoo Ship-Burial" and we shall soon have that on "The Graveney Boat". A fuller but by no means final account of the North Ferriby Boats is made available in the guide-book issued this month by the National Maritime Museum. Basil Greenhill's new book "Archaeology of the Boat" fills the gap in English left by the obsolescence of such an indispensable classic as James Hornell's "Water Transport". Unquestionably an important excavation has been Sean McGrail's of the Brigg "Raft" which has already been the subject of a preliminary published note. Notable activity in the area of construction and trials of a replica has been the work of Eric McKee and Sean McGrail with the Gokstad faering. Finally and by no means least, we have here at Greenwich an established Archaeological Research Centre with facilities to conduct research and conservation and also to guide others who may need help both in Britain and elsewhere.

It is timely, therefore, that we should to some extent take stock of the state of a number of aspects of boat archaeology, especially as it concerns North West Europe; expose ideas, some of which may be novel, to critical international review; draw on the experience of our international fellow-workers; and, we hope, emerge with improved equipment - in methods, terminology and skills - to meet the challenges of the future. Confession they say is good for the soul and I have no hesitation therefore in acknowledging that the past record in this country has been lamentable. I trust that our colleagues from abroad will recognise that our house is now in better order and join with us here in advancing our science internationally.

In setting the scene for our proceedings this week I intend to develop two complementary lines, the one material and the other abstract. I should, however, emphasise that my approach in each case will lean towards the practical and tangible rather than the theoretical. I have had personal experience of the temptation to theorise on the basis of limited evidence, a temptation which is as difficult to resist in Boat Archaeology as in any branch, not least because finds are rare and by their nature often dramatic. Nevertheless, a measure of controlled excitement is not entirely out of place since boats and ships are the largest and among the most complex movable structures attempted by man throughout his history and as such offer much insight into technical capability and organisation through the ages. This and the circumstances of discovery and excavation, which are seldom without drama and emergency, combine to make the study and reporting of episodes in maritime archaeology of unusual interest not only to the expert but to the general public also.

The first part of this paper is an attempt to order the regrettably sparse material available in Britain so as to present the opportunity open to us and the justification for applying resources to the study. The second contains some reflections on the method of approach which lies behind our policy for the development of a specialised facility in the National Maritime Museum and its work. This in turn has guided the committee in framing the programme for this symposium.

The basic needs which have induced man to build boats can be reduced to relatively few and in the main they have not varied much over the ages. I suggest these as the most important:-

Food-gathering: fishing, fowling, etc.

Travel and transport of goods by waterway as an alternative to land.

Crossing of water obstacles of whatever width: for purposes of travel, transport, exploration, etc.

War

Ceremonial and Religious observance

Pleasure

All but the last two figure in the material record of finds from Britain which fall into my own suggested category as properly of archaeological rather than historical concern. I personally define the latter as comprehending those covered by the sort of written records such as survive in this country from the latter years of the sixteenth century onwards.

Chronologically the earliest dated trace of boating activity in the British Isles is the early Mesolithic paddle from Star Carr (c. 7,500 b.c., i.e. in uncalibrated radiocarbon years). No boat was found near the settlement site despite deliberate extension of the excavation in the most likely direction, nor was the available timber in the area (Birch) thought to be suitable for construction of dugout boats. The inference was, therefore, that some form of skin-boat might have been used (Clark: Excavations at Star Carr Cambridge 1954). There is a dugout "dated stratigraphically" to the Mesolithic as well as paddles from Scotland, but the boat has not yet been securely dated. Indeed the only actual boat so far safely attributable to the Mesolithic appears to be the Drenthe dugout (van der Heide in his paper to the first of these Greenwich archaeological symposia, states that it "could be dated by radiocarbon at 8,000 years old). At this period Britain's land-links with the mainland had not been severed by rising post-glacial sea-levels. This is believed to have occurred at the end of the Boreal climatic phase perhaps about 5,500-5,000 b.c. At this point I should mention that I follow the convention of expressing calendar or true years as B.C., A.D. or B.P. (Before Present taken as 1950 A.D.) and radiocarbon years, which before the Christian era at least significantly understate the true age, as b.c., a.d. or b.p. There is not yet an universally accepted method of calibrating radiocarbon years with calendar years but the most recent British work on the subject is R.M. Clark's in Antiquity Vol. 49, 1975.

After the end of the Boreal phase we are faced with the undeniable fact that what was by then an archipelago was visited and settled by Neolithic farming people from the mainland well before 4,000 B.C. and that they could only have achieved this passage of the narrow seas with their domestic animals, cattle in particular, in boats of substantial carrying capacity and seaworthiness. We do not know for certain what these boats were, the earliest boats so far dated being the Branthwaite and Chapel Flat Dyke dugouts of c. 2,000 B.C., although there are several dugouts which have been ascribed to the Neolithic by association, e.g., that from the Erith marshes which contained a polished flint axe and a scraper.

For the earlier Bronze Age too we have clear indication of oversea contacts from the presence of "Beaker" elements from about the mid 3rd millennium, notably the connections between the south-west and Brittany which are generally accepted. More positive evidence for sea-traffic comes from the recent discovery by skin-divers of a hoard of early bronze axes off the Kentish coast. It is beyond argument that there was such traffic across the Channel and the Irish Sea at intervals throughout the Bronze Age and thereafter the Early Iron Age until the historical record begins. And yet, for all that we are confident that voyages were made, we know very little indeed for certain about the boats used in prehistoric times.

We can be sure of dugouts for many centuries and McGrail and Switsur's work emphasises that the radiocarbon-dated span shows them in use at least from 2,000 B.C. down to 1,000 A.D., long after more sophisticated methods of boat-building had become available. We are fortunate to have the only examples of Bronze Age planked boats known so far from Western Europe, the three North Ferriby fragments and the later Brigg "Raft". For other forms of prehistoric boat the sole evidence is documentary, Caesar's reference to skin-boats in Britain, an idea which his army copied in Spain. Whatever time-span may be assumed for the making of skin-boats in these islands, or elsewhere, in antiquity the oldest datable reference to their use in Britain is of the 1st century B.C. It is, of course, very likely that they were made in pre-metal ages; but no ancient examples have been found for certain, although there is a doubtfully authentic report of a "coracle" from a brickyard at South Ferriby.

For ships of the Roman period London River has been comparatively rich in its yield with the two finds from Blackfriars and New Guy's House, dated to the 2nd century A.D., and the rather later County Hall vessel lifted in 1910. From Saxon times Sutton Hoo has given up all its ghostly evidence and the timely discovery of the well-preserved remains of the Graveney boat provides material of the quality so badly needed for detailed and thorough technical study using modern methods. It will be surprising if we are not confronted before many years with medieval finds calling for "rescue" excavation and let it not be forgotten that the bottom of the hull of the Grace Dieu of 1418 lies in the Hamble River awaiting attention, if it ever be thought justifiable to apply major resources to the task.

In variety and time-range we are most fortunately placed in what has been discovered, though perhaps less so in the past standards of recording and reporting. This we trust has now been put right, as also has facility for conservation in those cases where it is practically and economically feasible. What, I suggest, should follow is the deliberate pursuit of a policy to seek out answers to questions posed so as to fill the gaps in our knowledge. Some of this can be tackled using finds and the evidence to be gained from them in a controlled fashion and to a planned programme. This is particularly true of experiments in use of materials, essays in reconstruction and the making and trial-testing of replicas. What cannot unfortunately be controlled is the discovery of additional material, especially examples of types from periods at present little known. I do, however, stress the fact that in the estuaries and waterways of the eastern half of England conditionsleading to preservation and possible discovery of ancient boats are unusually favourable. Happily, more people with some idea of what to look for are spending more time actually searching for ancient boats in the likelier places on the foreshores and banks of lakes, estuaries and rivers and I am confident that the opportunities for selection of material for excavation will be greatly increased. Nevertheless, the record of recent major finds is not unimpressive:-

1958	Guy's Hospital -	Romano-British
1962	Blackfriars -	Romano-British
1963	North Ferriby 3 -	Bronze Age
1965-7	Sutton Hoo - (re-excavation) -	Saxon
1970	Graveney -	Late Saxon
1974	Brigg Raft (re-excavation) -	Late Bronze Age

To these may be added over a score of dugouts in the last twenty-five years and a number of fragments of medieval vessels such as those from London, Dublin, etc.

It is by now a commonplace that the most formidable risk that attends a surviving relic of perishable material like an ancient boat is discovery itself. It is a measure of the improvement in circumstances in this country that the most recent excavation in the list above, incidentally in a controllable dry land situation, was carried out without undue pressure of time and led to complete recovery of the remains for conservation. The existence of an establishment, as at Greenwich, properly staffed and equipped however is, within the resources available, no more than a beginning rather than a solution. There remains the danger that because of inadequate assessment of priorities, or a series of emergencies created by accidental discoveries, even the most ambitious facilities could easily be overwhelmed. It is the more important, therefore, that work should be planned and programmes for research controlled. As in other branches of archaeology, it is a matter of asking the right questions and using the tools of excavation or other types of research to produce the answers. In this way compromise, short cuts and risks will be reduced and purposeful and relevant information be produced.

I therefore conclude this paper by attempting an analysis of the possible consequences of a discovery whether it be by accident or design. The steps as I see them are these:-

1. Find and ensure report to a body qualified to assess the importance of the find.
2. Identify and "police" for protection.
3. Record
4. Determine importance and establish priority for application of resources for future reseach by an organisation fully capable of the responsibilty.
5. Excavate and/or investigate - again by a properly qualified team.
6. Recover in total or partially for
   - examination, analysis, research
   - conservation
   - display
7. Interpret
   - on paper
   - with the help of reconstruction
   - by means of production of a replica
   - by practical trial
8. Publish reports at all appropriate stages.

A methodical approach of this kind may involve negative as well as positive decisions and it is especially important that a conscious decision NOT to proceed should be properly supported and fully understood. Inactivity for good reasons must be accompanied by adequate facilities for protection.

Activity must at all times be based on awareness of what is involved in access to resources: human, material and financial.

Our Symposium programme is designed to illuminate a number of aspects of an action programme such as that outlined and, in particular, those which may have received too little attention in the past or involve novel approaches. We have not included techniques of excavation which usually require to be improvised or developed specially according to circumstances. I believe that we shall have a fruitful and stimulating symposium.

3. A GLOSSARY OF BOAT ARCHAEOLOGY TERMS

J.E.G. McKee

Summary

With sailors coming fresh to archaeology and archaeologists first going to sea, words are not going to mean the same thing to the reader as the writer intended. Whereas archaeological terms have, for the most part, been closely defined and standardised, we lack an authorative English nautical dictionary.

Past attempts to compile a comprehensive work have failed due to aiming higher than the available support. To-day this means funds, which are harder than ever to find. Instead a shorter and more selective glossary would prove more practical and attainable. The paper suggests its scope in outline. Only headings covering a vessel's purpose, her shape, construction, components, propulsion and the relative position of these items are envisaged. Preference is given to the simpler useful words which combine well to describe more complex ones and so avoid terms of limited use.

Definitions which lend themselves to understanding and translation are desirable. The need for a multi-lingual version is admitted, but as being secondary to a countries unhindered progress in her own tongue.

Central control is essential if the work is to be completed and published. In addition to this, such an organisation would be able to settle terminological problems and pave the way for the more comprehensive dictionary.

The chestnut, which gives the advice to those considering marriage as, "Don't", probably also applies to those about to start a glossary. In both cases, however, it may be the only way of getting a desired goal. What is this goal ? In the case of the glossary, it is the need to use words that mean the same to the reader as the writer. This is as much a prerequisite for a scientist as marriage lines for a prudent maiden.

Proposals for innovations must answer the challenge, "What is wrong with what we've got?" There is not a simple answer, in that part of what we've got maybe good enough already. It seems that a Glossary of Terms for Nautical Archaeology would have two aspects. That is the part dealing with the archaeological terms used when examining, recording, evaluating and conserving a vessel's remains; and then the nautical part describing the terms used about the vessel herself. These two aspects have only been brought together recently; and thinking about the effect of this is, to some extent, the purpose of this symposium. While archaeology has had its own terminology for some time, the addition of a nautical side has modified old techniques and invented new ones. Seamen have always had their own argot, seldom exact, constantly changing. While one should have some sympathy for the grumbles of land archaeologists, I must urge the nautical archaeologist to master the correct use of sea-words.

May I deal with the archaeological part first? It seems to me that these terms have already been well defined, and if one has had and benefitted from a classical education, they are easy to use and understand. The few new terms can be integrated as they appear, and when Nautical Archaeology has settled down, there will be a case for an archaeological section. Certainly, this would benefit web-footed newcomers, like myself, who uses four letter words like 'boat' rather than five syllable ones like 'archaeology'. As there are many present well endowed to comment on the need for an archaeological section, I hope that they will do so in the discussion period.

I would now like to sit down and let you discuss the part of the glossary to do with vessels. But, I feel it is up to me to provide a few victims for you lions to tear apart. What nautical glossaries have we got? Though in the past there have been some excellent nautical dictionaries, which give valuable meanings for the terms then used, I know of no current English work that meets our needs. Captain Paasch's From Keel to Truck (4th Ed. 1908), in English, French and German is illustrated, but falls short as it is mainly about commerce over a narrow period. The same sort of criticism can be levelled at the purely military aspect of the 1937 Seamanship Manual, which is admirable, once the definitions scattered throughout the text have been excavated. Layton's Nautical Dictionary (Brown, Son and Ferguson), still in print and not costly, avoids bias as it was written by an officer of the Merchant Navy but was revised by one from the Royal Navy. It includes many of the older terms and is a good work-a-day reference.

A shelf full of dictionaries may well hold the correct definition, but this calls for individual decisions and risk of confusing the reader. The Concise Oxford Dictionary, a much abbreviated version of the New English Dictionary, offers reliable definitions of many nautical terms, but there is too little explanation. To get this one must turn to the New English Dictionary, which is a bit like signing on the London Philharmonic Orchestra to play the sea chanties on board a Thames Barge.

Feeling the need to regulate my own usage, I have been assembling a card index of boatbuilding terms; over three thousand of them and still growing fast. It is unlikely that any personal glossary can be completed or used if it was; not because the material is at fault, but because it will represent too narrow a precoccupation for general purposes.

If I am convinced that nautical archaeologists would benefit from an authorative glossary of terms, it is due to the frequency with which I have wished there was one. Any specialisation tends to weaken ones contact with normal usage. This doesn't matter until one picks up a pen or a book. This is the moment one is at a loss for a word or needs clarification of a meaning. For me, a glossary of nautical terms would need to cover descriptions of every aspect of shipping of all traditions and ages, how they were built for, and operated in, the pursuit of war and trade, fish and transport, pomp and pleasure. Such a glossary can either be concise with carefully chosen adaptable terms accurately defined, or it has to be very large to be truly definitive.

One of the declared aims at the foundation of the Society for Nautical Reseach was "to compile a complete and scholarly encyclopaedia or dictionary" (Rule 11, 3.). We know from L.G. Carr Laughton's article in the Mariner's Mirror (Vol. 10, p. 299), and the introduction by G.P.B. Naish to its reprint (Vol. 44, p. 190), that in spite of the fine intentions and profound scholarship of a dedicated few, this aim has, through lack of helpers, yet to be realised. We can learn from this that anything similar calls for sustained effort over a long period.

If in 1924 there were not enough scholars with independent means to carry through this project, I doubt if they could be found to-day. The only sure way to get continuity up to completion is to pay a competent staff; but what is the chance of finding funds for such an enterprise at present? This is not to say that the objective is not still desirable. It most certainly is, but it might be realistic to shelve it for the moment and set to work on a more concise glossary of useful terms for nautical archaeologists.

When a Chinaman learns to write, he has to memorise another ideograph for every new word he uses. This apparently unsurmountable task was simplified when the scholars drew up a list, known as the first hundred characters. These were so fundamental or common, that repetition fixed them in the mind. There then followed the second and third hundred characters and so on until; I suppose, the thirst for any more scholarship was slaked. I suggest that a worthwhile concise glossary could be put together in a not dissimilar manner.

Divided between a group of workers, three or four hundred words would give us a working nucleus, without creating too great a burden. This should at least indicate how much larger the concise glossary might have to be, and what sort of and how much work was still needed. While single terms can be defined by individual effort, a focal organ will be needed to apportion work, resolve difficulties and in due course, edit and publish the glossary. Without some form of central control it is unlikely that finality would be reached, and your views on the nature of this will be sought later.

To stimulate ideas, it is suggested that a concise glossary of terms should include terms under these headings:-

Purpose of vessel, words such as coaster, tender, rather than those like cog, fifie.

Description of form and ability, such as depth, stability, rather than clipper, ardent.

Building processes and tools, such as clench, auger rather than carvel, fabricate.

Parts of a vessel, such as deadwood, counter rather than skorvel, spirketting.

Propulsion methods and gear, such as clew and tholepin rather than raffee, thranite.

Position inside or in relation to a vessel, such as athwart, astern rather than windward, superior.

This list is far from exhaustive as yet. It could easily be extended in a number of ways; say by the inclusion of the terms used in commerce. But, my preference would be for the minimum number of terms consistent with sound usage. There is nothing objectionable about the words excluded above, other than they are sufficiently specialised that they ought to be put in a more comprehensive work than this is meant to be.

The definition of each term selected for inclusion would need to be substantiated from reliable sources, and there would have to be very good reasons for departing from the general sense given in the New English Dictionary. For clarity descriptions should use other than nautical terms whenever it is possible, and preference given to terms that combine well in descriptions over those whose use is limited. If the scope of the glossary is to be complete in itself, this will mean the inclusion of many quite common terms. Time and again, it is these words that are most often misused and misunderstood as they are the hardest to define.

Seamen are inconsistent and tend to use terms slackly, relying on context to convey meaning; but this is not good enough for written work. Take the word gunwale. Originally this meant the range of thicker strakes below the gunports; but a century ago it had come to mean the plank which covered the heads of the timbers between the drifts, or where the sheer rises to a higher line. In a boat, the term was used for the sheer strake, the inwale, the rubber or even the capping if it was sturdy enough. Turning to the Concise Oxford Dictionary one finds, "The upper edge of a ship's or boat's side". So it seems that, over the years, the gunwale is no longer part of a vessel, but is now a position on it. There are other common terms like Chine and Keelson which will need careful use until we have precise definition for them.

Several terms may be used for a single item; strake, stroke and streak being examples. If in each case the item meant is identical, one would expect the commonest form to be chosen as the standard for written use and translation. The others become dialectic variations, which should be recorded in some more detailed glossary. On the other hand, some variations may relate to an actual difference in form or act, for instance; bull, copp, and clog. The problem here is to decide whether a collective term has to be found for these three single tholepin oars or if all the terms are to be included.

Thought should be given to the need for a multi-lingual glossary developed internationally. Given the command of languages, it is hard to think of a better occupation; but to start with, keeping in step might hinder rather than help. If countries are free to make their own speed, while they follow a common pattern, I feel sure they will be able to align their efforts at some later date.

Though the idea of a comprehensive glossary was put on a shelf earlier, it was by no means abandoned. Much can be done to prepare for it, while helping nautical archaeologists at the same time. As an immediate step, a bibliography of sources of nautical terms is suggested. A revision of the lists of dictionaries given in the first two volumes of the Mariner's Mirror could be the way to start, in English and French anyway. This Museum's Select Bibliography for the Development of the Boat is another step in the same direction, not only for its valuable contents but for the clear indication that this is no small task. Some of the sources may be hard to locate, and if rare, access to them may not be easy. While this should not discourage research, it can slow down progress. It would help therefore if each source listed gave advice on how or where it might be seen.

To sum up, here are some points you might care to discuss.

Firstly, is a Glossary of Terms a requirement? If so, should it be:-

concise and available soon? or

comprehensive, taking longer? or both?

In any case, should archaeological terms be included? and is a list of sources of nautical terms needed?

Finally, what sort of organisation do you envisage to formulate and carry out decisions?

Though you may be ready to discuss these points, you may not be ready to make decisions right away. If there is sufficient merit in any of these proposals, is this something you would like to see taken further?

DISCUSSION

McGrail: We need an agreed working glossary defining the terms we now use, which may subsequently be expanded into the grand dictionary. There is so much to do in boat archaeology that we need land archaeologists with an interest in boats to work with us and not to be frightened by our technical terms. I suggest therefore that we use any common words that can be used without loss of precision. If cove is synonymous with groove let us use groove, because we will lose nothing, the land archaeologist will gain a lot, and we will gain by his active interst.

Wolle: A glossary is needed by both seaman and scholar.

Madden: This glossary should be fully illustrated.

Du Plat Taylor: Land archaeologists often misuse technical terms for say architecture as well as the sea, and a glossary would overcome this problem.

Wright: For material not covered by accepted naval architectural terminology we do need "new speak" - that is, we need to introduce new terms.

Greenhill: The trading ketch Shamrock, which this Museum and the National Trust are restoring, is perhaps closer in structure at the bottom-apart from the fact that she is not edge-joined-to the Ferriby boats than to most Victorian sailing ships. This is an example of why we need the clarification that Mr. Wright mentioned, so that we can describe the construction of these boats without ambiguity.

Christensen: I recommend what has been done in Norway: pick out a small and well defined area, one craft, one type of ship, the work of one craftsman, and define the terms used there.

Greenhill: One of the main problems in this country will be determining the preferred word for definition, when you have got such tremendous dialect variation in the regions.

Editorial Note. This topic was further considered in the discussion after Basil Greenhill's Summary.

4. DEGRADATION OF WOOD

J.F. Levy

Summary

The aspect of degrade of wood of most concern to the archaeologist is caused by bacteria and fungi. These organisms degrade wood without, apparently, altering the shape, size and outline of the object so long as it remains wet. The middle layer of the secondary cell wall is destroyed and with it the strength of the wood, which is no longer able to withstand the internal stresses set up as it dries out. This can lead to splitting, distortion and a change of shape which can render the object almost unrecognisable. The activity of bacteria and fungi and their effect on the wood cell walls is described.

Wood can be degraded in many ways. Fire, chemicals, mechanical wear and tear, and exposure to weathering out of doors can all bring about the partial or complete breakdown of the material. The most widespread degrade is, however, caused by the biological agencies which normally form part of the process whereby plant and animal remains are broken down to their constituent parts in the soil for the recycling of these materials into new plants and subsequently animals. These biological agencies take many forms: animal, from rodents and woodpeckers to marine borers; insects, including the termites; and plants, particularly the fungi and bacteria.

From the archaeologist's point of view the degrade caused by animals is sufficiently large and obvious to be seen, and the disfigurement caused may be so great as to render the object of doubtful value for conservation. Even the smallest wood-boring beetles make a gallery whose diameter is many times greater than all but the largest of the wood cells.

The fungi and bacteria, on the other hand, can cause considerable degrade without apparently altering the shape, size and outline of the object so long as it remains wet. It is only when the object begins to dry out that drying stresses can cause splitting, distortion and a change of shape, which can render the object almost unrecognisable.

This paper will, therefore, only be concerned with the type of degradation c caused by fungi and bacteria; their effect on the wood; the importance of the conditions of burial; the part or parts of the wood destroyed; and the reasons for the distortion on drying.

Wood as a material

The material, wood, consists of a limited number of basic units or cells which are formed in a regular manner as the tree grows. Most of the timber

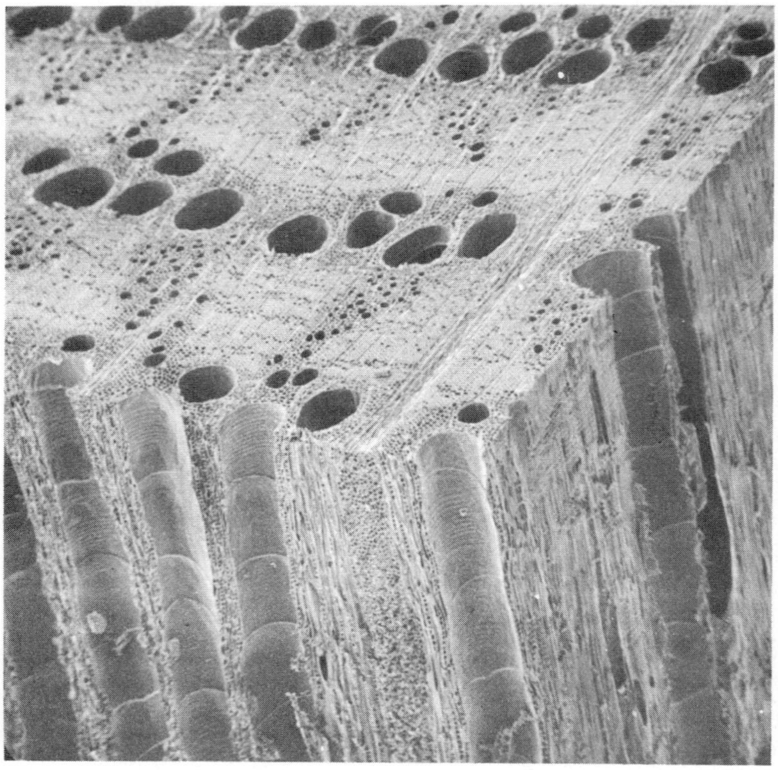

Fig. 4.1 Scanning electron micrograph of Oak (Quercus sp.) showing the wood cut in the transverse, tangential longitudinal and radial longitudinal planes (x 100).

Fig. 4.2 S.E.M. picture of Scots Pine (Pinus sylvestris) showing the wood cut in the transverse, tangential longitudinal and radial longitudinal planes (x 90).

of commerce has been cut from the stem of a forest tree. The arrangement of the cells within the timber will reflect the three-fold purpose of its formation, (a) mechanical strength to support the weight of the branches and leaves and to resist permanent damage when the wind blows, (b) a system of tubes or pipes to allow water from the soil to pass from the roots of the living tree to the leaves and (c) a system to enable the lateral transport of water and materials to take place and to store reserves of food to support growth and development whilst the buds are breaking and before the leaves are fully open.

The cells are characterised by the possession of a cell wall consisting of cellulose, hemicelluloses and lignin with other materials (known as extractives) added in varying amounts at a later stage of development. Cellulose is the main structural component, the long chain molecules becoming aggregated together to form long part-crystalline structures known as microfibrils. These microfibrils are set down in closely packed helices, somewhat analogous to the winding of a hollow ball of string, but each helix is cross-linked with its neighbours.

The microfibril is thought to contain hemicellulose in its outer regions and these hemicellulose molecules may, in some way, act as a key between the structural cellulose and the amorphous lignin whose precise position and function is not entirely clear, but may be regarded as part of the matrix of the wall. The orientation of the microfibrils with respect to the long axis of the cell has an important bearing on the strength properties of that cell, which will differ according to whether the stress acts along or across the long axis of the cell. Wood is thus an anisotropic material, that is to say, its properties will vary according to the orientation of the axis of the cell or, in other words, the grain direction.

Since the strength and properties of wood depend on the organisation of the cell wall, it follows that the thickness of the wall and the proportion of a unit volume of wood occupied by cell wall material will have a marked effect on the properties of wood. Balsa can weigh as little as 10 lb/ft^3 (160 kg/m^3) when air dry, whilst lignum vitae is eight times as heavy at 84 lb/ft^3 (1250 Kg/m^3). The former has thin cell walls and much air space per unit volume of wood, whilst the latter has very thick walls and very little air space per unit volume.

Fungi and bacteria act on the components of the cell walls. Some attack and destroy all the components, others only attack one or two of the components, whilst others only destroy a particular part of the wall in a particular way, Bravery (1971 and 1975), Kirk (1975), Liese and Greaves (1975), Levy (1974), Levy, Millbank, Dwyer and Baines (1974).

Types of fungi and bacteria

The fungi associated with degrade of wood may be separated into five groups.

Moulds
Stainers
Soft rots
Brown rots
White rots

Moulds feed on the carbohydrates, fats and proteins and general detritus exposed when cells are cut open as wood is sawn or planed. They rarely, if ever, penetrate beneath the surface, and then only through holes in the cell walls since they do not normally possess the enzyme systems capable of breaking down the wall material. They serve to indicate that the wood is in a suitable condition to support fungal attack.

Stainers are those fungi which colonise the ray cells in sapwood, feeding on the stored reserve food materials laid down by the tree. The fine strands (or hyphae) of the fungi become pigmented at maturity and this dark colour can disfigure the sapwood of most species of timber. In some softwood species the bright blue discolouration gives rise to the term "blue stain". In wood they cause very little structural degrade but are able to penetrate through cell walls from the lumen of one cell to the lumen of the next, sometimes crossing many cells. The hypha on reaching the cell wall bulges at the tip and penetrates through the wall, with a much reduced diameter to emerge into the lumen of the next cell where the hypha assumes its normal size. This strange, characteristic, but largely unexplained behaviour, has given rise to much speculation, and may well be effected partly by localised enzyme action and partly by some sort of mechanical means.

The three types of rot all cause degrade of wood by destroying some or all of the cell wall. The end point in each case is greatly reduced strength properties and collapse of the wood structure. The differences in micromorphology have been well illustrated by Bravery (1971).

The soft rot fungi penetrate into the wall in a manner somewhat resembling the stainers. However, instead of penetrating through the wall they only go part of the way and in the thickest layer of the wall (known as the middle layer of the secondary wall, or the S_2 layer) the hypha appears to turn at right angles and branch at the bend in the opposite direction to form what has been called the T-branch. The two ends of the cross of the T extend their growth along the orientation of the cellulose micrfibrils in this layer of the wall. As they grow, the hypha behind expands in girth and enzyme is released along its length to form a cavity in the wall round the hypha with pointed ends. A chain of such cavities is formed, characteristically forming a helix within the wall parallel to the orientation of the helix of the cellulose microfibrils. The reason for cavity formation is not fully understood, but it must be due to an interaction between the enzyme system of the fungus and the arrangements of the constituent parts of the cell wall. The cellulose and hemicellulose components are destroyed, but the activity of the enzymes towards the lignin is a matter of some speculation. The significance of this type of wood rotting fungus has only been established within the last twenty five years and it is the centre of much research activity.

The brown rot fungi are capable of destroying both cellulose and hemicelluloses but lignin is not broken down, only modified. The hyphae lie along the cell wall inner surface and enzymes diffuse from the fungus into the wall and appear to permeate all the wall layers. The reactive components are broken down and the soluble sugars diffuse back to the hypha. The action of these fungi can be fast and devasting and amongst their number are many of the well known wood-rotting fungi, such as Dry Rot (Serpula lacrymans) and the Cellar Fungas (Coniophora puteana).

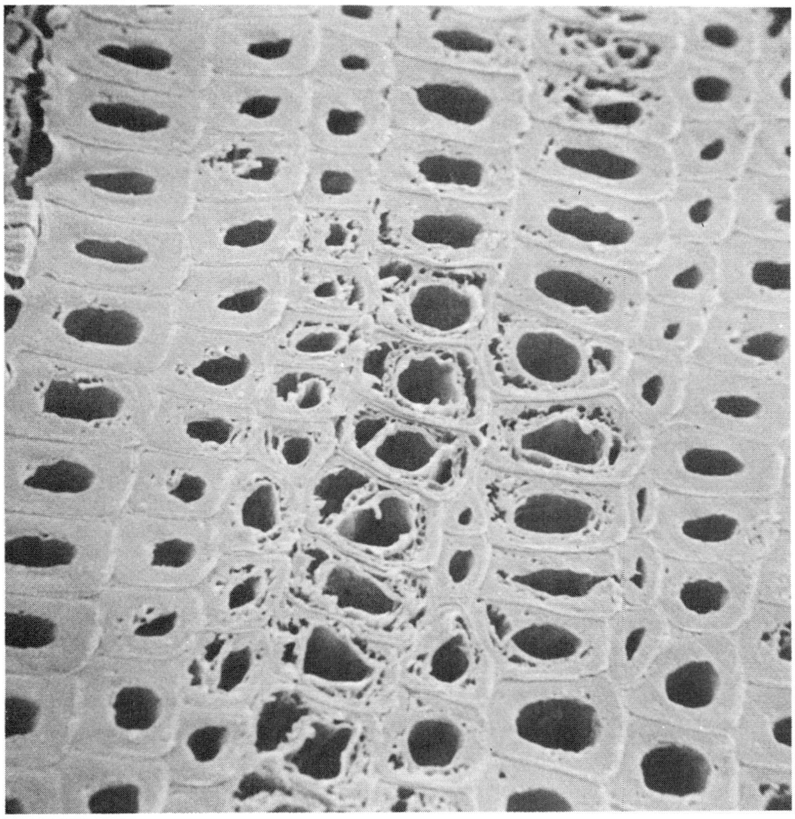

Fig. 4.3 S.E.M. picture of Scota Pine (Pinus sylvestris) showing the wood cut in the transverse plane. Note the cells (tracheids) to the right and left of the picture have intact cell walls. The cells in the centre have been actively attacked by a soft rot fungus and the middle layer (S_2 layer) of the cell walls almost completely destroyed (x 750).

The white rot fungi attack and destroy all the wall components, but they do so in a manner different from the other two wood-rotting fungi. The branched hyphae lie along the cell wall inner surface and the enzymes secreted from them erode a groove round them. The groove widens and deepens and the wall is slowly eroded in its entirety from the inner surface.

The effect of bacteria on wood has only been studied to any extent within the last ten years. They have been shown to increase the permeability of logs floating or stored in water by destroying the thin membranes covering the pits in the cell walls. This also occurs in timber in ground contact and has the effect of opening up the structure to enable fungi to penetrate into the wood in depth. It also enables gaseous diffusion to take place and may thus change a region that was anaerobic and capable of supporting bacterial activity, into a more aerobic situation and capable of supporting fungal decay.

Bacteria have been shown capable of destroying the wood cell wall under certain conditions (usually of long term burial), and to cause a destruction of the wall layers similar to the effect of the soft rot fungi (Liese and Greaves, 1974). They are known to be early colonisers of wood. Nitrogen fixing bacteria have been shown to penetrate into wood and nitrogen fixation has been demonstrated in wood in ground contact, in wood in water cooling towers and in window joinery. They clearly have a considerable part to play in the complex of organisms causing wood decay.

Wood in ground contact

Soil is usually considered to be a reservoir of water and it is certainly a reservoir of micro-organisms. Wood lying on the soil surface or buried in the ground is likely to be in contact with a complex of micro-organisms, some of which will be capable of penetrating into the wood, colonising it and then either destroying the wood or preparing the way for its destruction by other organisms. The processes of colonisation are complex and the successions and association of organisms are only beginning to be elucidated Käärik (1975), Levy (1975 a and b). A number of points are however becoming clearer and it is well known that not all species of wood behave in the same way, some exhibit a greater durability to decay than others. Recent work suggests a number of other effects.

1. The moisture content of the wood may be of paramount importance in determining which organisms will colonise. An anaerobic situation will support a quite different mixture of organisms to an aerobic one.

2. The moisture content and the degree of aeration will change as decay proceeds.

3. Charcoal shows a very high degree of durability to degrade by biological agencies, and much of the cellular detail is retained intact in situations where wood itself would be totally destroyed.

Site conditions are therefore critical in determining the degree of preservation of the cell wall structure in wood buried in the ground. It is not beyond the bounds of possibility that long term burial in slowly moving water could produce an acid or alkaline chemical breakdown of wood. Certainly fossil evidence is a clear indication that petrification can occur with the carbon of living organisms being replaced by silica. However the evidence for long term breakdown of wood by acids or alkalis is largely non-existent whilst there is much evidence of biological degrade, which can often be seen to have stopped at a certain point, presumably by a change of some sort in the condition of burial.

The archaeologist is thus faced with the prospect of finding a wooden object usually in a very wet condition. The size and shape may appear to be unaltered apart from a softening, almost a plasticising of the wood. Can it be removed from the wet condition in which it was found and dried? The answer is "yes", but only if care is taken to examine the condition of the wood. If any of the wood-rotting micro-organisms have been only moderately active, there may be little visible evidence without micropscopic examination. What is required is to know what has happened to the cell wall layers. Have some or any of them been destroyed? If so, then inevitably stresses will be set up in drying which may cause great distortion of the object. Two courses are open.

(a) Solvent drying to remove the surface tension effects of water drying out by replacing the water with a solvent whose surface tension effects leave the remaining parts of the wall in situ without drying stresses being built up. The resulting sample may be the right shape, but it will often be very brittle and fragile and may well have changed colour.

(b) Impregnation with materials capable of filling the spaces left behind by the destruction of the cell wall components so that on drying the remaining parts of the wall are supported by impregnating materials and shrinkage is resisted because the spaces are all filled.

Archaeology, so far as wood is concerned, deals with those wooden objects that have survived complete degradation by biological agencies because the condition of burial has prevented, delayed or stopped the activity of the organisms involved. Where partial destruction has occurred considerable distortion is likely to take place on drying the wood unless action is taken to prevent this happening.

REFERENCES

Bravery, A.F., 1971, 'The application of scanning electron microscopy in the study of timber decay'. J. Inst. Wood Sci. 5 (6), 13-19.

Bravery, A.F., 1975. 'Micromorphology of decay in preservative treated wood'. In Biological Transformation of Wood by Microorganisms, (Ed. W. Liese), Springer-Verlag, Berlin, 129-144.

Käärik, A., 1975. 'Succession of micro-organisms during wood decay'. In Biological Transformation of Wood by Micro-organisms, (Ed. W. Liese), Springer-Verlag, Berlin, 39-51.

Kirk, T.K., 1975. 'Chemistry of lignin degradation by wood destroying fungi'. In Biological Transformation of Wood by Micro-organisms, (Ed. W. Liese), Springer-Verlag, Berlin, 153-164.

Liese, W., and Greaves, H., 1975. 'Micromorphology of bacterial attack'. in Biological Transformation of Wood by Micro-organisms, (Ed. W. Liese), Springer-Verlag, Berlin, 74-88.

Levy, J.F., 1974. Fungi in Wood. Scanning Electron Microscopy/74 11, 462-468.

Levy, J.F., 1975a. 'Colonisation of wood by fungi'. In Biological Transformation of Wood by Micro-organisms, (Ed. W. Liese), Springer-Verlag, Berlin, 16-23.

Levy, J.F., (1975b). 'Bacteria associated with wood in ground contact'. in Biological Transformation of Wood by Micro-organisms, (Ed. W. Liese), Springer-Verlag, Berlin, 64-73.

Levy, J.F., Millbank, J.W., Dwyer, G., Baines, E.F., 1974. 'The role of bacteria in wood decay'. Record of the 1974 Annual Convention of the British Wood Preserving Association, 3-13.

5. CONSERVATION OF OLD WATER-LOGGED WOOD

J. de Jong

Summary

In the Lake IJssel area many shipwrecks have been found and excavated. The conservation of old waterlogged wood, if large objects are to be treated, still causes a number of problems. Using the existing treatment methods with polyethylene glycol it was found that little deteriorated european oak, was very difficult to conserve. Based on data from literature and the observed impregnation results a division of european oak into 3 classes is given. The division is based on the degree of degradation of the wood. The wood of two of these classes can be treated by existing water-PEG processes.

INTRODUCTION

During the seventeenth century a plan was made in Holland to enclose the Zuiderzee. The aim was to establish a better protection of the land against the sea. It was not, however, until 1918 that the Dutch Government, after a large flood in 1916 and a period of food scarcity during World War 1, decided to have the Zuiderzee separated from the North Sea by means of a barrier dam, and to construct 5 polders in the fresh water lake, Lake IJssel, so formed.

The first polder, the Wieringermeerpolder, was drained in 1930, and the barrier dam was completed in 1932. The first polder was therefore constructed in a marine environment.

The other polders have been reclaimed from a fresh water lake. The second polder, the Noordoostpolder (North-east-polder), was drained in 1942, Oostelijk Flevoland (Eastern Flevoland) in 1957, and Zuidelijk Flevoland (Southerm Flevoland) in 1968. Zuidelijk Flevoland is still under development.

Figure 5.1 gives an impression of the area of reclaimed land in Lake IJssel.

During reclamation, development and farming operations in the new polders many finds of archeological and historical value have been made. Remains of neolithic settlements, dated between 3650 and 3350 B.C. have been excavated. There are also finds of a more recent date. The latter finds are mainly shipwrecks or parts of ships dating from the Middle Ages to recent times, and crashed airplanes from World War 2. Besides being an important

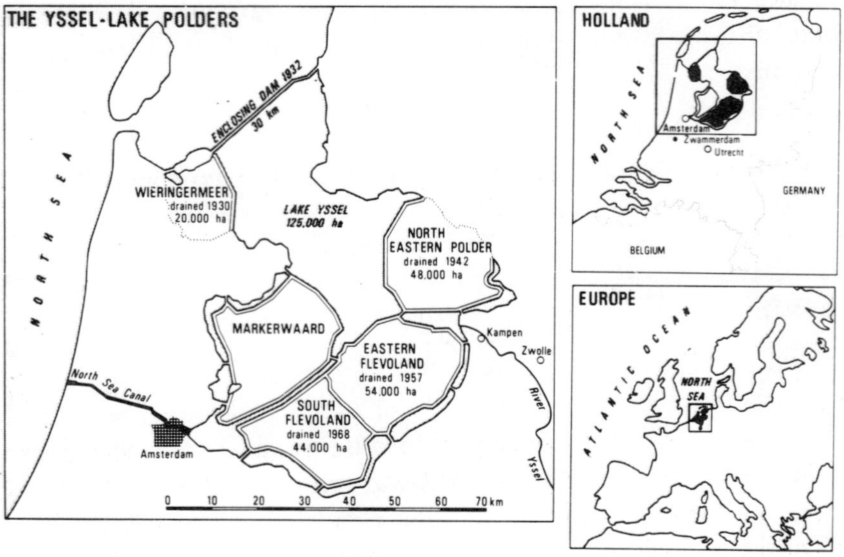

Fig. 5.1 The Lake IJssel land reclamation and development project.

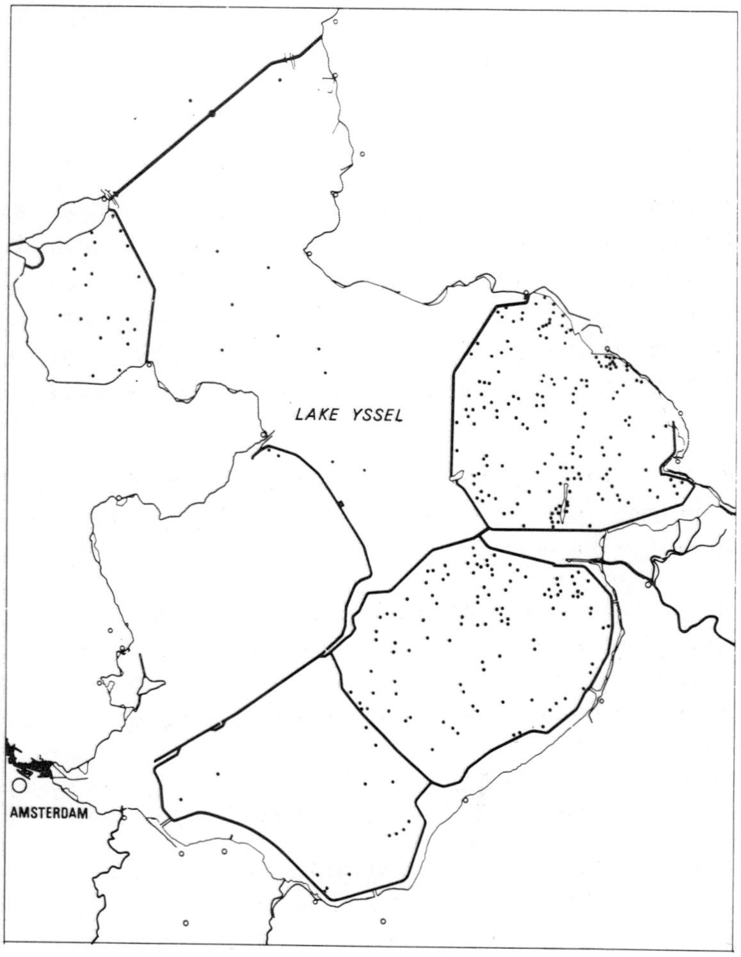

Fig. 5.2 The location of the shipwrecks found in the Lake IJssel area.

fishing area, Lake IJssel was in previous centuries a centre for national and international shipping, and it is therefore not to be wondered at that the bottom has so many shipwrecks. Up till now over 300 of them have been found (Fig. 5.2).

A number of these shipwrecks have been excavated to study the history of the trade of the shipbuilding and life on board. In several cases the entire wreck, large parts of it, or the ship's inventories appeared to be of such an interest and value that it was decided to add them to the collection of the Museum of Maritime Archaeology at Ketelhaven.

If it is decided to preserve the excavated remains of the wrecked ships it is necessary to apply a certain technique of conservation of wood to prevent shrinkage, cracking and distortion of the ship's remains, which often happens when old waterlogged wood is allowed to dry.

THE PROPERTIES OF OLD WATERLOGGED WOOD

If wood is kept in a very moist or wet environment it will absorb water and eventually become waterlogged. After a certain amount of time changes will occur in the wood structure, and in the physical and chemical properties of the wood. These changes adversely affect the suitability of the wood for construction purposes, and are mainly caused by microbiological attack on the wood. Physical-chemical processes also contribute to the breakdown of the wood.

How far wood will desintegrate depends on the wood species, its growth history, the kind of environment during the period the wood has been used, and the conditions and the length of the period after its use. Important factors in this respect are mechanical damage, type of medium, humidity, temperature, pH and redox potential [1,2]. Decay of waterlogged wood may occur under both aerobic and anaerobic conditions. The amount of oxygen in watersaturated soil, in which most of the shipwrecks are found in the Netherlands, is small, hence the disintegration will be mainly anaerobic. The microbiological decay causes more damage than physical-chemical decay in both aerobic and anaerobic conditions.

This becomes clear when finds made of a wood species which is very resistent to microbiological attack, and finds from an environment where microbiological attack could not occur (e.g. the Norwegian Viking ships and the Egyptian mummy coffins) are compared with decayed old waterlogged wood

In many cases deterioration of the wood may start before it became waterlogged. Attack by rot causing fungi, and teredos may be taken as examples. Finally the period between the moment the wood became waterlogged, and was exposed to decay causing phenomena, and the moment of excavation will play a role.

The process of deterioration of wood in wet, anaerobic conditions is visualised as follows [5]:

At the outset, the readily water soluble compounds in the wood will diffuse into the surrounding medium, followed by readily hydrolised compounds such as the pectines and pentosans. Then a microbiological degradation of the more stable hemicellulose and cellulose will follow. Finally only lignine remains, which can also be decomposed slowly by micro-organisms in an anaerobic environment. Both fungi and bacteria cause the microbiological breakdown of wood and both use extra-cellular enzymes to hydrolise hemicellulose, cellulose and lignine. In the case of a fungal breakdown degradation spreads throughout the capillary system of the wood at least as far as the extra-cellular enzymes can penetrate into this system [3,4].

The result of the attack on the wood and the decay of the cell material is a considerable loss of original strength. When drying, old waterlogged wood shows a very strong shrinkage, warping, cracking, and sometimes a complete disintegration; cell-collapse can also often be observed. The contracting capillary forces of the evaporating water exceed the strength of the capillary system of the deteriorated wood. As a result of the decay of cell material in the wood the size of the submicroscopic capillaries increases, causing a decrease in the specific surface of the wood. The cellulose breakdown also results in a decrease of adsorbtive sites on the wood, resulting in a decreased swelling capacity of dried old waterlogged wood as compared with new wood. In practice, dried and shrunken old waterlogged wood swells only to a very limited extent when again in contact with water. Cell-collapse is mostly irreversible. Noack[6,7] recorded 9-10% radial and 20% tangential shrinkage in little decayed wood of the Bremer Kogge (the description little decayed is based on chemical composition and water content in the waterlogged state). The longitudinal shrinkage was almost zero. These data are based on drying from waterlogged to ovendry (approx. 105°C). Figure 5.3 indicates what the results are of such a shrinkage process.

From investigations conducted on the wood of the Zwammerdam and Utrecht ships it was learned that even higher degrees of shrinkage may occur. These are, combined with data recorded in literature, listed in Table 1. The water content as given in this Table is based on the oven dry weight (105°C) of the wood. The more deteriorated the wood is, the higher the water content will be for any given species of wood. The breakdown of the wood also causes considerable changes in the chemical composition of the wood. This is shown in Table 2 by giving the results as obtained for the Wasa, the Bremer Kogge [7], and the investigations conducted on the wood of the Zwammerdam and Utrecht ships. All data given are based on the ovendry weight of the wood material.

Table 1 shows that in many cases old decayed waterlogged wood shows after drying degrees of shrinkage considerably higher than those of dried new waterlogged wood. Old and deteriorated waterlogged wood can only be allowed to dry in a few special cases when the wood is very slightly degraded.

The data of the Tables 1 and 2 together show that even in cases in which almost no chemical changes are found, there is yet a considerable increase in the degree of shrinkage as compared to that of new waterlogged wood. Increased degradation of the wood means increased radial and tangential shrinkage. In highly degraded waterlogged wood a longitudinal shrinkage will also occur. The shrinkage in this direction may well be over 10%. When an outer layer of highly degraded woods is overlying a less degraded inner core, the highly degraded layer and the inner core will shrink in all

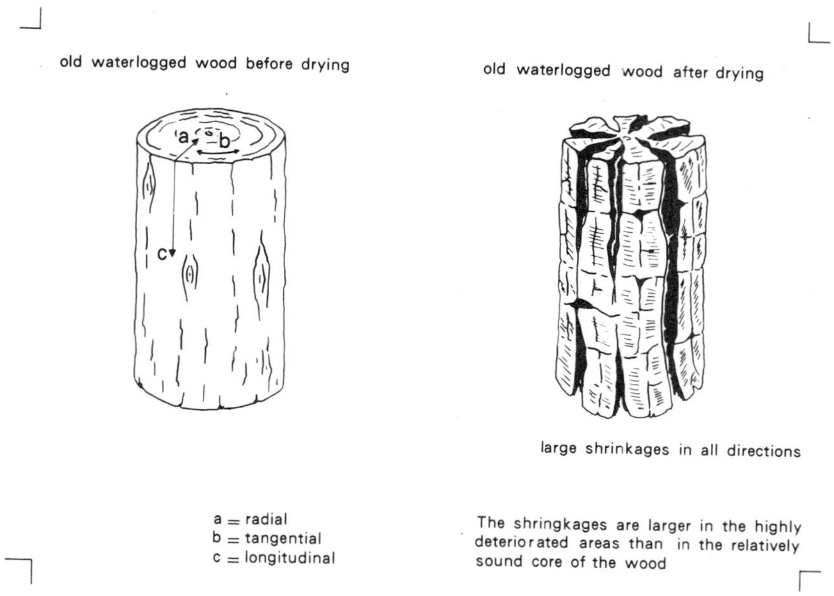

Fig. 5.3 The three main axes in wood and the damage due to shrinkage after drying a piece of old waterlogged wood.

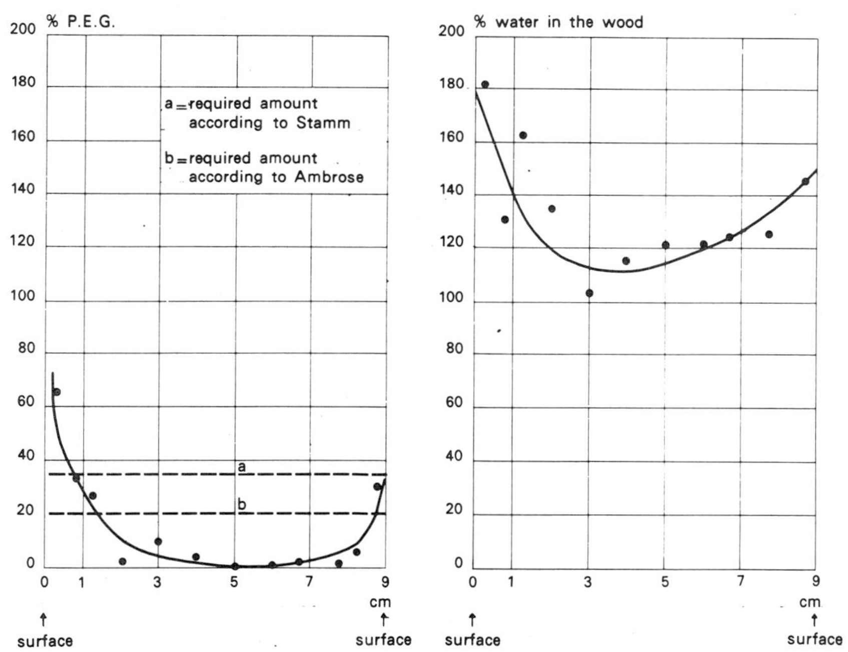

Fig. 5.4 The distribution of PEG 4000 perpendicular to the longitudinal direction in a piece of European oak. Data based on the dry wood material. The conservation process is described in the text.

Table 1. Shrinkages percentages in new and old waterlogged wood when dried at 105°C. Data partly based on data from Noack.[7]

Wood	Origin	% water content	% shrinkage when dried from waterlogged to ovendry (105°C)		
			radial	tangential	longitudinal
Oak heartwood	new	± 120	4.8	10.6	0.4
Oak sapwood	new	± 145	4.0	9.4	0.4
Beech	new	± 115	± 6	12	0.3
Oak heartwood	Bremer Kogge	120–145	9–10	20	-
Oak heartwood	Utrecht ships	± 100	8–9 (8)*	17–22 (19)	0–2 (1)
Oak heartwood	Zwammerdam ships	100–120	12–14 (13)	17–23 (21)	0–2 (0, 5)
Oak heartwood	Utrecht ships	375–425	5–13 (10)	26–52 (33)	9–11 (10)
Oak heartwood	Zwammerdam ships	375–400	8–16 (11)	38–56 (47)	10–17 (13)
Oak sapwood	Bremer Kogge	600	18–19	54–64	10–13
Oak sapwood	Utrecht ships	670–1000	12–16 (14)	41–55 (47)	10–16 (12)
Silver spruce	Zwammerdam	450	11–15 (13)	12–49 (26)	5–7 (6)
Beech	Bremer Kogge	590–610	8–9	65–70	15–17

* The numbers between the brackets are the mean shrinkage values as calculated from 4 to 7 experiments.

Table 2. The main chemical constituents in new and old waterlogged wood. Data for the Wasa and the Bremer Kogge from Noack[6,7] and Mühlethaler.[5] The data for the Zwammerdam and Utrecht ships were determined by the Forest Products Research Institute T.N.O. Delft, the Netherlands

Wood	Origin	Content in % of ovendry wood					
		Water	Cellulose	Hemi-cellulose	Lignine	Ash	Fe
Oak heartwood	new	± 120	± 65	-	± 25	± 0.3	-
	new	-	± 50	-	32	0.4	-
Oak sapwood	new	-	± 50	-	32	0.6	-
Beech	new	± 115	± 66	-	± 23	0.8	-
Softwood*	new	-	40-50	20-22	28-30	-	-
Hardwood*	new	-	40-50	26-28	20-22	-	-
Oak heartwood	Wasa	± 150	67	-	29	3.4	-
Oak heartwood	Wasa	± 800	26	-	62	5.6	-
Oak heartwood	Bremer Kogge	-	46	-	36	5.3	-
Oak heartwood	Utrecht ships	131	46.0	8.1	25.1	2.0	0.8
Oak heartwood	Utrecht ships	455	5.5	9.4	50.6	8.0	0.5
Oak heartwood	Zwammerdam ships	98	47.3	11.1	23.1	1.4	0.7
Oak heartwood	Zwammerdam ships	494	5.5	16.9	51.3	7.6	0.5
Oak sapwood	Bremer Kogge	600	17	-	62	7.7	
Oak sapwood	Utrecht ships	760	6.8	3.1	64.1	10.1	0.4
Beech	Bremmer Kogge	590-610	31		64	3.2	
Silver spruce	Zwammerdam	463	11.0	1.6	69.3	2.4	0.1

* after K. Seifert

Table 3. The classification of wood species according to durability. The division is based on an outdoor exposure of untreated heartwood in contact with a humid soil and in a temporate climate.

Class		Lifetime (years)	Wood species (examples)
I	very durable	25	afromosia, afzelia, azobé, iroko, merbau, teak, wengé
II	durable	25-15	Californian redwood, red cedar, white cedar, European oak,* Am. mahogany, sipo mahogany, red meranty (heavy)
III	moderately durable	15-10	Oregon pine, pitch pine, larch, red meranty (light), sapelé mahogany, European walnut
IV	little durable	10- 5	fir, European pine, parana pine, Am. red oak, maple, elm
V	very little durable	5	balsa, birch, beech, ash, poplar, alder, lime, willow

* at the boundary of class II and III

directions during drying, but the outer layer will shrink considerably more than the inner core. These differences in shrinkage, cause the higher degraded outer layer to chip off from the inner core. Even in cases with only a partially chipping off of this outer layer almost all details of the texture of the surface of the wood will be lost.

To prevent these adverse affects it is necessary to conserve the material. The aim of this conservation is, as far as possible to stabilise the dimensions of the find. This means that the goal of any conservation process is to obtain an equilibrium between wood and environment, preventing the high degree of shrinkage. In some cases, especially with highly degraded wood, an improvement of the mechanical properties of the material is desirable or necessary. As can be seen from Table 1 even new waterlogged wood shrinks when drying, mainly in the radial and tangential directions, and some shrinkage will almost prove inevitable during any conservation process. The acceptable degree of shrinkage depends largely on the nature of the object and its use after conservation. If for example the object is fairly small and has a very fine and delicate surface texture (e.g. sculpture) and will be used for further research (archaeological or historical) shrinkage should be as small as possible. If, however, the surface of the wood is rather unfinished, and the finds will be used after conservation to give an impression of the entire constuction (as is the case with many old ships) some shrinkage and surface defects can be accepted.

A particular case of wood deterioration has to be expected if wood is found in land reclaimed from sea or lakes. During the time that these finds were covered by the sea or lake sediments the wood of them was attacked by anaerobic microbiological and physico-chemical processes. After the reclamation of the land, causing a lowering of the ground water level, the degradation of the wood will be continued by aerobic processes and hence deterioration is speeded up [9].

An indication of the lifetime of wood under such circumstances can be obtained from the durability classes which have been developed for new wood. This division of wood species into several durability classes is based on the lifetime of untreated heartwood, during an outdoor exposure and in contact with a humid soil in a temporate climate (Table 3). European oak heartwood is, according to this division, placed on the boundary of the classes 11 and 111. The expected lifetime will be about 15 years and for small objects even less.

METHODS OF CONSERVATION AND THEIR RESULTS

A review on the existing conservation techniques and the results obtained is given by De Jong [9].

Two approaches can be discerned in the conservation of old waterlogged wood [10]. The first approach is to attempt to apply a coating to the surface and to the wood material close to the surface. The coating has to be impermeable to water and water vapour to a very high degree. Unless very thick coatings are applied this requirement will never be met, and drying will occur and cracks readily form. Hence a coating techniques will not prevent shrinkage. The result is merely a retardation of the establishment of an equilibruim between the wood and its environment, and shrinkage corresponding to the quality of the wood and the environmental conditions will finally occur.

Only in a very few cases, where for example a tiny deteriorated surface layer overlies a sound and almost undegraded inner core, can surface treatment give satisfactory results. Such was the case for the Kentmere boat where good conservation was said to have been achieved [1] by impregnation and consolidation of the surface of the wood with an epoxide resin. Near Lelystad a 19th century ship has been excavated and left to dry in the open air. The wood was in a very good condition and no other damage than a local slight surface cracking occured after a few years.

The second way of approach is to impregnate the wood as thoroughly as possible, the aim being to treat both the micropscopic and the submicroscopic capillary system of the wood. Water present in the old waterlogged wood is substituted by non volatile compounds. If such a compound is strongly absorbed into the cell walls it can take over the function of swelling agent from the water and force the latter to leave the wood. Hence the wood, though now containing less water, retains its swollen state. Especially ethyleneglycol and polymers based on this compound are suitable for this purpose [12, 13, 14, 15]. These polythyleneglycols are capable, when absorbed into the cell walls, of acting as a swelling agent like water. Unlike water these compounds are non-volatile and remain under dry circumstances, in the wood, once they have been introduced. So, with these compounds good and lasting dimensional stabilisation can be achieved. Several methods for the impregnation of polyethyleneglycols into old waterlogged wood have been described in literature and will be discussed later. If the hard, wax-like polyethyleneglycol (PEG) grades are used and the capillary system is filled with these compounds, an improvement of mechanical strength can also be realised.

When treating new wood excellent results are as a general rule, obtained when, about 35% of polyethyleneglycol - based on the dry wood material - is present in the wood [14]. When, after or during the introduction of such an amount of PEG in the wood, there is sufficient time to allow for a diffusion of PEG from the cell lumia into the cell walls, these cell walls will remain swollen when the wood dries out and is kept at low relative humidities. It is sometimes necessary to raise this diffusion from the macrocapillary into the microcapillary system of the cell walls by storing the wood, after the impregnation with PEG, in non drying conditions. From the investigations of Ambrose [16] it appeared that the dimensions of old waterlogged wood found in swamps in New Guinea were stabilized when the wood contained about 20% PEG with a molecular weight of 400.

Based on the experience obtained in treating new wood conservators started to use polyethyleneglycols for the conservation of old waterlogged wood [17, 18, 19, 20, 21, 22].

In 1973, using the bath method described by Barkman[17] the conservation of old waterlogged wood was started in the workshop of the Museum of Maritime Archaeology at Ketelhaven. PEG 4000 was used as a conservation agent. The entire impregnation time was 18 months, the objects were then stored in the impregnation liquid at 60°C for one month more. After removing the objects from the bath they were dried in an environment of initially 70-80% and finally 50-60% relative humidity. During this drying shrinkage, crack formation and warping occurred in a number of cases. The main weight loss of the objects was approximately 30%.

The cracks, which developed mainly parallel to the fibre direction, varied in width from a few millimeters up to some centimeters. Round objects with the centre of the tree or branch present especially developed these fissures. In some cases the picture characteristic of an insufficient impregnation with PEG was found. Near end planes, more or less perpendicular to the fibre direction, there is sufficient PEG present and no shrinkage occurs during drying, while other parts show shrinkage. These phenomena have also been observed by Brorson Christensen [23].

These disappointments led to investigations into the actual distribution of PEG in objects which appeared badly and well conserved. To obtain information about this PEG distribution pattern, samples were taken with a diameter of 2.5 cm and mainly perpendicular to the fibre direction of the wood. By dividing the samples in several slices and extracting the PEG with water, average values of the PEG present in different layers could be obtained. Initially the extracted PEG was determined tititrimetrically but since this is a time consuming method, a spectrophotometric method was developed. [9]

In Figure 5.4 an example of a PEG distribution curve is given. The amount of PEG 4000 expressed as a percentage of the dry wood material is plotted against the distance from the wood surface. The measured water content of the wood determined from the extracted samples is also given. It is assumed that no shrinkage has occurred during the conservation process.

It was remarkable that oak almost exclusively caused the shrinkage problems. During investigations oak appeared to have retained its highly impermeable structure, and almost no penetration of PEG 4000 into the wood was observed. Brorson Christensen [23] described these phenomena. The water content of this wood was found to vary between 100% and 185%.

We found that the success of a conservation process also depends on the size of the objects. This is in no way remarkable as will be explained later. In small test samples of little deteriorated oak heartwood of dimensions 5 x 5 x 10 cm (tangential x radial x longitudinal) a good penetration of PEG 4000 was observed, whereas in larger objects such as anchorshanks of comparable wood quality (comparable water content) a far less satisfactory impregnation was found, and in large objects such as knees and hull planking the impregnation was even worse. This clearly illustrates the fact that results obtained with small test samples can be strongly misleading when conclusions are drawn about the applicability of any process for the treatment of large objects.

The explanation of these phenomena is based on two facts. First the mass-transfer of PEG from the solution into the wood is a function of the area available for this mass-transfer. Since in small samples the surface to volume ratio in higher than for large samples, the mass-transfer per unit of volume will be higher for the small samples than for the large objects. Secondly the transport of PEG and other dissolved materials in the wood proceeds with different speeds in the three main directions in the wood. The transport in longitudinal direction is considerable faster than in the two other directions [24, 25, 26]. The consequence is that in small objects, with surfaces perpendicular to the fibre direction at relatively short distance from each other, the degree of impregnation will largely be determined by the "fast"

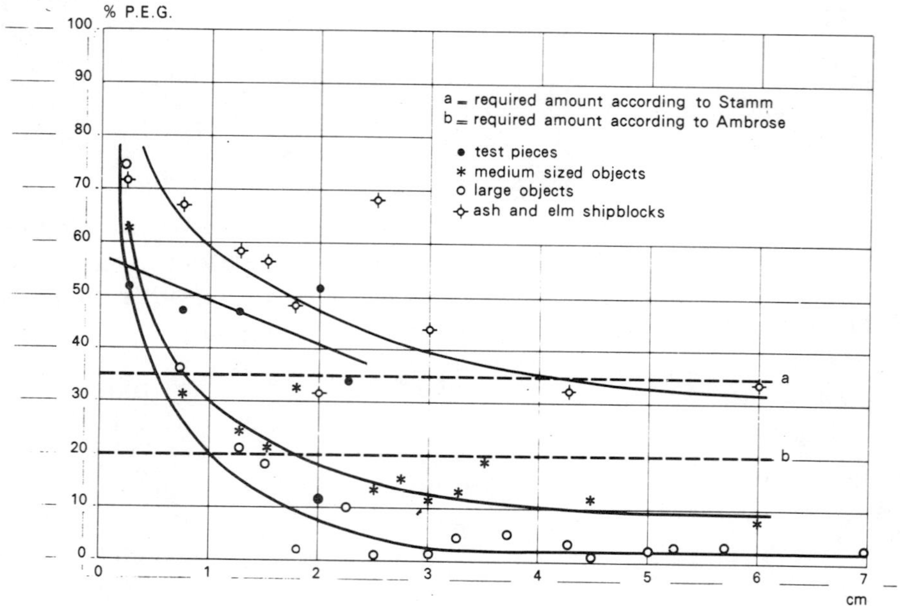

Fig. 5.5 Graphs showing the difference in impregnation and hence conservations for objects of the same wood species and quality (European oak containing less than 185% water when waterlogged), but of different size. Results for small test samples, medium sized objects (e.g. anchor stocks), large objects (e.g. hull planking and heavy knees) and rigging blocks made of ash and elm.

transport through those surfaces. For large objects this "fast" transport will be too slow to compensate for the much larger distances between the surfaces perpendicular to the fibre direction. From these considerations one can clearly see that no conclusions about the conservation of old waterlogged wood finds of large dimensions can be based on the results for small test samples. How the penetration of PEG 4000 in objects of different sizes but of the same wood species (European oak) and comparable state of decay (water content 120-185%), after the same impregnation process with PEG 4000, may differ is shown in figure 5. With the exception of small test samples, for good dimensional stabilization, the required amount of 35% PEG (with respect to the dry wood) is only found near the surface of the objects. The same holds if the 20% level as given by Ambrose is used.

It is remarkable that the surface concentration is the highest in the large samples. When using the impregnation technique a high surface concentration may be expected, since the conservation is started with a low PEG concentration, which is increased gradually during the process. After the removal from the bath drying starts, and water will initially evaporate from the wood surface. The evaporating water will be replenished from the interior of the wood. The result is a net flux of water saturated with PEG from the interior towards the surface, where the water evaporates and the PEG precipitates. This effect will be the largest in objects having the lowest surface to volume ratio, hence generally in the large objects.

Besides the size of the objects the wood species proved to be important, European oak is very difficult to penetrate and hence to conserve but other wood species can be impregnated much more easily. From our investigations and those reported in the literature, it can be concluded that old waterlogged wood can be successfully conserved with polyethyleneglycols provided there is a thorough knowledge of the wood to be treated. Knowledge about the different stages of deterioration present in the find and possible conservation techniques are essential. Further fundamental research for methods of improving the speed of penetration of PEG into some wood species, e.g. European oak is needed. The problems of treating slightly deteriorated old waterlogged European oak also led Brorson Christensen (23) to search for new methods and develope the method of dehydrating the wood with tertiary butylalcohol (tba) followed by impregnation with a solution of PEG 4000 in tba. After impregnation the tba is removed from the wood by freeze-drying. This method is used for the treatment of large objects in Denmark. Results are as yet not available. Based on this principle the following experiment was carried out in our laboratory, Wood samples in which highly degraded oak sapwood, deteriorated oak heartwood, and sound heartwood were present were dehydrated in 88% tertiarybutylalcohol by axeotropic distillation of the alcohol water mixture (9). The samples had wet weights from 0.5 to 30 kg and dimensions from about 5 x 10 x 10cm up to 100 x 15 x 20 cm (radial x tangential x longitudinal). Dehydration took two months. After dehydration the wood was impregnated with a solution of PEG 4000 in 88% tba at 50-55°C. The concentration was raised in 10% steps up to 50% in a period of four weeks, after which the concentration was maintained for another four weeks at 50%. Occasionally PEG was added to maintain the concentration of the solution at 50%, because of diffusion of PEG into the wood. The final PEG concentration in the bath liquid was 52%.

After this impregnation the wood was removed from the bath and cooled. The PEG crytallises from the solution and therefore little or no transport of PEG will occur to the outside of the wood while the solvent evaporates. After two weeks the samples were heated for 7 days up to 65°C to melt the PEG. The aim of this is to diffuse the PEG present in the macrocapillary system of the wood into the cellwalls. After this treatment, which has the advantage that freeze drying can be omitted and that a second heat treatment can be given to ensure the absorption of PEG into the cellwalls, it was found that even in samples in which highly degraded sapwood, deteriorated heartwood and sound heartwood were present, no shrinkage damage occurred, only some small fissures along the fibre direction developed. As a result of the heat treatment the wood lost the familiar waxy surface of PEG treated wood.

The shrinkages have been measured at several stages of the treatment, but since the final shrinkages are the most important summarising will be limited to the latter. These values are determined one month after the heat treatment. During this period the samples were stored in the laboratory at 18-20°C and 55-65% r.h.

The following shrinkage values were found for degraded heartwood and sound heartwood (it was impossible to isolate the values for the sapwood since this layer was too thin and too soft to permit measurements in the wet state with sufficient accuracy):

	% shrinkage radial	% shrinkage tangential	% shrinkage longitudinal
little deteriorated oak heartwood	3.0 - 3.7	2.9 - 4.1	-*
highly deteriorated oak heartwood	1.8 - 2.2	2.1 - 9.6	1.2 - 1.7

* this value could not be measured with sufficient accuracy but was less than 1%.

This method has now been investigated into more details. It proved successful for samples of European oak, ash, elm.

POSSIBILITIES OF CONSERVING THE SHIPS FOUND AT ZWAMMERDAM AND AT UTRECHT

In February 1973 the IJsselmeerpolders Development Authority established a board of Dutch experts in the field of wood conservation whose task was to advise on the conservation of the large finds of waterlogged wood found in the Lake IJssel polders. In the first months of 1973 the Zwammerdam ships, dating from Roman times, were found and it was decided that the IJsselmeer polders Development Authority should excavate and salvage them. It was also decided that the conservation and restoration of the finds should be done in the workshop of the Museum of Maritime Archaeology at Ketelhaven. Some of the ships will be exhibited there after conservation. In Utrecht in 1974 two medieval ships were found and it was decided again that the IJsselmeerpolders Development Authority should salvage and conserve these finds. It will be clear that the attention of the advisory board has been focussed almost entirely on the possibilities of conserving these finds.

The condition of the wood was too bad and the damage sustained by the wood in the soil too heavy to enable salvaging and restoration of the ships as a whole. So it was decided to take the timber of the ships apart, and put the separated parts on new boards upon which the old waterlogged wood could be transported, cleaned and conserved. At the excavation site and during the transport to the museum workshop the wood was kept wet. At the workshop the wood was stored in water, first in temporary basins outside, then after cleaning in a number of concrete tanks. To prevent the growth of microorganisms thymol was added to the water (5). However this chemical proved during summer to be insufficient, and it was decided that the water should be renewed regularly.

Since the ships had been taken apart in separate pieces and since much of the wood, especially of the Zwammerdam ships, was highly deteriorated, it was decided to conserve the wood first and to restore the ships afterwards.

Any process for the conservation of large finds of old waterlogged wood should fulfill the following requirements.

- The process should be safe. This means that it must be reasonably certain that a good result can be expected, or the entire process should be reversible.
- The conservation process should result in an as 'uniform' as possible distribution of the impregnated chemicals throughout the wood. The movement of the chemicals in the wood after conservation should be as little as possible.
- The chemicals should exert no stress upon the wood either during application, or afterwards.
- The chemicals should not react with large dimensional changes in response to changes in the surrounding medium.
- The chemicals should not be growth media for microorganisms. Finally
- Finally the process to be used has to be hazardless, - the workshop is located at a polder dike - and require as little labour as possible. It has to be possible to automate the process. The costs of the conservation have to be as low as possible.

Since six ships were found at Zwammerdam and two at Utrecht and all these ships will have to be conserved, it is impossible to evaluate the applicability of all possible processes for the conservation of old waterlogged wood. The amount of wood was too large to consider using conservations techniques that still are only suited for a laboratory bench. Four possible processes for the conservation of the ships have been studied before a choice was made:

1. Freeze-drying followed by a final treatment of the wood.
2. Impregnation with glycolmethacrylate followed by a thermal polymerisation of this monomer. (28)
3. Dehydration of the wood with a volatile, water soluable solvent followed by a final treatment of the dehydrated wood.
4. Impregnation of the wood with a solution of polyethyleneglycol in water.

Freeze-drying requires large investments in both equipment and energy while there is as yet no experience with the treatment of very large pieces of timber available. In addition a two step process is involved, since a final treatment of the freeze dried wood is necessary. This makes the conservation more complex, requires labour, and increases costs, The conservation of wood with glycolmethacrylate though very attractive in theory - the monomer is watersoluable and can be polymerized thermally without the addition of a catalyst - is still in development and cannot be used for the treatment of large or valuable objects. The dehydration of large quantities of wood with solvents such as methanol,acetone or tertiary butylalcohol requre safety precautions. This fact and the consideration that a two stage process is involved, does not make this process likely to be suitable for the conservation of this large amount of old waterlogged wood. Although the conservation of old waterlogged wood with polyethyleneglycols has not always been successful and a number of drawbacks can be listed, this method was selected as the best for the treatment of the ships found at Zwammerdam and Utrecht. The following considerations led to this conclusion.

- The experience obtained in conserving large quantities of old waterlogged wood is limited to the conservation with PEG (Roskilde ships, Wasa), and there are publications describing the practical problems encountered. Since both the Zwammerdam and the Utrecht ships were taken apart during excavation, the bath method seems to be the best technique. The ships can subsequently be reconstructed with the conserved wood. The conservation of a previously reconstructed ship has the disadvantage that, expecially with the highly deteriorated timber of the Zwammerdam ships, damage will increase and a very large bath will be needed for the conservation. Another solution is the spraying method, but this will result in longer impregnation times than for the bath method.
- The PEG bath method is a technically simple process and had the lowest estimated process costs.
- The process is safe with respect to the wood to be conserved and is easily reversed.
- The process is hazardless from an environmental point of view.

A boat, 13.5 m long and 2.6 m wide was found in Krefeld Germany, and the possibilities for excavation and conservation studied. In that case it was also decided to salvage individual elements of the boat. These separate parts will be conserved, by polyethyleneglycol followed by reconstruction (29). This way of solving the problems of excavation and conservation shows a striking resemblance to the methods used by the Museum of Maritime Archaeology.

Using baths with PEG 4000 at 60°C Brorson Christensen (23) met difficulties when conserving decayed European oak. Barkman too, though initially starting with PEG 4000 for bath treatment of oak timbers and sculptures from the Wasa, afterwards used PEG 1500 for the bath process. Ambrose (16) indicates that the use of PEG of low molecular weight, requires comparitively low relative humidities to be maintained after conservation to prevent liquification and leaching out of PEG, When PEG 1000 is used, relative humidity should be less than 40%. In our situation, where we have high relative humidities in the museum, we choose PEG 4000 since liquification and leaching out of PEG occured with objects treated with PEG 1500. When using PEG 4000 each piece of timber should be selected for a specific method. A number of classifications of European oak are given in literature (23, 30, 16).

Based on these classifications and our own investigations the following scheme for the treatment of old waterlogged European oak has been made and is now used:

I Wood containing over 400% water. This wood has lost almost all of its hard core.

II Wood containing 185-400% water. A hard core is present but is comparatively small.

III Wood containing less than 185% water. A hard, little-deteriorated core is present below a thin deteriorated surface layer.

Wood belonging to the classes I and II can be conserved with PEG 4000; large objects consisting of class III wood cannot successfully be conserved with solutions of PEG 4000 in water.

Class I wood can be conserved by means of the warm bath method with PEG 4000, while for class II wood the cold bath method with PEG 4000 can be used.

Treatment of class III wood with lower molecular weight PEG is possible, but requires many years (Wasa spraying with PEG 1500 for 15 years; Bremer Kogge treatment with PEG 1000 for an expected period of 20 to 30 years). Conservation methods which are less time consuming still have to be developed, and the research in this field should be stimulated.

Possible treatment for Class III wood.

In some cases, when the wood is very little deteriorated just leaving it to dry may give good results. This has been done for a 19th century wreck found near Lelystad.

The ship was excavated and left to dry in the open air where it is still lying. Except for a local, slight surface cracking, no problems arose, as figure 6 may illustrate. We have recently obtained more experience using this method. A number of large objects and ship segments all consisting of very little deteriorated European oak and fir were dried. Almost all results were good. Controlled experiments with wood from a 17th century ship showed that after a fast drying under laboratory conditions the shrinkages in this wood are comparable to those occurring in new wood upon drying from a waterlogged condition. The observed mean shrinkage data are given in Table 4.

It will be clear that in many cases such shrinkages can be accepted, so simple controlled drying of old wood can be used for some Class III wood. After drying a surface coating can be applied.

If the wood is too much degraded to permit drying, conservation is still necessary. Using water-PEG impregnations, even when low molecular weight types PEG are used, long impregnation times are needed. When large amounts of timber are to be conserved one soon feels the need of faster processes. On

Table 4

The observed shrinkage data in very little degraded
European oak and fir as compared to new wood

Species	% shrinkage air dry l*	t*	oven dry l	t	new wood oven dry l	t
Oak heartwood	0.3	9.2	0.4	12.1	0.4	10.6
Fir heartwood	0.2	4.2	0.4	6.3	-	7.7

*l longitudinal shrinkage
*t tangential shrinkage

a laboratory scale the tertiary butyl alcohol PEG 4000 method has proved itself successful. It gives good results as can be seen from Figs. 5.7 and 5.8, and is less time consuming. A pilot plant has been constructed based on the tba-PEG process enabling the treatment of samples up to 3 m length. Results will probably be available in late 1977. If they are satisfactory the method can be used for the treatment of those class III timbers which cannot be allowed to dry.

Fig. 5.6 The wreck of a 19th century ship, found near Lelystad. The wreck was dried in the open air and except for a local slight surface cracking no problems arose.

Fig. 5.7 Two rigging blocks made of elm. The dark one is treated with PEG 4000 in water, the bright one using the tba-PEG 4000 method.

Fig. 5.8 Pieces of an oak tree, dating from the 2nd century B.C., in wet condition, air dried and after conservation using the tba-PEG 4000 method. The hole in the conserved sample was made during the sampling for PEG distribution.

REFERENCES

1. Barkman, L., The preservation of the Wasa. Wasastudier 5, Statens Sjöhistoriska Museum Stockholm, (1965).

2. Boutelje, L.B., and Kiessling, H., 'On water-stored oak timber and its decay by fungi and bacteria'. Archiv. Mikrobiol. 49 (3), 305-314, (1964).

3. Stamm, A.J., Wood and Cellulose Science. Ronald Press Company, New York, (1964).

4. Hashimoto, Y. and K. Nisizawa, Advances in enzymic hydrolysis of cellulose and related materials, edited by E. T. Reese. Pergamon Press London, (1963).

5. Mühlethaler, B., Conservation of waterlogged wood and leather ICOM Travaux et publications X1, Ed. Eyrolles, Paris (1973).

6. Noack, D., 'Der gegenwartige Stand der Dimensionsstabilisierung von Holz und Schlussfolgerungen für die Konservierung der Bremer Kogge'. Bremisches Jahrbuch 50, 43-72, (1965).

7. Noack, D., Die Bremer Hanse Kogge. Fund. Konservierung, Forschung p. 127-156. Friedrich Röver, Bremen 1969.

8. Elburg, v.d. J., Private comm.

9. de Jong, J., The conservation of old waterlogged wood from shipwrecks found in the Netherlands. In: Biodeterioration Investigation Techniques Interscience. 1977.

10. Risi, J., and Arseneau, D.F., 'Dimensional stabilization of wood'. Forest Products Journal, 210-213, (1957).

11. Werner, A.E.A., Technical notes 218. CIBA Ltd. Duxford, Cambridge, England.

12. Stamm, A.J., 'Dimensional stabilisation of wood with carbowaxes'. Forest Prod. J. 6, 201-204, (1956).

13. Stamm, A. J., 'Effect of polyethyleneglycol on the dimensional stability of wood'. Forest Prod. J. 9, 375-381 (1959).

14. Stamm, A. J., 'Factors affecting the bulking and dimensional stabilization of wood with polyethyleneglycols'. Forest Prod. J. 14, 403-408 (1964).

15. Heslinga, F. J. M., 'The impregnation and conservation of deteriorated oak using ethyleneglycol and polyethyleneglycol'. Museologia No. 1-X-1973 pp. 24-39.

16. Ambrose, W., The treatment of swamp degraded wood by freeze-drying. Report to the ICOM, Madrid, 1972.

17. Barkman, L., 'The preservation of the warship Wasa'. in Oddy WA (ed) Problems in the conservation of Waterlogged Wood. Maritime Monograph 16 (1975) Greenwich.

18. Lefeve, R., 'Conservation treatment with polyethyleneglycol of a wooden Roman bucket excavated at Wemmel'. Bull. Inst. Royal du Patrimone Artistique, 4, 97-108, (1961).

19. Organ, R.M., 'Carbowax and other materials in the treatment of waterlogged paleolithic wood'. Studies in conservation 4, 96-105, (1959).

20. Seborg, R.M., and Interverarity, R.B., 'The preservation of old waterlogged wood by treatment with polyethleneglycol'. Science 136, 649-650 (1962).

21. Seborg, R.M., and Interverarity, R.B., 'Conservation of 200 years old waterlogged boats with polyethyleneglycol'. Studies in conservation 7, 111-119 (1962).

22. Organ, R.M., 'Polythyleneglycols in the treatment of waterlogged wood: some precautions'. Studies in conservation 5, 161-162, (1960).

23. Brorson Christensen, B., The conservation of waterlogged wood in the National Museum of Denmark. Copenhagen, 1970.

24. Behr, E.A., Briggs, D.R., and Kaufert, F.H., 'Diffusion of dissolved materials through wood'. J. Phys. Chem. 57, 476-480, (1953).

25. Kumar, S., 'Diffusion through wood. Diffusion of boric acid'. Holzforschung und Holzverwertung 25, 21-24, (1973).

26. Stamm, A.J., Wood and Cellulose Science. Ronald Press Company, New York, 1964.

27. Munnikendam, R.A., Private comm.

28. Pirling, R., and Buchwald, G., 'Ein Schiff aus karolingischer: Zeit und seine konservierung'. Naturwissenschaften 61, 396-398 (1974).

29. Moren, R.E., and Centerwall, B., 'The use of polyglycols in the stabilizing and preservation of wood'. Modo. Technical Information no. 64, (1960)

6. ASPECTS OF WATERLOGGED WOOD CONSERVATION

C.W. Gregson

Summary

Though wood is preserved in a water-saturated environment, the structure is weakened by leaching and microbiological decay. European Oak (Quercus Sp.) shows a marked pattern of differential degradation, a soft surface surrounding a "hard core" of relatively sound wood. If allowed to dry, waterlogged wood suffers irrevocable shrinkage; it must be kept wet during excavation and stored under water after lifting.

There are two main methods of conservation: replacement of the water for a liquid of low surface tension which can evaporate without causing cell collapse; or bulking with a water soluble substance wuch as Polyethylene Glycol, which remains within the wood on evaporation of the water. Large structures are usually treated with Polyethlene Glycol, either by spraying or immersion in heated baths.

Modifications are required to impregnate "hard core" oak; use of a very slow rate of impregnation, with low molecular weight grades of Polyethelene Glycol.

Wood in contact with the soil is rapidly decayed by the action of insects, bacteria and fungi: the first step in humus formation essential for maintenance of soil fertility and the growth of green plants and consequently animals. If wood is covered by a constantly wet soil, the rate of decay is greatly reduced due to the lower density of micro-organisms capable of functioning in the absence of air. These conditions pertain to estuarine, flood plain or heavy clay soils, and it is from these environments that the majority of ancient wooden boats have been excavated in North West Europe. Several other factors influence preservation; wood species, acidity or alkalinity of the soil, (Table 1) and temperature. The Viking ships of Gokstad[1] and Oseberg[2] were remarkably well preserved due to their burial in a soil with a high clay content, a wet anaerobic environment. By contrast, the Sutton Hoo ship[3] buried in well drained sandy soil, a moist aerobic environment, had completely decayed leaving only an impression and the deposition of metal fastenings to establish its form and method of construction.

Waterlogged wood often appears sound except for being darkened and superficially soft, but deterioration has altered its chemical composition and structure. The main elements in the composition of wood are: cellulose hemi-cellulose and lignin.

Table 1. Likelihood of wood preservation, buried under different environmental conditions

Environmental condition	Aerated	Anaerobic	Waterlogged	Action of fungi	Action of bacteria	Likely condition of wood
Acidic soil	X	-	-	Active	Less active	Decomposed
Acidic soil	-	X	-	Less active	Very restricted	Preservation likely
Acidic soil	-	-	X	Very restricted	Very restricted	Preservation very likely
Basic soil	X	-	-	Very active	Very active	Decomposition
Basic soil	-	X	-	Less active	Less active	Preservation possible
Bogs, alkaline	-	-	X	Minimal	Minimal except methane producing forms	Preservation likely but in very weak condition
Bogs, acidic	-	-	X	Minimal	Minimal except sulphate reducing	Preservation very likely especially those woods with high tannin content, chestnut, oak.

Cellulose is made up of pairs of glucose units to form long chain macro-molecules, these chains bond together making micro-fibrils, which, wound like a ball of string make up the layers of the wood cell wall. It is the nearly axial orientation of the fibrils of the thickest cell wall layer which contribute to the expansion and contraction of wood when exposed to different moisture gradients; for the water molecules attach themselves to the sides of the fibrils causing transverse dimensional changes. Hemi-cellulose does not possess the chain forming property of cellulose but acts as a matrix into which the cellulose fibrils are set, and has a much lower resistance to chemical or organic attack. Lignin, which is very resistant to decay, lies amongst the cellulosic fibrils and is most concentrated between the wood cells. Fungi and bacteria both produce enzymes which can break down cellulose and hemi-cellulose elements to their basic units for absorption into the micro-organisms metabolic process. Lignin is also affected but far more slowly. After attack, the cell walls are weakened and the porosity of the wood increased by enlargement of the microscopic passageways, this consequently decreases the microscopic internal surface area of the wood cell, both factors affecting the drying characteristics of waterlogged wood. Fresh wood contains "free water" within the pores for conduction of nutrients, and "bound" water within the micro-fibril structure. On drying, shrinkage does not occur during the loss of free water, but begins with the loss of bound water, that is, at a water content of about 30%, the so called "fibre saturation point". In fresh wood shrinkage varies in the three major planes: in the fibre direction, longitudinally, it is small, averaging less than 0.1%; in the tangential plane, at a tangent to the annual rings, it ranges from about 5% to 10%; and in the radial direction from 3% to 6%; these values increasing with the density of the wood. In fresh wood, loss of water during seasoning does not effect the capacity of wood to regain its green dimensions on rewetting. However, loss of water from degraded waterlogged wood causes irrevocable changes; for the water within the wood is now supporting the weakened structure and is linked with exposed terminal hydroxyl goups on the remaining cellulose chains. On evaporation, the surface tension of the water causes collapse of the fragile cell walls which cannot be restored to their original dimensions by rewetting; this is attributed to the low number of sites available for water absorption due to the decrease in internal surface area, and the joining of the exposed hydroxyl groups to each other. The wood splits longitudinally and transversely with cell collapse to become permanently distorted.

The shrinkage characteristics of waterlogged wood vary with the extent of decay, and wood in this condition can be divided into two main groups: those which are evenly degraded throughout, and those where a degraded surface covers a hard core of relatively sound wood. Evenly degraded wood collapses completely, whilst the greater shrinkage of degraded surface material around a hard core causes splitting, cupping and separation at the interface of the differentially degraded zones. Partial decay can come about by a short period of burial or a long period of burial in a favourable preservation environment. The most likely species to produce this pattern of decay is European Oak, which even in an unfavourable preservation environment shows a strong resistance to degradation. But the central hard core is not unaffected; averages of ten samples from the Swedish warship "Wasa" showed compressive strength in the direction of the grain to be 57% that of fresh wood, flexural strength 58%, hardness perpendicular to the direction of the grain 63%, and impact strength to be 26% that of fresh wood. [4]

Passive conservation

Passive conservation entails maintaining the wood in a wet condition during excavation, transportation, and pre-treatment storage, so preventing drying and shrinkage. The dependence of waterlogged wood on the water contained within its structure, dictates the form of excavation in that all work must be carried out in wet conditions. Each excavation site presents its own special problems excavation techniques being modified to circumstances. The Viking ships of Roskilde Fjord deliberately sunk as block ships, were isolated from the surrounding water of the fjord by a coffer dam which was gradually pumped dry. When the fjord bed was reached, water sprays kept the wooden remains wet while work progressed. [5] These craft were fragmented due to failure of metal fastenings and the weight of stones used to stabilise the blockade, so most of the component parts could be lifted individually.

Even when the form of a craft has been retained by infilling clay or sand, it may need to be dismantled. For existing conservation facilities and extraction equipment will limit lifting as a "whole", as will the loss of mechanical strength of the wood itself, and suction of underlying clay. A memorable exception was the raising of the Swedish warship "Wasa" from Stockholm Harbour by hawser and floating pontoons. Large sections of well preserved craft have also been raised, complete, from the reclaimed polders of the Zuider Zee. [6]

The time available for excavation also influences the method chosen for lifting wooden remains. The rescue excavation of the Graveney Boat by the National Maritime Museum and British Museum in 1970 which was completed in two weeks, did not allow time for reinforcing the shell of the boat for lifting as a whole. Each strake was separated from its neighbour and lifted with the help of marine ply strips inserted beneath each piece. [7, 8] By contrast, another land site excavation by the National Maritime Museum, that of the 6th B.C. Brigg 'Raft', demanded a different technique. [9] The raft, of six planks stitched together, with cross battens through projecting cleats, had to be raised with as little disturbance of the underlying soil as possible. It lay on blue-clay and covered by the 19th century backfill, near the junction of the present River Ancholme and the New Cut, at Brigg Lincolnshire. After mechanical clearing of the site, the remainder of the overburden was cleared manually from boards laid across the excavation trench. Exposed wood was immediately covered with strips of plastic foam saturated by periodic spraying. Opaque black polythene covers prevented solar gain during the day and water loss at night. Due to the fragmentary nature of the planks it was possible to separate them into cleat units, some 1.2 m long; some double cleats longitudinally and some adjacent cleats laterally were also lifted as one unit. Heavy duty polythene worked under each cleat enabled each section to be slid on to a slave bearer, which was then removed from the excavation trench and slid again on to a transportation pallet covered with a large polythene sheet folded to form an envelope around the wood and the layers of overlying water-saturated plastic foam. So it was possible to lift each section with the minimum disturbance of its component parts and without contaminating the underlying clay. The transport pallets served a dual purpose, being subsequently used for the underwater storage of the excavated material.

As a rule the volume of wood exceeds the capacity of treatment equipment, and storage for several years can be anticipated; this is a common problem and the use of ancilliary materials with a high water resistance is essential. Very long term storage is feasible if an artifact is kept under water; sledge shafts and wooden posts from the Oseberg [2] were stored for 50 years (1904-54) in water and formalin before a suitable method of treatment was devised.

Active Conservation

The aims of active conservation are removal of the water within the wood and, by bulking or otherwise strengthening the wood, retention of the ' as excavated ' dimensions. It is also essential to minimise the mechanical damage that can come about by the expansion and contraction of wood in response to changes in the relative humidity. When dealing with composite structures which need to be separated for treatment, the consolidant should allow for any reshaping necessary during re-assembly. It is also desirable that consolidants should be reversible to allow for re-conservation should more effective techniques be devised; and finally the wood should retain a wood-like appearance.

The means of exchanging the supporting water in waterlogged wood for a more stable supportive substance break down into two main groups:

1. Replacing the water by a liquid which by virtue of its low surface tension can be allowed to evaporate without causing the collapse of the wood structure. (e.g. alcohol-ether-resin; all freeze drying methods). Impregnation with a consolidant after drying is often necessary.

2. Replacing the water by a liquid which can be solidified in situ. (e.g. "Alum", "Acetone-Rosin" and Polyethylene Glycol PEG).

Alcohol Ether Resin

This is an extension of the standard botanical method of dehydrating organic material. Contained water is displaced by increasing concentrations of methylated spirit, which in turn is exchanged for di-ethyl ether which has an extremely low surface tension and can be allowed to evaporate without causing distortion. If the wood is highly degraded and fragile a consolidant such as dammar resin is dissolved in the final ether bath and remains within the wood cells after evaporation of the solvent. This is an excellent method, for the wood returns almost to its natural colour and tone, often revealing coatings such as tar which are indistinguishable from the deep brown black colour of wood in the waterlogged state. But it can be used only with small artifacts; the extremely low flash point and anaesthetic properties of ether preclude its use on a large scale.

Freeze Drying Methods

By freezing waterlogged wood, the water within it can be sublimed from its solid phase by warming under vacuum. Two of the most serious problems encountered are ruptures caused by initial freezing, and post drying shrinkage; both have been largely overcome by soaking the wood in a weak solution of Polyethylene Glycol 400 [10] a water soluble wax. Polyethylene Glycol has terminal hydroxyl groups which join with those terminal hydroxyl groups exposed in the cellulose chain so reducing shrinkage; also the expansion of the ice on freezing is countered by the contraction of the Polyethlene Glycol. Above 4°C the PEG melts and is absorbed by the wood. This method is an improvement on exchanging the water for tertiary butyl alcohol, which freezes at 25°C so obviating expensive refrigeration equipment, but lacks the capacity of satisfying exposed hydroxyl groups and reducing shrinkage. Good results have been obtained with freeze drying and fine detail is well preserved, but the wood is light and fragile and may suffer from fine transverse cracking. It is often necessary to consolidate after drying. Small evenly degraded wooden objects generally yield the best results.

Alum

Introducing a substance in liquid form into the wood which can be solidified in situ is an extensively used technique. From the middle of the 19th century the salt alum had been used successfully, the object being boiled in a saturated solution of the salt which drystalizes within the wood structure on cooling. Subsequent modifications in the industrial preparation of the salt involved the addition of glycerine which leads to the breakdown of the salts crystalline structure, and its use in now discontinued.[12]

Small evenly degraded wooden artifacts can be treated relatively easily. Large partially degraded timbers present considerable problems, especially the most commonly used boat building wood, European Oak, in which the marked pattern of incremental decay results in a firm "hard core" within the softer more extensively decayed surface layers. Insufficient volume of consolidant within this hard core will allow it to shrink within the strengthened outer layers causing hollowing or other distortions.

Acetone-Rosin

What actually prevents the penetration of consolidating liquids into oak wood which has a "hard core" is still not clearly understood. It is possibly due to insoluble sulphides or insoluble gallates and tannates formed from reaction of the resin acids in the wood with iron from the surrounding environment. The recently devised acetone-rosin method involves cleaning obstructions from the wood structure prior to treatment and has been sucessfully used to treat well preserved wood. Pretreatment with dilute hydrochloric acid, dissolves water insoluble salts, which can be washed away; then acetone is used to dissolve the resins and resin acids. Rosin (colophany) which is soluble in acetone proved to be the most successful consolidant, forming supporting crystals within the wood on evaporation of the acetone. A low level of shrinkage has been reported with this technique.[13]

Polyethylene Glycol (PEG)

The use of Polyethylene Glycol (PEG) for the conservation of waterlogged wood was introduced by Moren and Centerwall in 1952.[14] Polyethylene Glycol, a product of petroleum is manufactured in several grades; lower molecular weights are liquid, the higher molecular weights hard wax-like solids. It is the higher grades which have the widest application in conservation due to their strength and low hygroscopicity. Contained water within the wood is replaced by a hot (60°C) aqueous solution of PEG; by maintaining a constant volume the percentage of PEG is increased by controlled evaporation of the water. On cooling, the PEG hardens to support the cell voids.

It has been found that the molecules of the higher grades are too large to penetrate the hard core often found in European Oak. Christensen discovered during experiments for treating the Skuldelev shipwrecks, that the higher grades filtered into the decayed surface layers, surrounded the hard core and drew the water from it so causing it to collapse with consequent surface distortion. To overcome this the method of treatment was modified, depending on the classification of the wood into one of three groups. Those timbers with insignificant hard core (strakes) substantial hard core (keel, ribs); considerable hard core (stems). For wood with little hard core the basic impregnation cycle has extended from seven to twenty four months, during which time a 10% solution of PEG 4000 at 60°C was increased to 100%. Hard core speciments were treated with a cold 25% solution of PEG 4000 which was increased to 50% over twelve months. Slow drying with periodic immersion in the 50% solution increased the surface concentration within the wood to 60%. Results were described as good.[12]

Barkman in conserving the 'Wasa' also found that reducing the rate of increase in concentration improved penetration. Even better results were achieved by use of PEG 1500 which has a smaller molecular size, the increased hygroscopicity of this grade requires a carefully controlled environment to prevent 'weeping'.

Very large structures such as the 'Wasa'[4] can only be treated by spraying, a controlled environment being maintained to prevent drying during breaks in the spraying cycle. Artifacts which can be dismantled are best treated by immersion in heated baths. The latter method will be used by the National Maritime Museum for the conservation of excavated wooden craft. For the consolidation of timber with 'hard core' a slow impregnation with the medium grade PEG 1540 will be followed by impregnation with the harder PEG 4000, for increased strength and reduction in hygroscopcity. Porous wood will be treated with PEG 4000 alone, with a very slow rate of increase in concentration.

Table 2. Proposed system of impregnation with Polyethylene Glycol

Grade of PEG	Oak wood: condition	Temp. °C	Initial concentration	% (W/V) increase	Final concentration
4000	Porous	60	0.0%	1/10% daily 1 year 1/5% daily remainder	60%
4000	Resistant	60	0.0%	1/10% daily	60%
1540	Very resistant	60	0.0%	1/10% daily	60%

Despite the time required for treatment with PEG, often in excess of 24 months for resistant timbers, and the darkening of treated wood by the release of colouring matter, the use of this material has many advantages. It is non-toxic, and no hazards are involved in large scale use. Once begun, the process is easily monitored and controlled. And during post conservation re-assembly further qualities of PEG can be exploited, not only in the case of joining broken fragments, but the shaping of planking. Distortion in the ground, of the overall form of a craft, by the weight of overburden for example, can be remedied by rewarming the PEG impregnated wood until it is pliable, and then reshaping. This is an important factor in the accurate reconstruction of ancient wooden craft.

REFERENCES

1. Brøgger, A.W., and Shetelig, A., 1971. The Viking Ships. Their ancestry and evolution. - C. Hurst & Co. Ltd, London.
2. Rosenquist, A.M., 1959. "The stabilization of wood found in the Viking Ship of Oseberg" Part 1. Studies in Conservation Vol. 4 No.1: 13-21. Part 2. Studies in Conservation. Vol. 4 No. 2: 62-72.
3. Bruce-Mitford, R.L.S., 1975. The Sutton Hoo Ship-Burial. Vol. 1. British Museum Publications Ltd.
4. Barkman, L., 1975. "The preservation of the Warship Wasa" in Problems in Conservation of Waterlogged Wood. Oddy A. (ed). National Maritime Monographs and Reports, No. 16. pp. 65-105.
5. Olsen, O., and Crumlin-Pederson, O., 1967. "The Skuldelev Ships". Acta Archaeologia. Vol. 38: 73-174. Copenhagen.
6. Van Der Heide, G.D., 1974. Scheepsarcheologie Scheepsopyravingen in Nederland en elders in de wereld. Strengholt Hotstede Oud-Bussen, Flevoland, 41, Noarden.
7. Evans, A.C., and Fenwick, V.H., 1971. "The Graveney Boat" Antiquity 45: 89-96.
8. Oddy, W.A., and Van Geersdaele, P.C., 1972. "The Recovery of the Graveney Boat". Studies in Conservation 17 No. 1: 30-38.
9. McGrail, S., 1975. "The Brigg Raft re-excavated" Lincs History and Archaeology 10: 5-15.
10. Ambrose, W.R., 1971. "Freeze drying of swamp degraded wood". Conservation of Wooden Objects (2nd Ed) pp. 53-57. The International Institute for Conservation of Historic and Artistic Works, London.
11. Rosenquist, A.M., 1975. "Experiments on the conservation of waterlogged wood and leather". in Problems in Conservation of Waterlogged Wood. Oddy, W.A. (Ed.) National Maritime Museum Monographs and Reports. No. 16: 9-24.

12. Christensen, B.B., 1970. <u>The Conservation of Waterlogged Wood in The National Museum of Denmark</u>. National Museum of Denmark, Copenhagen.

13. McKerrell, H., Roger, E., and Versanyi, A., 1972. "The Acetone/Rosin method for conservation of waterlogged wood". <u>Studies in Conservation</u> 17: 111-125.

14. Centerwall, B., and Moren, B., 'The Use of Polyglycols in the stabilization and preservation of wood.' Modo Kemi (UK) Ltd, 25 Haymarket, W.C. 1. <u>Technical Information Leaflet, No. 64</u>.

7. NEW POSSIBILITIES IN CHEMICAL ENGINEERING FOR CONSERVING STRUCTURES OF WATERLOGGED WOOD

J. Roelfzema

Summary

In the FOIL COCOON system investigated, the whole wood construction is wrapped twice: first in a porous layer of carefully selected material, about half inch thick; next in a sheet, built of polypropylene strips, about four feet wide, carefully connected by an air tight sealing system.

The top of the inside of this wrapping system, is connected to a waterjet vacuum pump system, maintaining automatically about 80% vacuum, thus pressing both the wrapping-stiff-porus against the wood and the outside layers against the inside ones.

Through the foil, the porous layer is then connected, to the store of the appropriate chemical liquid kept at 60°C. This liquid is circulated by a pump.

Satisfactory results obtained so far; for further investigation more man power and money would be needed.

A few engineering students from the Amsterdam Technical College accepted two years ago as a baccalaureous project, the design and construction on a trial installation for an engineering method for the preservation of large parts of waterlogged wood. This method may be applicable to wooden ship constuctions or, may be, even whole ships. Without considering the chemical process chosen, the aim was to investigate the possibility of submerging large pieces of timber in a heated liquid, without constructing a tank or container.

The investigation was supervised by the "Centraal Laboratorium" Amsterdam (Dir. Ir. J. Lodewijks). The project was sponsored by the Dutch Ministry (CRM) and Royal Dutch Shell Oil Co. The work of the students, after finishing their initial part of the project, was taken over by members of the institute for medical revalidation under Dr. F.J.M. Heslinga (Vrije Universiteit Amsterdam). Active help was received from the staff of the Amsterdam Maritime Museum; chemical industries, Phillips Duphar, DSM a.o.

The aim of the investigation was to achieve submersion by building around the wooden construction, two layers of flexible material. The first, with a thickness of about half an inch, having stiffness in the direction perpendicular to the surface of the layer, but being porous in the directions parallel to the surface: Fig. 7.1.

The second layer covering the first, enclosed the whole construction in an airtight foil, built from long sheets about three feet wide. Like the inside surfaces of housewalls covered with wall paper, the whole ship construction is now wrapped in the two layers, forming a "FOIL COCOON SYSTEM" both inside and outside; Fig. 7.1.

On the top of the construction, through the plastic foil, the porous layer is then connected (line 1) with a system of waterjet vacuum pumps, discharging the air and waterdamp from the porous layer, hence pressing both layers on the wood by the atmospheric air pressure, leaving the thickness of the porous layer intact due to its vertical stiffness. Next, other outside connections (lines 2) to the porous layer are made through the outside wrapping foil. Two on the bottom of the construction, leading to a fluid store tank (with heating coil) hence allowing the fluid to enter the porous layer when vacuum is built up on the top. The other connections (lines 2) are on top of the structure, working as a fluid discharge, hence introducing a top level of the fluid, to be maintained by the amount of vacuum realised by the waterjet pumps, which are automatically controlled by a level indicator and switch control. The top discharge connection leading back to the store tank, and a circulation pump installed in this circuit, makes fluid circulation possible together with the thermostatically controlled temperature level.

Waterdamp or drops are collected in a tank containing circulation water for the pressure pump feeding the waterjet vacuum pumps, making the whole system basically a closed circuit; see Flow Sheet Fig. 7.2.

Photographs in Fig. 4 give an impression of the installation with measuring and control equipment. An engineering scheme based on these principles calls for some remarks describing its limitation.

1. Vapour temperature under vacuum drops, since the water in the circulator starts boiling. Fig. 7.3 shows the drop of boiling temperature against pressure above water level, and shows that at about 80% vacuum, (0.2 atm. pressure) water boils at about 60 °C. This means, that if that temperature is chosen as acceptable for the liquid for diffusion into the waterlogged wood, the height of the cocoon-wrapped structure never can be higher than 80% of 33 feet, making about 26 feet. If higher, the plastic foil on the bottom of the structure would not be suctioned against the structure, undergoing pressure from the liquid column above, and collapse. There seem to be possibilities in the application of a "water lock device", possibly enabling several heights of 26 feet above each other, but experiments in this field have not yet been done, so for the time being, a Maximum column height of 26 feet exists, although the temperature of 60°C seems to be in accordance with the wishes of chemists, regarding preferred diffusion temperature.

2. Air leaking and foil sealing accuracy.

Much attention was given to this problem, being of crucial importance for the whole system. We found good possibilities in selected sealing systems of polypropylene foil, the latter maintaining sufficient strength at the temperature demanded. On the other hand, the stiff porous layer underneath the plastic foil should be smooth enough to avoid possible leaks penetrating the outside foil. After elaborate search, useful materials appeared, allowing the vast three dimensional distortion necessary.

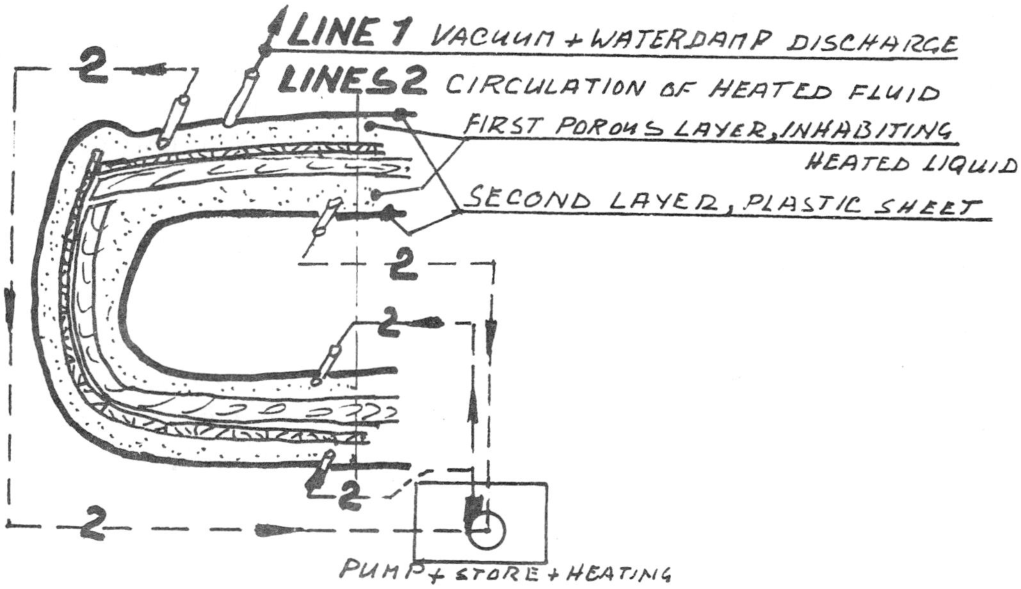

Fig. 7.1 Principles of the "Vacuum Foil Cocoon" wrapping system with heated fuel circulation.

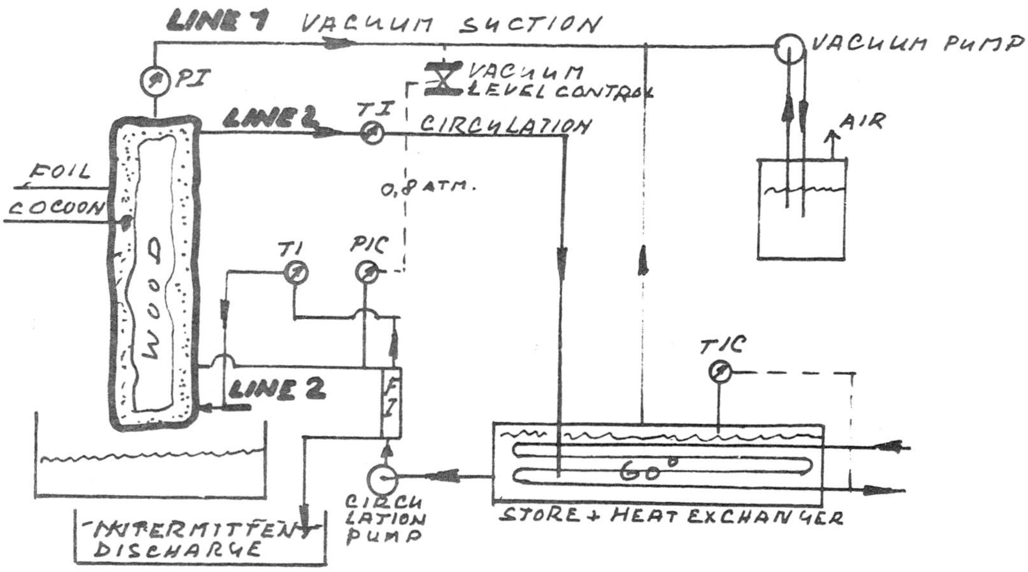

Fig. 7.2 Flow sheet of the pilot plant.

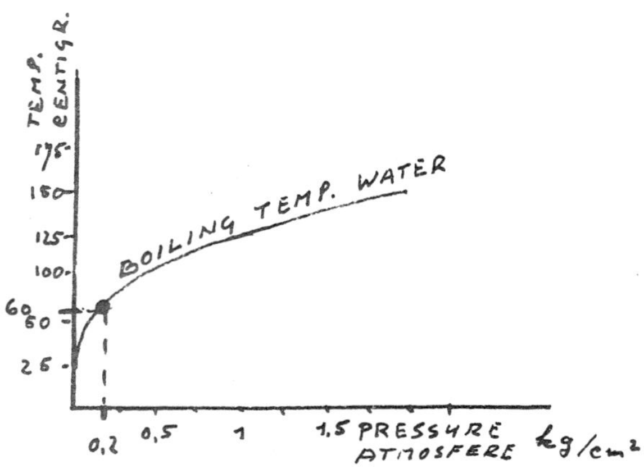

Fig. 7.3 The boiling point of water related to atmospheric pressure.

Fig. 7.4 The Vacuum cocoon testing installation.

3. A good control system to maintain the vacuum is needed, otherwise collapse of the foil would occur, causing liquid leakage. A mercury level pressure meter, with variable electric contact was used.

4. Fluid circulation through the porous layer should be equally divided all over the surface of the layer. As only slow circulation is needed, "dead corners" will be inevitable, causing local temperature drop. This was overcome by emptying the whole system at intervals, letting all the liquid into a tank underneath the level of the structure. As no special chemistry was considered in our experiments, water was used as circulating fluid.

5. Building a cocoon around parts or structures will not always be possible. If it is, amongst the advantages are: "large" even "very large" parts can be submerged; distortion through shrinkage can be limited, when several parts in a structure support each other; in many cases a cocoon might be highly cost-saving by avoiding large tanks with vast quantities of (possibly dangerous) liquids.

CONCLUSION

Within the restrictions mentioned, the problems around the foil cocoon system appear to have acceptable solutions. Further investigation in the direct practical direction would be useful; the enthusiasm and inventivity of the mainly voluntary research workers, is however overwhelming the financial resources. Consideration of this method by those who are interested in the preservation of big waterlogged constructions could lead to more coordination and possibilities.

DISCUSSION

Oddy: Can bacteria be used to increase the porosity of wood, by controlled attack on the cell structure to allow PEG or other consolidants to penetrate more readily?

Levy: For sap wood, yes; for heart wood, as far as I know, the answer is no.

Turner: Does the type of burial environment determine the preservation of wood.

Levy: Yes, it is the major factor. Where the burial conditions are dry you get very sound preservation. In wet aerated conditions you can get active fungal decay; in anaerobic conditions where the wood is completely waterlogged and at some depth then there is very little decay.

de Jong: We have a number of wrecks found just after pumping out the water from the Zuider Zee polders which were in a very sound condition. Ten to fifteen years later, when we were able to excavate the ships, we found that aeration of the soil and the lowering of the ground water table caused the sections of wreck above the water table to be completely destroyed. In anaerobic conditions you can expect the wood the preserve fairly well, but when it becomes aerobic degredation is speeded up enormously.

Crumlin-Pedersen: Our problem is that European oak was probably the best shipbuildings material which was at hand for centuries, and thus, whereas there are very good methods developed for other species of wood where you can get penetration, with European oak there is no penetration except for the first centimetres of the surface. Today we still have a number of important ships built mainly of oak awaiting conservation. Does Dr. de Jong have any cost figures comparing conservation by PEG with conservations by butanol? In round terms our costs were £5 per litre volume of wood for PEG and £7 for butanol. Secondly, we have developed our methods on the basis of the north European climate, but even here we require proper air conditioning systems when we display conserved timber. In tropical conditions this requirement is too costly to meet. Could we find another element to put into the timber which would not require all that air conditioning?

de Jong: The PEG method costs about 2,000 Dutch guilders, that is £450 per cubic metre of wood treated. We only have experience on a limited laboratory scale with the tertiary butyl alcohol method which we think will be 50 per cent more expensive. One should first investigate the condition of the wood and then decide upon the specific conservation needed. For example with the Zuiderzee timber the result of controlled drying can be fairly good for Class III timber, and that treatment involves a lot less money.

Oddy: It had been suggested elsewhere that reservoirs and lakes could be used to store excavated waterlogged material, but there is the problem that you are changing the conditions, the water in these lakes will be aerated and thus you will get decay unless you can add chemicals to prevent the growth of micro-organisms.

As for the control of the atmosphere in display and storage areas, people have often said that the lower the molecular weight of polyethylene glycol the better the conservation because you get better penetretion. But then there is the problem that as you lower the molecular weight of PEG it becomes more and more hygroscopic so air conditioning will be required in future.

Levy: Getting the water out without destruction of the material is the major problem. If you can do that you might be able to use, say, a commercial impregnation plant to put in one of the organic solvent type wood preservatives in which would kill off any fungi that remain, but although it would impregnate what wood remains it would not be miscible with water; PEG is miscible with water and you can replace the water slowly. Anything that involves handling or expensive chemicals is going to put the price up; just holding a boat in a lake or a reservoir for some time puts a cost on it, because it has got to be transported there and taken out and examined from time to time. Classifying your timber as suggested by Dr. de Jong so that you use the appropriate method would seem to be a way of minimising costs. Oak is a major difficulty, it has a very broad ray which is not strong and collapses, and sometimes the main water transport system in the tree may get plugged with ingrowth into the vessels so that the permeability is very much less; this particularly applies to the timber used for shipbuilding because it is probably chosen specially for its impermeability. That increases the problem of removal of water or getting any other material into the wood.

de Jong: We did several experiments trying to exchange the water for organic liquids but we found that in many cases there was a shrinkage of the wood which was too much to be acceptable. What shrinkages do others think can be accepted during conservation, because the wood when excavated will be in an expanded condition?

Johnston: Can we find a way, where we work from the inside out by injection of some sort, without destroying the fibres that have been damaged inside?

Levy: To get to the inside in that sense you would have to drill into a plank, and put some diffusable material in, and in wood preservation for instance, this is a method of treating a beam in situ. The big problem for archaeological conservation is how are you going to get your material inside for it to diffuse through. There are sap displacement methods, probably not feasible with oak, where the sap in a tree trunk is displaced by a head of pressure at the butt end pushing the preservative fluid in and the sap out. If you can get the plank and orientate the direction correctly, so that you have got the flow right, you might be able to replace the water that way in the sap wood, but you would only be able to do that I should think with Dr. de Jong's class 3 timber and perhaps some of the class 2.

de Jong: With class 3 oak transport of liquid within the timber is very slow. When using water and water dissoved chemicals holes would have to be drilled very close together; for example to complete within 2 years you would need holes every 6 cm.

Oddy: On a small scale we can conserve European Oak successfully by either the alcohol/ether/resin method or by McKerrell's recent adaptation of this using acetone for dehydration, and then resin dissolved in acetone for impregnation. There are serious problems in using this on anything bigger than a laboratory scale. McKerrel is trying to do this but it may well be expensive and the safety procedures and restriction which would be placed on a plant of this sort would probably be outside the capabilities of most Museums.

Crumlin-Pedersen: Dr. de Jong asked what shrinkage was acceptable to archaeologists. For edge joined boats the shrinkage should be less than 2 per cent. This is less than the natural shrinkage of fresh wood, but Viking ships for example were built of Green Oak, recently felled, where the moisture content is above fibre saturation point. There is no extra swelling from the extra amount of water in waterlogged wood of this type. As long as you stay above the saturation point of the fibres you are at the same dimensions as at the moment of building the vessel and therefore the condition you excavate it in is the condition that you want to maintain during conservation. The larger non/edged-joined ships have frames which do not tend to shrink longitudinally, and as the planking shrinks so gaps are left between the planks. Aesthetic considerations could determine how much shrinkage can be accepted here.

For the workability of timber after conservation. Methods should be developed to allow archaeologists to put the pieces together and re-shape them. Our recent experiences with the butanol glycol method is that it gives in general a very fine result with less shrinkage than other methods, but it is hard to re-shape the wood, even if we put it in a bath of hot glycol we cannot bend it back. This problem in fact affected a most useful piece of the Skuldelev ships, the stem of Ship No. 3, so we cannot replace it in the ship.

Varrall: Generally you find that all three of Dr. de Jong's classes of timber degradation are present in one piece of timber. The system used to conserve the hard inner core may not be the best to preserve the outer surface detail where the wood will be soft.

de Jong: The classification is useful to enable each piece of timber to be given its appropriate treatment, and thus one is able to reduce the conservation costs.

McGrail: Are the water contents and the shrinkage for, say, the Bremen Cog or for the Zwammerdam boats in Dr. de Jong's Table 1 a mean value for the whole ship or are they for one specific timber in that ship?

de Jong: They are mean values of 4 to 7 samples which were dried experimentally.

Wright: What is the present state of the art of conservation, and are we optimistic or pesimistic?

de Jong: We are rather optimistic about possibilities of conserving large structures or large amounts of timber, but there are many problems to be solved.

Levy: I don't see any reason for pessimism at this stage. It depends on the number of people, the amount of money, the intellectual capacity of the people working on it, and above all luck. But recording before conservation is essential, even if we do get accurate conservation you must still record dimensions and measurements and the conditions under which the find was originally discovered.

Editorial Note:

National Maritime Museum Monograph No. 16 is relevant to this discussion:

Oddy, W.A., (ed.) 1975, Problems in the Conservation of Waterlogged Wood.

8. ETHNOGRAPHY AND LIVING TRADITION

Olof Hasslöf

Summary

This paper considers problems concerning different methods and kinds of source material used to explain the development of the ship-building technique. Four chief methods have developed. First: Investigations in archives and literature, that is written sources. Second: Iconographical analysis of pictures, sculptures and the like. Third: Archaeological excavations and underwater explorations of finds from the earth and the sea bed. Fourth: Ethnographical field inquiries about ships and shipbuilders in their workshops.

It stresses the importance of the ethnographical source material, which is living in itself and so animates even the silent objects, the meagre texts and the mysterious pictures from long ago.

Attention is called to the concept of living tradition and it is emphasized that it must be scrutinized with regard to the six different means of communication - or tradition media, through which evidence of people and their activities are transmitted through time, space and social groups.

The Organizing Committee for this Symposium has invited me to give an introductory lecture on Ethnography and Living Tradition. This invitation is, I take it, a sign of interest in the viewpoints on source material and research methods which our Scandinavian maritime working group has set out in the book Ships and Shipyards, Sailors and Fishermen. I should therefore like to take as my starting point the papers published in this book.

As the terms archaeology, ethnography and tradition are interpreted in rather different ways by different groups, I shall begin with a few remarks on the meaning we place on these terms.

Archaeology comes from the Greek word archaios, old, and logos, study. So it means the study of the past. In modern terminology it comprises research into the development of human culture as revealed by remains excavated from the earth or salvaged from the sea bed. Thus archaeology uses quite different source material from that studied in historical research, which follows cultural developments mainly or exclusively through written records.

Ethnography, from the Greek ethnos, people, and graphein, to write, strictly speaking means the description of peoples. In the International Dictionary of Regional European Ethnology and Folklore it is defined as "the description of peoples, especially primitive peoples, a description based on observation and noting down of cultural facts in the field".

So the archaeologist digs in the ground or dives to the bottom of the sea to find his source material, and the historian finds his material in archives and libraries.

The ethnographer, however, goes to the people themselves in their natural environments, on land and at sea, in villages and towns, at home and in their families, at their places of work and their businesses, in their team groups or institutions, in order to collect and record his source material. He uses a variety of methods and aids. He takes notes, he makes tape recordings, he measures, draws, photgraphs and films. He is given presents and buys objects and documentary material for ethnographical museums and research institutes. In short, he uses all the means of communication, all the tradition media which people use in their daily work and their social life. His finds are of many different kinds, but in spite of this diversity the varied source material is commonly placed under the heading of traditions. Naturally, lumping everything together under one heading is hardly satisfactory when the source material is to be used as a basis for scientific historical and sociological research.

There have been long discussions aimed at clarifying and defining more exactly the nature of tradition. In the dictionary from which I quoted, which was published in 1960, there is a collection of the most important statements on this subject. There is also the comment that "some authorities understand ethnography as a sub-discipline under anthropology or ethnology". In fact these two terms are the names of fields of research more or less similar to ethnography. But there is one difference, namely that they emphasise the latter part of the word, logos, study, rather than graphein, to describe. This underlines the fact that they not only describe phenomena but also, and above all, interpret and explain them by comparative studies and the analysis of traditions.

The differences between anthropology and ethnology are not such that we need to examine them here and now. It is sufficient to refer the student to the bibliography listed in the dictionary.

Since 1960 there has been a continuing debate on this question. Our team has to the best of its ability contributed a number of papers whose fundamental principles are outlined in our book under the title The concept of living tradition. I must of necessity restrict myself here to certain main points and for the rest refer you to the article itself.

The word tradition comes from the Latin verb tradere, which means to hand over or transmit. On this basic meaning we can formulate a definition that covers three main points which are vital to our discussion. They are the contents of tradition, the bearers of tradition and the tradition media. The definition is as follows:

1. What is handed over consists of evidence about human beings, their ways of living and their activities.

2. The handing over - the tradition - takes place between individuals, groups

and institutions; between districts and countries, across continents and oceans; between generations, over decades and centuries.

3. The evidence is transferred by means of various kinds of communication - tradition media - such as the spoken and written word, pictures and sculptures, actions and behaviour, objects and equipment (artifacts), social groups and institutions.

The tradition media are by nature very dissimilar. Each and every one of them can for this reason pass on - transmit - only that kind of evidence which is suited to its nature. Because of this we have to distinguish at least six different kinds of trandition. They are: 1 Oral, 2 Written, 3 Iconographical, 4 Manual or visual-motor, 5 Objects, and 6 Social or Institutional tradition.

In the article The concept of living tradition I have described each of these types of tradition in detail and as time is short I must refer you to this analysis. I also emphasize in my article a fact that we are all aware of, namely that those who pass on traditions master to varying extents and degrees the various forms of expression, that is to say, the tradition media. Another point we all acknowledge is that every person who transmits a tradition, whether consciously or unconsciously, passes on only what he himself finds worthy of transmission or what it is within his intersts to transmit. Evidence about people and their activities will therefore tend to be sifted and coloured in many ways, depending on the varying kind of trandition and on those who pass it on.

On top of this there is the influence that the interpreters - that is, the research workers and writers in the fields of history, culture and sociology - impose on the evidence they collect. As E.H. Carr, Lynn White and other modern scholars have pointed out, interpreters have for many years made use of one-sided, arbitrarily chosen material which they have described and interpreted in line with prevailing doctrines, and in doing so have used terms whose meaning has not been clearly defined.

As we have come to an understanding of this problem, and as new technical aids have been invented, more and more kinds of source material have been brought into use. Archaeologists brought in material things to add to their written documents. Ethnographers turned their attention to living traditions.

On a closer examination these prove to be of widely differing kinds, not only in form but also in content. They also differ in their distribution both among various primitive peoples and among various groups in our own society.

I suppose that language is the tradition medium that most clearly divides us into different tradition groups. As you can all hear, English is not my mother tongue, and I sometimes find it difficult to make myself understood in every detail in an English-speaking community. The same is true when people from different social classes or occupations have to communicate with each other. Let me illustrate this point with a story told by the famous Swedish writer Albert Engström. It comes from the islands outside Stockholm - the Stockholme skärgard, as it is called. A fisherman there took in a paying guest for the summer. This guest was a landlubber, but he wanted to taste the joys of the ocean wave, so he decided to take sailing lessons. The fisherman began the first lesson: "Supposing I'm on a starboard tack, and want to ...." "Stop. Stop. What on earth are starboard tacks?" "Hm.. If you're so bloody uneducated you don't know what a starboard tack is, there's no point in talking to you." And that was the end of that lesson.

In the thirties those of us who were working at the Gothenburg Museum were often faced with problems of this kind. During the summers we were out on archaeological and maritime ethnological field expeditions in the province of Bohuslän and the surrounding districts along the Skagerrak and the North Sea. Then we spent the winter analysing the material we had collected and researching in archives and libraries. We soon discovered that we had to communicate with our friends out in the field with different tradition media from those we used with our learned colleagues in the world of books. We found, too, that living, oral and manual traditions were best for explaining the fundamental principles of the boat and shipbuilders, their procedures, methods of contruction and ways of thinking. And they varied enormously between the large and the small shipyards, and between the clinker and the carvel builders. What was self-evident gospel truth among the former could not even be thought of among the latter. In fact it was just these contradictions that gave us our most salutary lessons in the art of understanding and interpreting the evidence of objects, drawings and other tradition media. It was these contradictions that drew our attention to the heart of the problem in shipbuilding research, the fundamental differences between shell and skeleton constructions, between clinker and carvel building.

These differences have, of course, been mentioned in a few books on ship-building, but they have not been described in detail and have certainly never been accurately defined.

We published some of our experience and ideas in a few short preliminary essays in Swedish journals. A couple of these were translated into English in Folkliv, 1958 and Mariners Mirror, 1963 and attracted the attention of international boat archaeologists who, with the help of new technical aids, had succeeded in investigating a whole series of wrecked ships. Their finds helped to throw new light on fields which had previously been sadly neglected.

In a valuable article on Ancient wrecks and the archaeology of ships in The International Journal of Nautical Archaeology and Underwater Exploration, 1972, Mr. Lucien Basch has presented a lucid survey of the research into the history of ship-building, principally in the Mediterranean area during antiquity. He includes a welcome discussion of the concepts of shell and skeleton construction, clinker and carvel building, which most scholars and writers up to then had treated far too superficially. It is important to get these terms even more precisely defined. Fortunately the discussion has been taken up by Mr. Arne-Emil Christensen in a most readable article in the same journal in 1973.

I have no time here, unfortunately, to go deeper into these problems, but I would mention that I have touched upon them in several earlier essays, and more recently in the article Main Principles in the Technology of Ship-Building, which Mr. Basch has evidently not read. There are, in fact, many problems involved, such as various methods of construction and their variants and hybrid forms; the use of different methods of fastening the planks together, to the frame and to the keel, the use of models, templates, instruments, drawings and other aids; not to speak of the various kinds of tools, equipment, customs and business organisation, all of which play very important roles. Other speakers, in this symposium I am sure, will deal with such matters in other contexts.

My function at this conference is mainly to remind you of the part that ethnography and living traditions can play in maritime archaeological research. I have pointed out that our team has listed six such traditions, and that those which are transmitted orally, manually or by visual-motor means particularly warrant the description living and deserve specially close study. May I also point out again that it is these traditions that have provided the clearest answers to the problems that Mr. Basch has raised. But they do have one weakness, a considerable one. The spoken word dies rapidly. Hand movements and patterns of behaviour also disappear. But they are repeated in more or less stereotype forms. And what is more, they form the basis of or are manifested, become tangible and take concrete form in objects and things - shipwrecks, drawings, written documents, organisations and institutions and so on. And these live on through the centuries. The ephemeral words, movements and turns of the hand are preserved in them. Dead and voiceless things bear witness of their creators, and how they spoke and worked, lived and believed. They, too, are the bearers of living traditions, fixing them in time and space. They provide us with fixed points for our research and accounts of cultural development, in which the art of ship-building has played a highly significant role.

Fig. 8.1 Shell construction. Clinker building with overlapping planks.

a. The keel and the lower part of the stem are hewn out of the root end of a spruce.
b. The keel is roughly outlined in the forest.
c. The keel is shaped with an axe in the workshop.
d. The upper part of the stem is nailed to the lower part (in one part with the keel).
e. The first plank is put in position.
f. The second plank is put in position.
g. The third plank is put in position.
h. All the planks are in position. The boat shell is 'bordfyllt' (completed).
i. Wood for frames is chosen in the forest.
j. A frame is hewn in the workshop.
k. A frame is fitted to the hull.

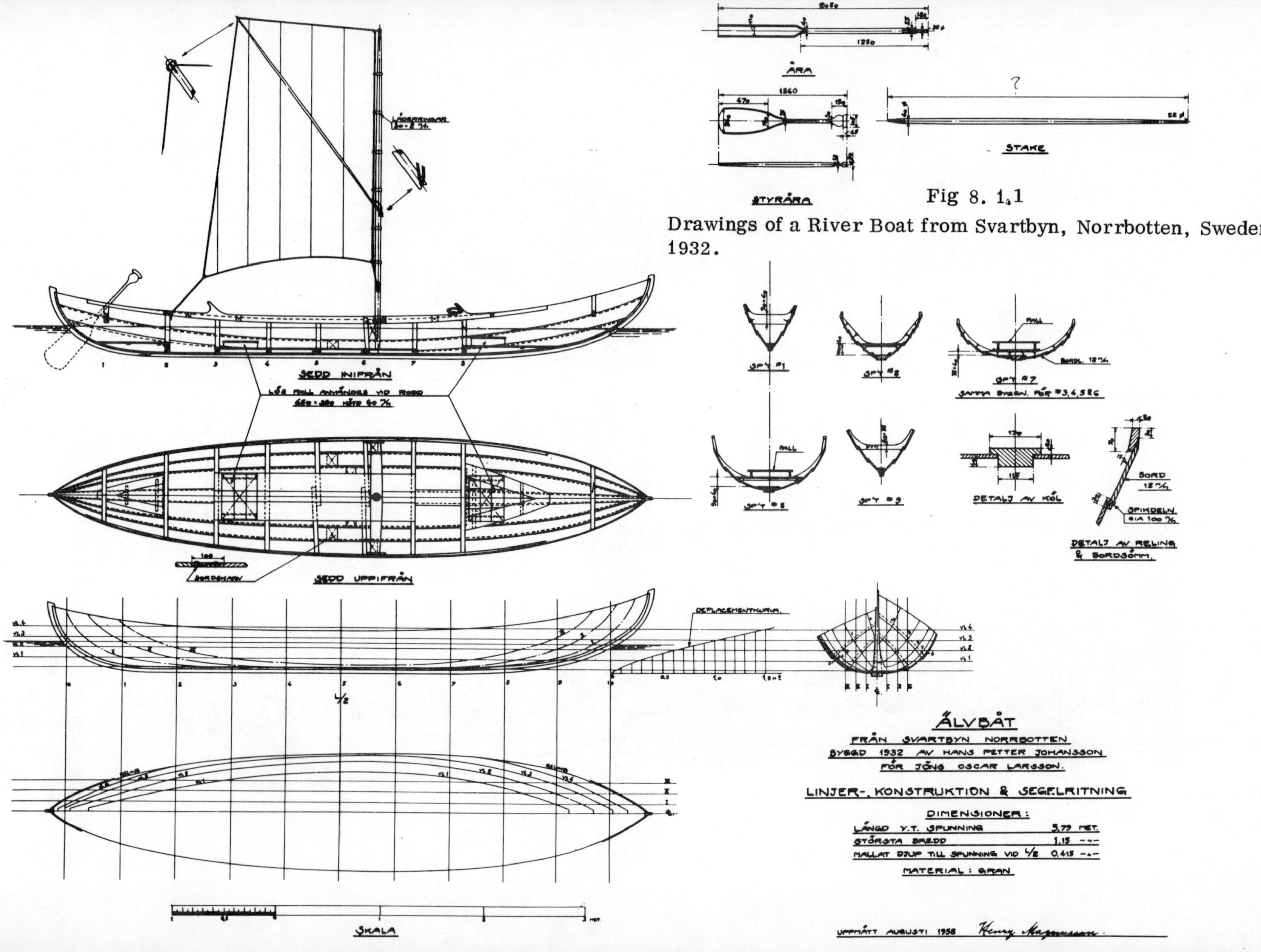

Fig 8. 1.1

Drawings of a River Boat from Svartbyn, Norrbotten, Sweden 1932.

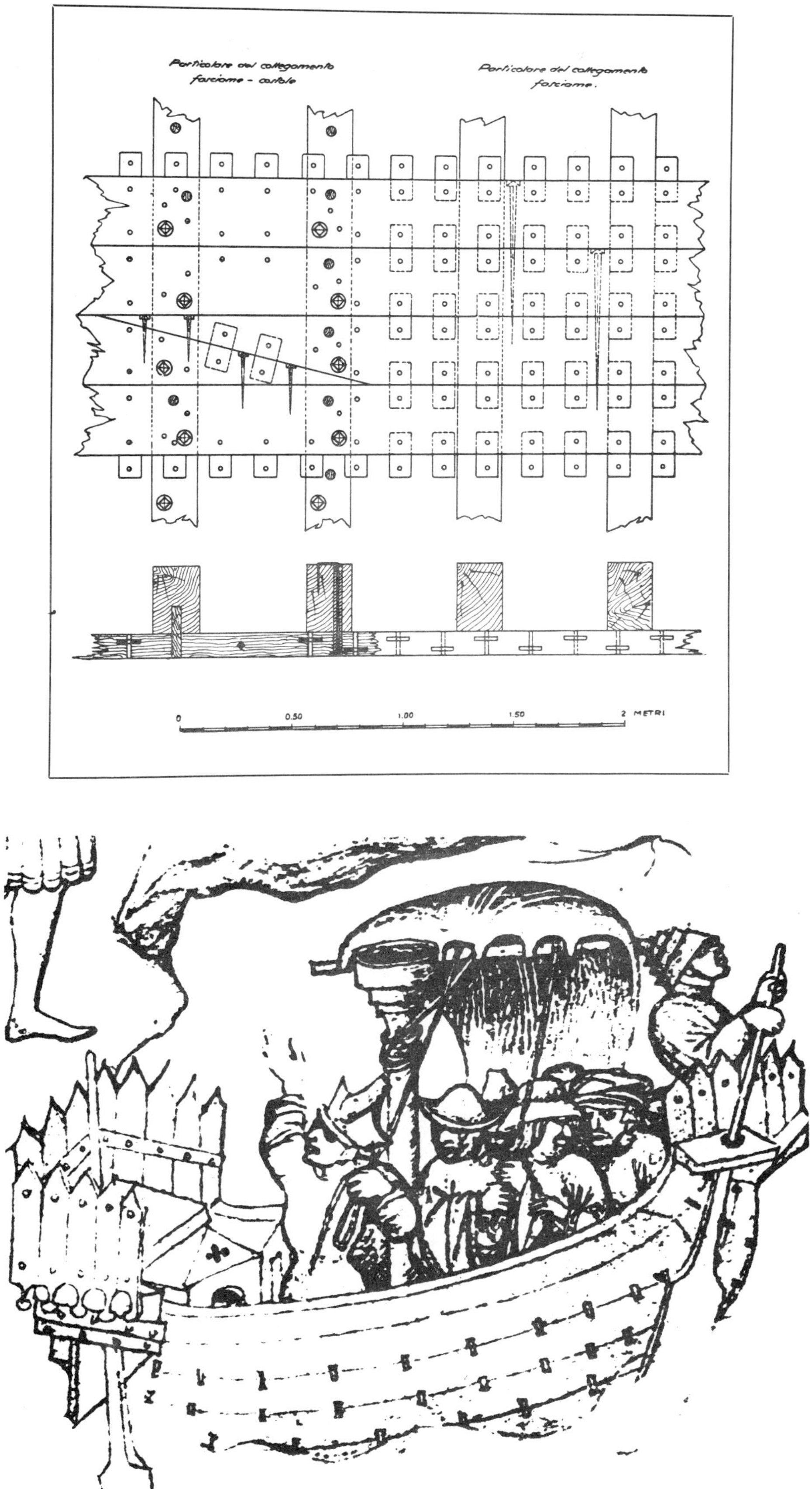

Fig. 8.2 Shell construction. Boats built with smooth sides, planks joined edge to edge.

a. Planking with tenons and mortises. Nemi ships First century A.D. After Ucelli.

b. Planking with dovetail tenons. Flemish Manuscript 1410. British Museum.

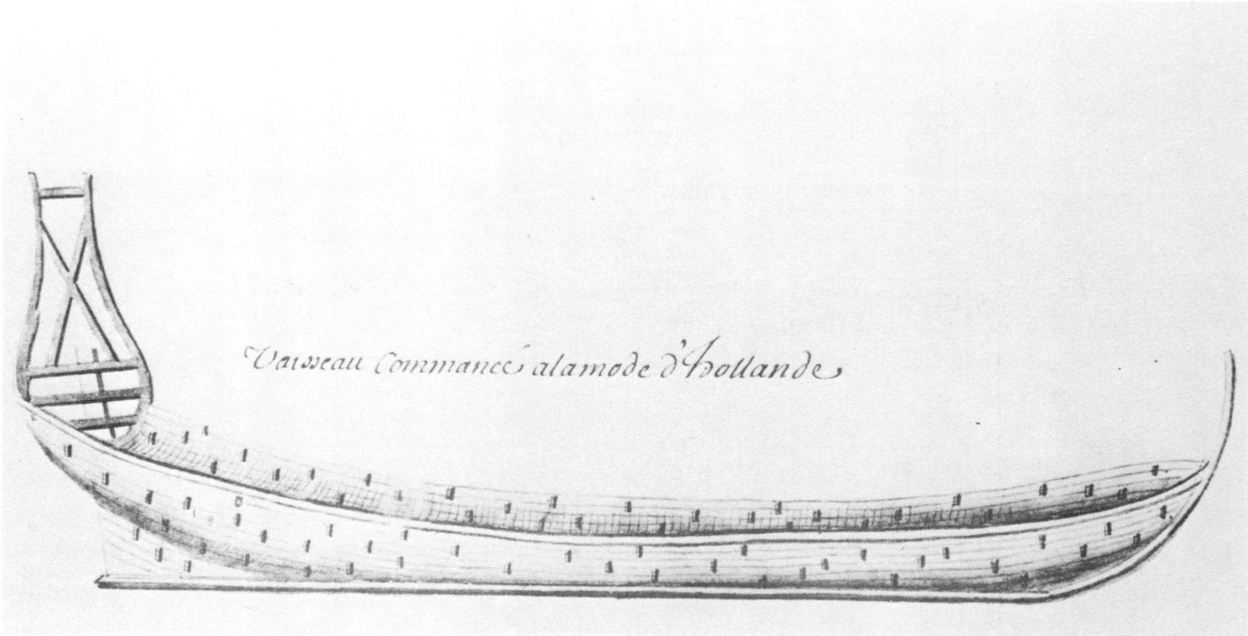

Fig. 8.2 Shell construction. Boats built with smooth sides, planks joined edge to edge.

c. Planking temporarily held together with clamps. Amsterdam 1670. Drawing by Sieur Arnould. Bibl. Nat. Paris.

d. Frisian yawl in Stavoren 1955. Zuidersee Museum.

Fig. 8.3 Shell and Skeleton Construction combined.
The hull below the waterline is clinker built. Above the waterline it is carvel built on a rib-frame. A galeas under construction 1938. Kongsviken, Bohuslän, Sweden.

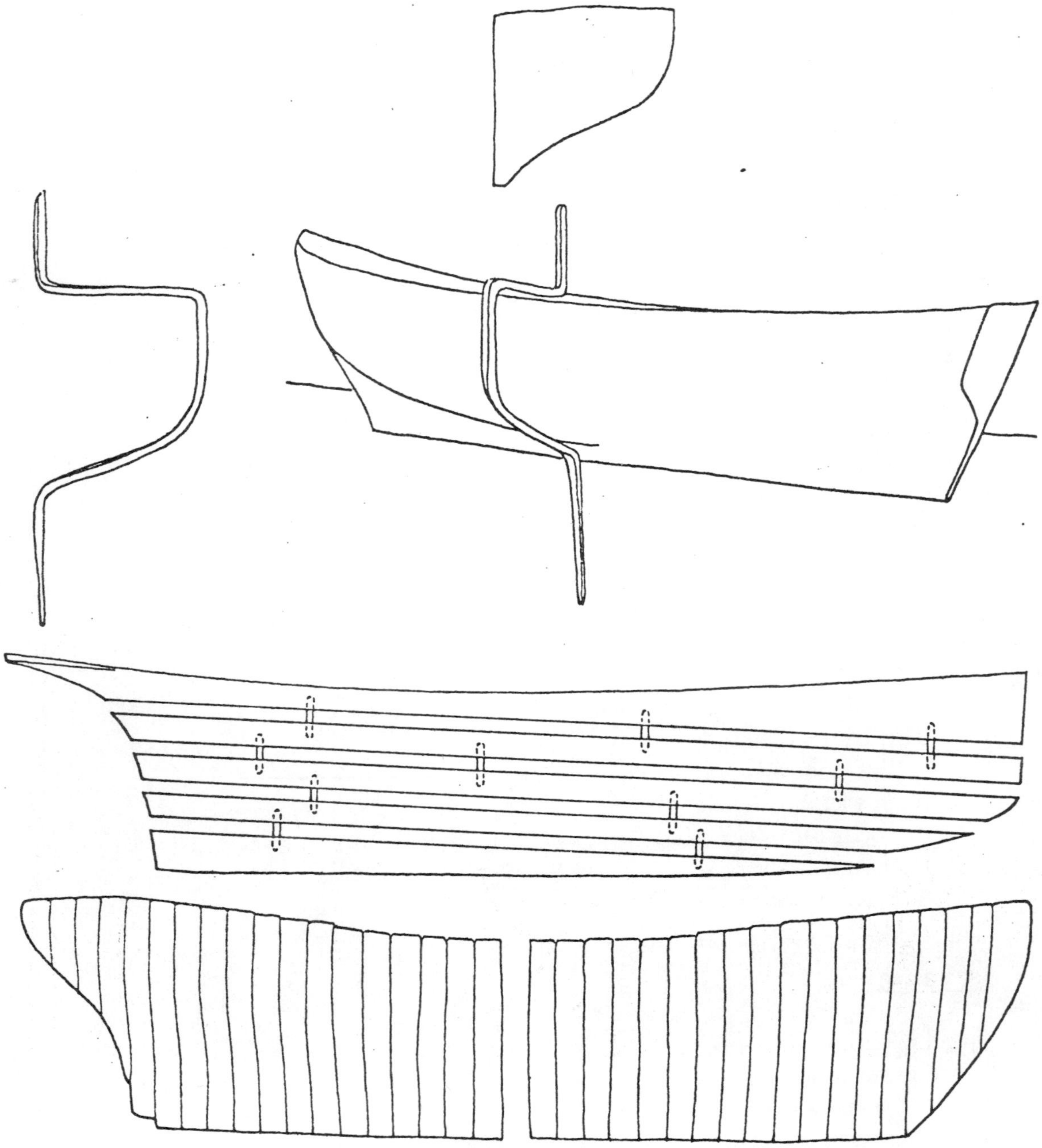

Fig. 8.4 Skeleton construction. - Three types of half model.
Most of the shipbuilders who built the middle-sized and smaller sailing vessels could not make ships drawings. They built their carvel ships from block models, which the shipwright made in collaboration with experienced seamen incorporating alterations and additions they considered advantageous. Some shipwrights cut such a model out of a solid block and sawed it vertically into the same number of slices as there were to be construction ribs. The ribs are then modelled on these slices. Instead of sawing the model up, another method is to use a light lead rod to make impressions that will give the shape of the ribs. Other builders pin together into a block horizontal layers of wood whose thickness is shaped to the waterlines of the planned ship. The half-model is then cut out of the pinned block and the slices are taken out and measured to give the shapes and sizes of the actual ribs.

9. THE XAVEGA BOAT
A CASE STUDY IN THE INTEGRATION OF ARCHAEOLOGICAL AND ETHNOLOGICAL DATA

Octavio Lixa Filgueiras

Summary

Division of Portugal by the Douro:

- to the North, early Scandinavian/Germanic (river craft) and Breton (inshore fisheries) influences;
- to the South, Mediterranean influences.

The Saveiro (the most significant boat in the South region, fishing with the xavega seine) as main vector of the oldest Mediterranean influence:

- Mesopotamian origin - striking identity (form and technical) with ancient Ur craft still surviving in the Lower Euphrates,
- The Ur to Ugarit (via Euphrates, Aleppo, Alalakh ?) trade routes as possible vehicle to the Mediterranean,
- Iconographic evidence (Cretan seals, Hagia Triada painting, possibly Thera murals) of its presence in the Aegean,
- Documented westwards progression,
- Survival of similar craft in Italy.

The problem of the ethnic origin of the communities fishing with the saveiro:

- Their peculiarity,
- Co-incidence of the areas of distribution of these boats and the main areas of refuge of South Iberian peoples after the fall of Tartessos,
- The open question of Cretan influence in Tartessos.
- The above proposed sources for the saveiro as a new factor for the re-evaluation of these arguments.

Due, perhaps, to the different environmental conditions prevailing North and South of the Douro, this river has traditionally marked the border between different ethnic or tribal groups and contrasting cultures: Galicians/ Lusitanians, Bracari/Turduli,[1] Christians/Muslems, ...

Such a division is still apparent today in the distribution of our vernacular boats,[2] but Strabo had already reported the navigation of the Douro by leather craft[2] (the use of such craft is often invoked as an important criterion in distinguishing North European from Mediterranean cultures) bringing to mind the possible connections between Galicians and Oestrimnii, of whom Avienus sang:

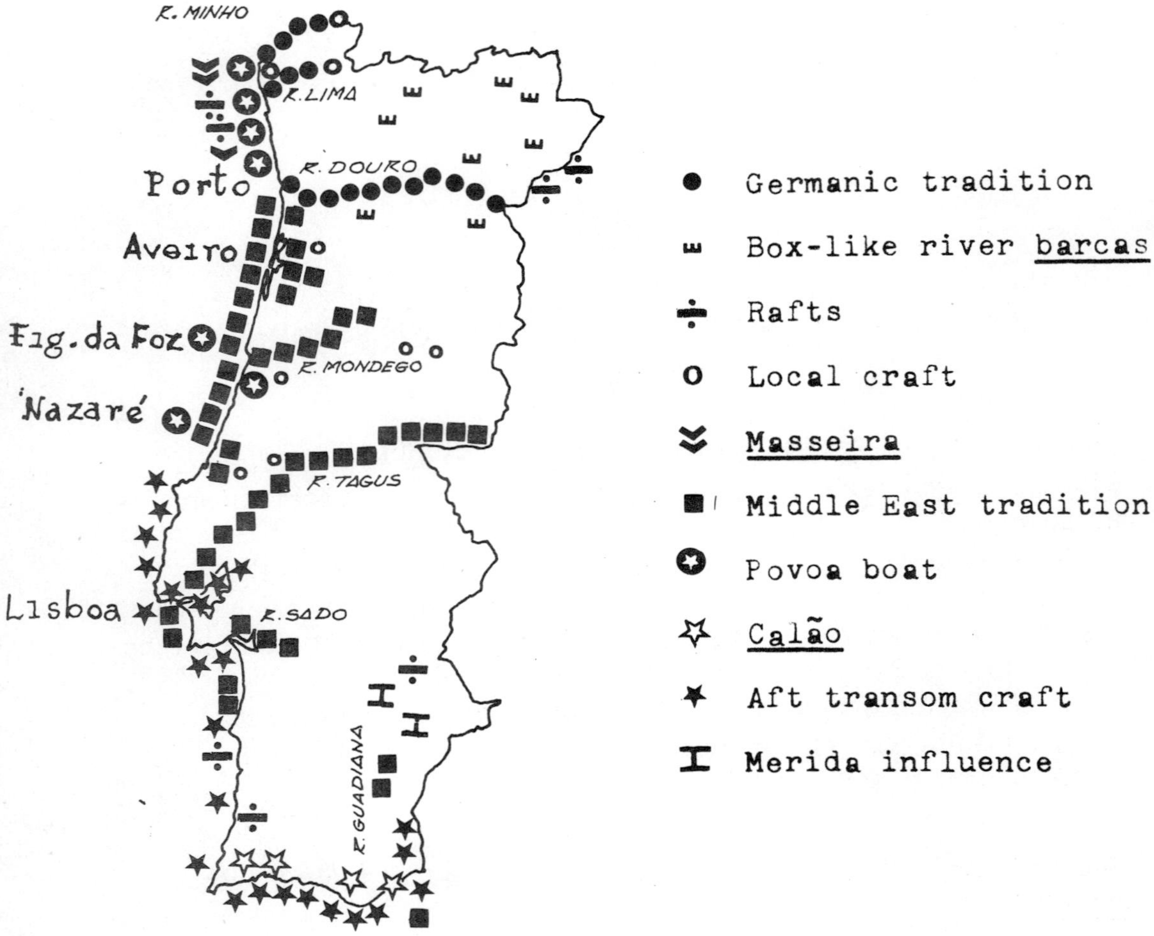

Fig. 9.1 Portugal: types of craft.

Fig. 9.2 The rabelo for transporting the Port wine casks.

Fig. 9.3 The flat-bottomed rabelo construction sequence: clinker-built plus shell technique.

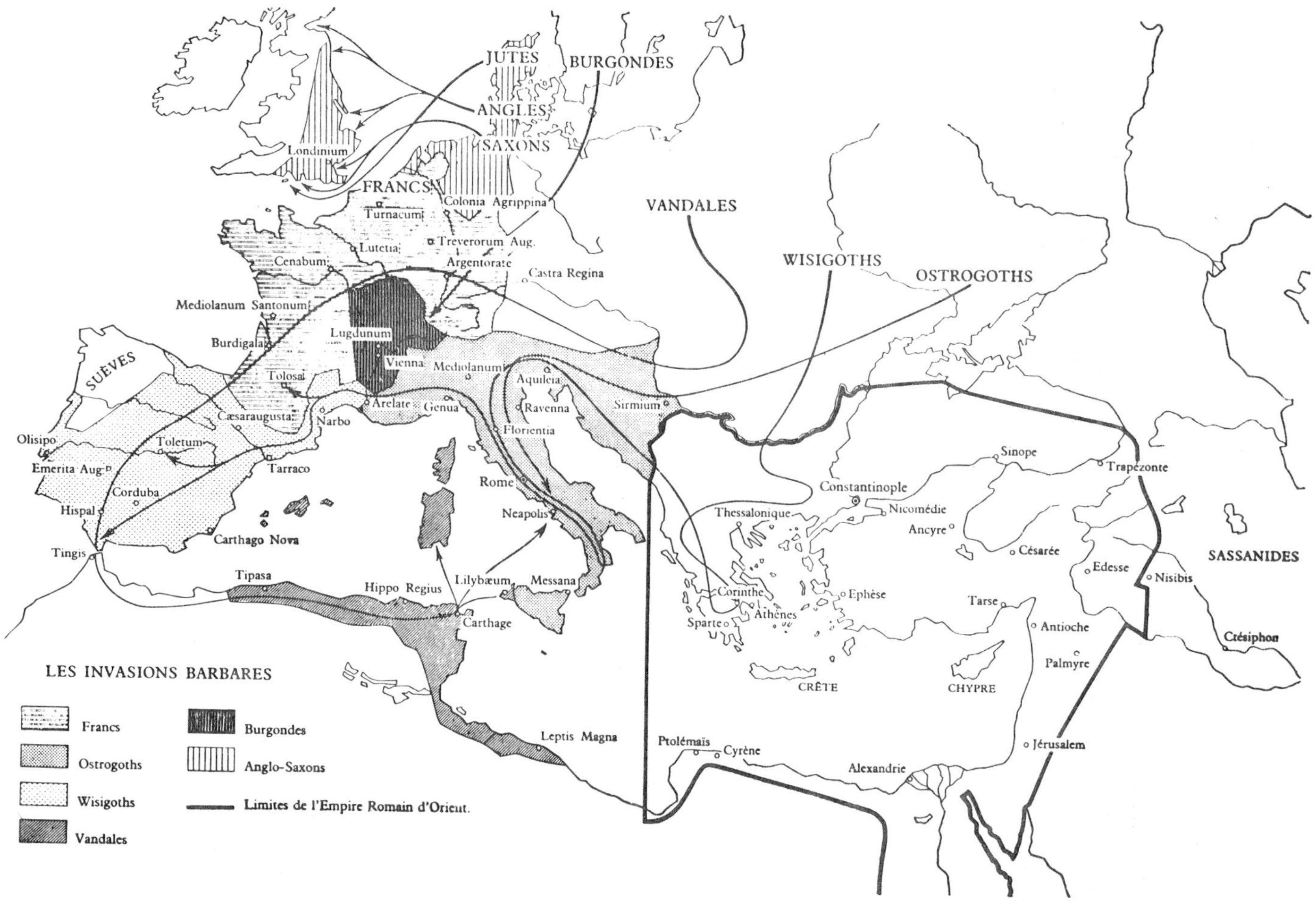

Fig. 9.4 The great migrations in Europe: V-VII centuries. (from Dawson, 1960 Le Moyen Age et les Origines de L'Europe Paris, 1969).

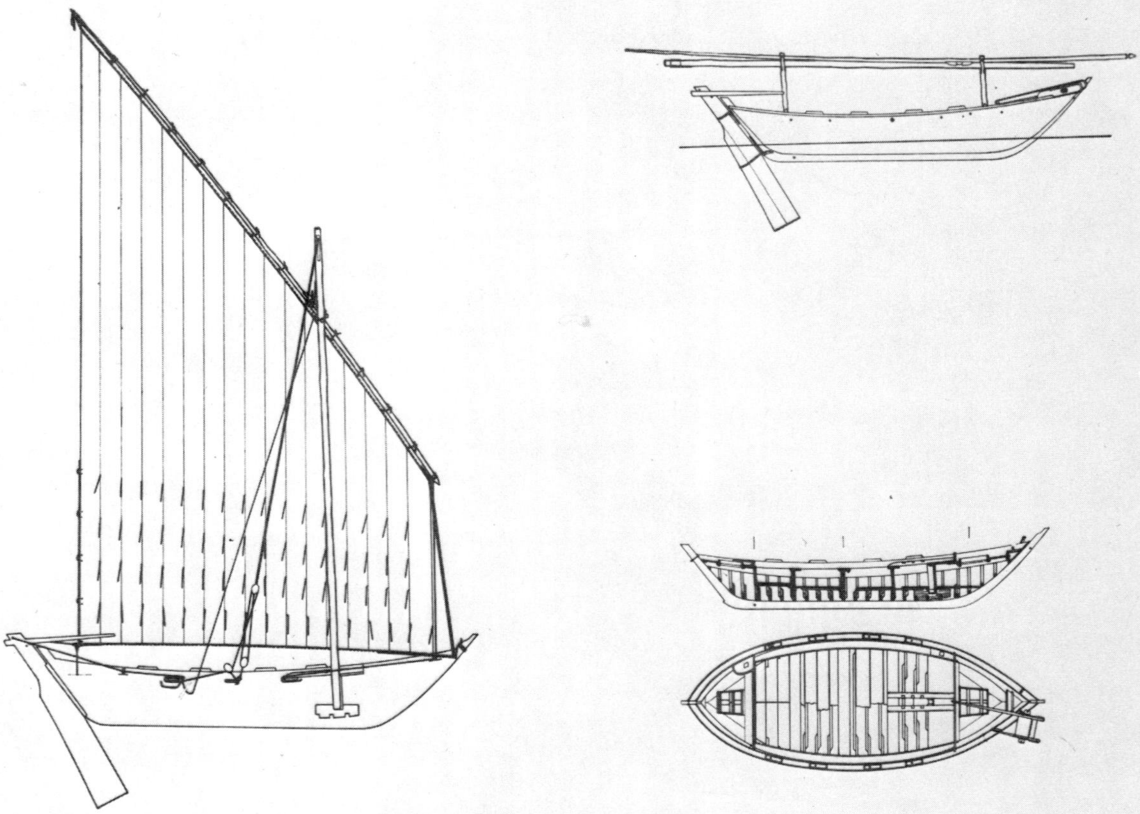

Fig. 9.5 The Póvoa lancha reminds us of the Breton sinagot. She is a 'carvel-built plus skeleton technique' vessel.

Fig. 9.6 The saveiro, used to carry the xávega seine.

Fig. 9.7 The saveiro's resemblance to the silver model of an ancient Ur boat (top: from Woolley, 1950 / below from De Castro, 1943).

"........................
They know not to fit with pine
Their keels, not with fir, as use is,
They shape their boats; but strange to say
They fit their vessels with united skins
And often traverse the deep in a hide.
................................" [3]

Even today it is possible to distinguish two cultural lines North of the Douro, both of Nordic tradition: an earlier Scandinavian Germanic influence shown by the building techniques of the boats sailing along the main rivers (Minho, Lima, Douro) and a later acquisition of inshore fishing craft, possibly derived from ancient Danish models. These may have arrived in Northern Portugal via Britanny and Galicia (once the Norman raids subsided).

The most interesting feature South of the Douro is the appearance of fishing boats of the Mesopotamian plank-canoe family. Having a privileged position in the central Western coast, this family of boats is characteristic of the Aveiro lagoon and the Mondego, but a few less important models can still be found in the Tagus, the Sado, and the Guadiana.

South of the Nazaré/S. Martinho do Porto area the traditional fishing communities are strongly acculturated due to the proximity of the two important harbours of Lisbon and Setubal and the improved fishing systems dominant in Algarve (Fig. 9.1).

Although we are mainly concerned here with boats typical of this central area, it is relevant to introduce two of the Northern models: the spectacular rabelo, that until quite recently carried the bulk of Port wine casks from the vineyards to the cellars in Gaia, across the river from Oporto, and the Povoa de Varzim lancha.

The construction of the flat-bottomed rabelo (Fig. 9.2) leaves no doubt as to her North European origins: clinker built[4] following the shell technique sequence (Fig. 9.3). The introduction of this North European feature may be understood, even if not demonstrated, when we recall the long occupation (V to VII centuries) of the Peninsular Northwest by the Suevs[5] (Fig. 9.4) and their other parallel legacies, e.g. ploughs.[6]

The Povoa lancha (Fig. 9.5) reminds us of the Breton sinagot, mainly by the shapes of her hull and sail, as well as by her building system - carvel plus skeleton-technique. The somatic features of the Povoa fisherman and some of his cultural ideosyncrasies (endogamy, family siglae, transmission through the youngest son, ...) find their counterparts in Britanny and Denmark; this is no co-incidence as the re-organisation of our Northwestern fisheries after the Viking raids was certainly done under strong North Atlantic influences.[7]

The saveiro (Fig. 9.6), used in the central area to carry the xavega seine, introduces us to another and contrasting world:[8] watching her sturdy, moon-shaped hull being hauled ashore by oxen we cannot remain indifferent to her extreme ressemblance to the Baghdad silver model of an ancient Ur boat [9] (Fig. 9.7). This resemblance has, however, a far deeper meaning, as shown by other convergent events. To start with, such formal identity is not unique and it is possible to relate other Portuguese boats in the plank-canoe family to old Ur vessels, of which the boats of the Lower Euphrates Ma'dan are local direct survivers.[10] The Aveiro moliceiro (Fig. 9.8), a seaweed gathering vessel, also finds her replica in two of the British Museum models of Ur boats (Fig. 9), as the S. Martinho do Porto masseira seems to correspond feature by feature, to the shartoush of the Euphrates (Fig. 9.10). Even the barca serrana (Fig. 9.11), although she has no direct replica in Mesopotamia, is an important member of the family, as will be seen later when evaluating building techniques.[11]

The existence of similar craft in the Mediterranean area as, for instance, the gondola,[12] the rascona[13] and some Rumanian boats[14] witness the westward progressions of this family of vessels through the Mediterranean, having possibly followed the Ugarit and/or the Al-Alalak/El-Mina trade routes (Fig. 24) to reach the Syrian coast.

Amongst the many contradicting theories that have been forwarded about the origins of the saveiro, only two have enjoyed some audience: while some authors favour the hypothesis of a Phoenician vector others prefer to give to the the Megalithic "missionaries" (whose Eastern origins have been advanced by many scholars) the honours of having introduced that craft into Portugal. This later thesis stands up well to strictly technological scrutinizing: the saveiro, as her sister craft, results from adapting into wood construction the shape of their remote ancestor, the pitch coated reed canoe, a primitive boat well within the technical capacity of a stone implemented industry such as the Megalithic.[15] A similar technological conversion took place amongst the Ma'dan, as can be seen on photographs of their boats[16] (Fig. 9.12). Such uncongenial adaptation of the skeleton technique to the shape of this type of Portuguese boats is fully testified by the importance attached in the building sequence to the central bottom plank, that, for construction purposes, replaces a real keel. This pecularity is more relevant on the barca serrana, where the rest of the bottom is laid after the side planking (Fig. 9.13), while on the saveiro the outer planks of the bottom, touching the bilge, are laid immediately after the central plank (Fig. 9.14). We can perceive here vestigial traces of the Mediterranean Gabarit de St. Joseph fashion.[17] This theory of the Megalithic carriers[18] does not, however, resist a broader approach as it is difficult to explain why no similar type of craft is portrayed on the rock engravings of the North West of the Peninsula[19] (Fig. 9.15), and no graphic representation of this type of vessel, prior to the Iron Age, appears in the North European areas of Megalithic expansion. The only known images of crescent-shaped boats curiously appear in Scandinavia, a marginal Megalithic area. The Scandinavian petrogliphs (Fig. 9.16) have, however, been interpreted as a product of Roman influence:

"Lindqvist (...) suggested that the ships developed from the figure of the crescent moon found on Roman stones along with the disk of the sun, and although no simple moon-shaped vessels appear on the Gotland stones, there is an example from Austrheim, Nordfjord in Norway, which Shetelig would date to the Roman period..."[20]

In any case we cannot ignore the purely symbolic motivation of many representations of boats amongst Indo-European populations, as the case might be in the Scandinavian examples.

Fig. 9.8 The moliceiro, seaweed gathering vessel.

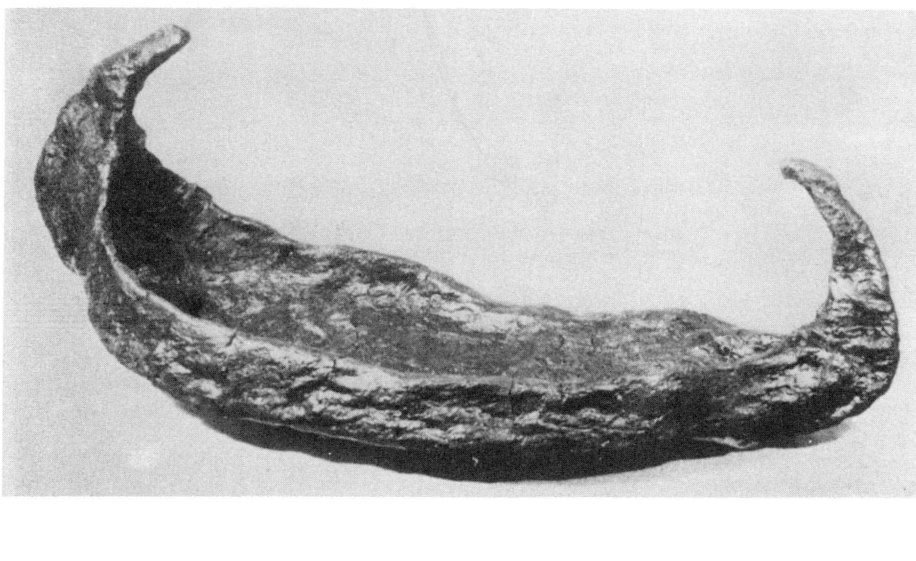

Fig. 9.9 Another ancient Ur boat (from Sumerian Art Illustrated, Plate XIIIb, British Museum, 1969).

Fig. 9.10 The S. Martinho masseira (top) seems to correspond feature by feature to the shartoush of the Euphrates (below - from Poujade, 1946 p. 201).

Fig. 9.11 The barca serrana of the Mondego river (top - photograph by Vasco Cunha) and another Ur model (below - by kind permission of the Baghdad Museum). Cfr. the seal number 51 of plate 20 following.

Fig. 9.12 The technical conversion to Mediterranean patterns observed amongst the Ma'dan boatbuilding: the zaima (top - from Thesiger, Feb. 4/56) made of giant reeds; the tarada (middle - from Thesiger, Feb. 19/55) shows the ribs with interior planking of the craft; and (below - from Thesiger, 1958 p. 212) marshmen waterproofing a canoe - compare the same type of ribs as those of the barca serrana, the mercantel, etc.

Fig. 9.13 The barca serrana building sequence: the central board of the flat-bottom plays the role of a real keel in this adaptation of the skeleton-technique.

Fig. 9.14 The wide hull of the saveiro explains, perhaps, why the outer bottom planks are laid before the ribs therefore helping to keep them in place (photo repr. by kind permission of David Goddard).

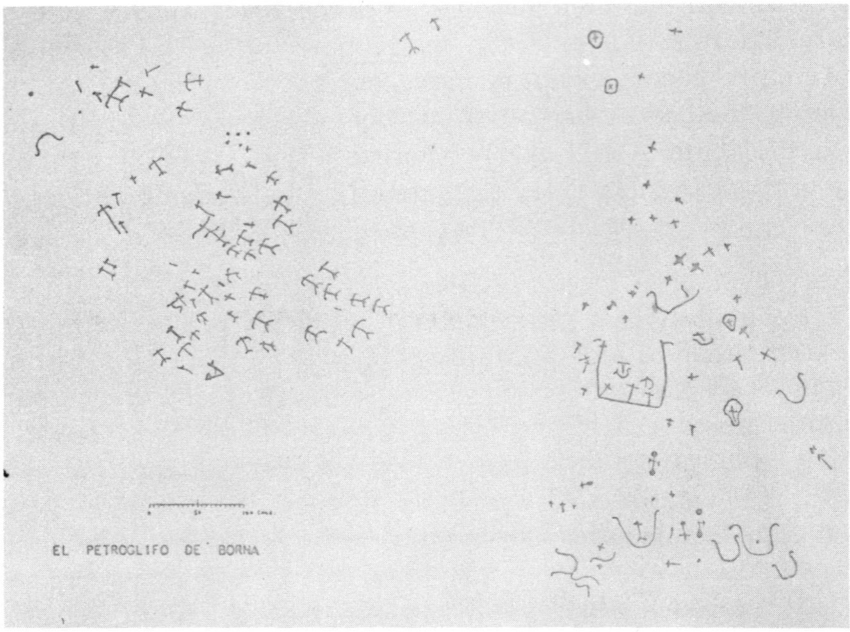

Fig. 9.15 The petrogliph of Berna (from Alonso Romero, 1974, p. 25).

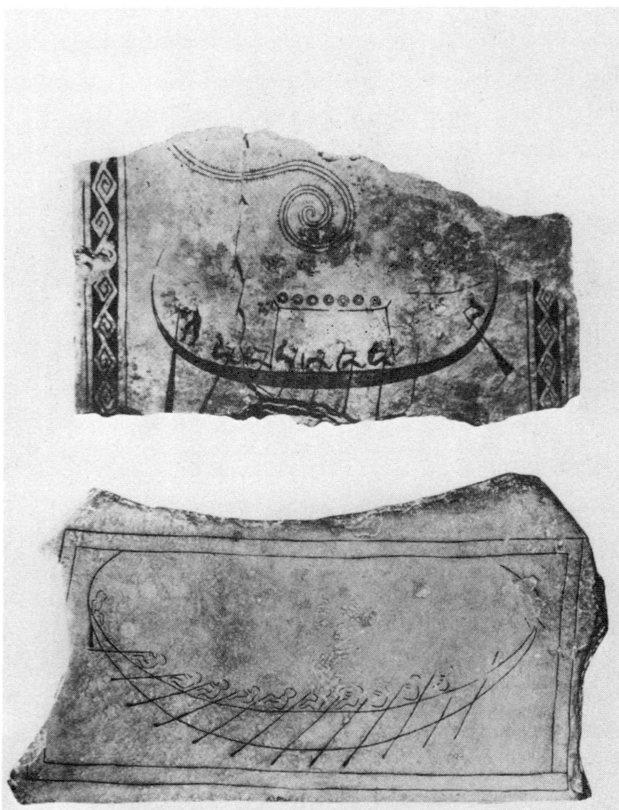

Fig. 9.16 The petrogliphs of Stenkyrka, Gotland (top - from Gelling and Davidson, 1969) and of Häggeby (below - from Brøgger and Shetelig, 1959).

As for the Phoenicians, it is again difficult to reconcile their importation of the saveiro with the historical notice of their use in Peninsular waters of an entirely different type of craft, the hippos[21] (Fig. 9.17) apart, of course, from their more recent ordinary merchant-men. The application of painted magic eyes to the bows of some of our boats (Fig. 9.18) and the use of stone anchors in the South of Portugal - Algarve - (Fig. 9.19), by themselves, provide only very weak subsidiary arguments to the Phoenician origin of the saveiro.

I believe that it will be far more rewarding to review Schulten's theory on the Cretan origin of the pre-Tartessian settlements in Southern Spain:[22] from the first steps of local metallurgy c. 2700 B.C., to the arrival c. 1100 B.C. of the Phoenicians in Gadir (close to Schulten's proposed location for Tartessus) a number of important events were taking place in the Aegean area. These most certainly had their sequel on the Spanish coast, a matter that nautical archaeology may be instrumental in clarifying. We can refer, amongst other things, to the deep change in the type of craft represented on Aegean seals around 2200/2000 B.C.[23] (Fig. 9.20), when the long ram-headed boats gave place inter alia to wide moon-shaped vessels: this change coincides with the emergence of Cretan thalassocracy and subsequent establishment of their trade on Peninsular shores. Clelia Laviosa has advanced the hypothesis that those moon-shaped models were originally Egyptian and were adopted at an early stage, by the Phoenicians who, according to the same author, controlled Mediterranean sea trade from as early as the 15th century B.C.[24] It is very difficult, however, to fit this hypothesis into some fundamental facts and, long before Laviosa, Hornell had already remarked that certain boats "with sharply up-turned ends appear to represent types of wooden vessels introduced into Egypt from foreign sources by purchase and imitation or through invasion; among these may be classed the straight-bottomed boat with high vertical ends carved on the Gebel-e-Araq knife handle".[25] This argument is strongly reinforced by Björn Landström's reproduction of two remarkable documents designated as Egyptian boat of Mesopotamian type and silver model from the II Intermediate Period (Fig. 9.21). In any case the obvious lack of interest of the Egyptian for seafaring,[26] discourages any attempt to overestimate their influence to the detriment of other peoples that effectively developed some kind of sea power. All this brings new strength to Schulten's statement: "...those vases, necklaces and bracelets (found in SE Spain) were certainly not brought here by the Egyptians...It would be safer to invoke the Cretans who had very large boats even before 2200 B.C. as can be inferred from the sail boats appearing in Early Cycladic stoneware and in Early Minoan III (2200 to 2000 B.C.)[27] seals. Similar Egyptian vases and ornaments were found in Crete and it seems that Spanish copper daggers dating from the III millennium B.C. were also found on that island. It has also been suggested that some silver vases found in Troy had come from Spain".[28] To be fair, it must be said that Laviosa correctly relates the small marble model boat from Hagia Triada to the vessel in the Tiryns ring.[29] Moreover, the points of similarity between the saveiro and the boat painted on the Hagia Triada tomb (1500 B.C.) (Fig. 9.22) are too obvious to need any further development here. Incidentally, this painting leaves no doubt as to the painted nature of the oblique bars appearing on the hulls of the boats of Minoan seals. Finally, the Thera murals, with their splendid collection of boats, will cer-

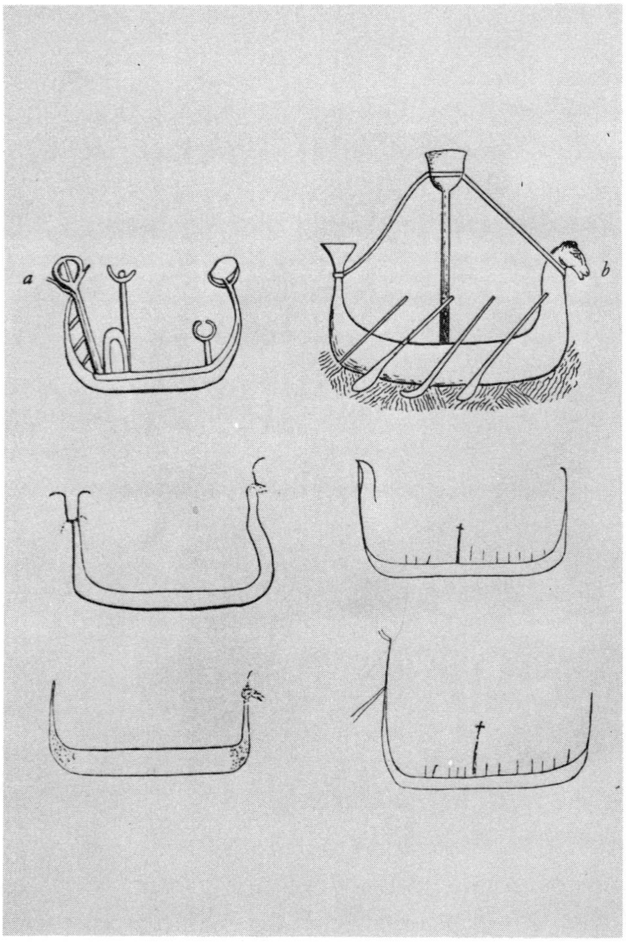

Fig. 9.17 An hippos compared with the foreign ship (a) carved on the Gebel-el-Araq knife and other foreign ships from petrogliphs of the Eastern desert of Egypt (after Winkler, repr. by Hornell, 1946, p. 6).

Fig. 9.18 Caparica: magic eye painted on the bow of a local craft.

Fig. 9.19 Baleeira (Sagres/Algarve): stone anchors used in conventional vessels.

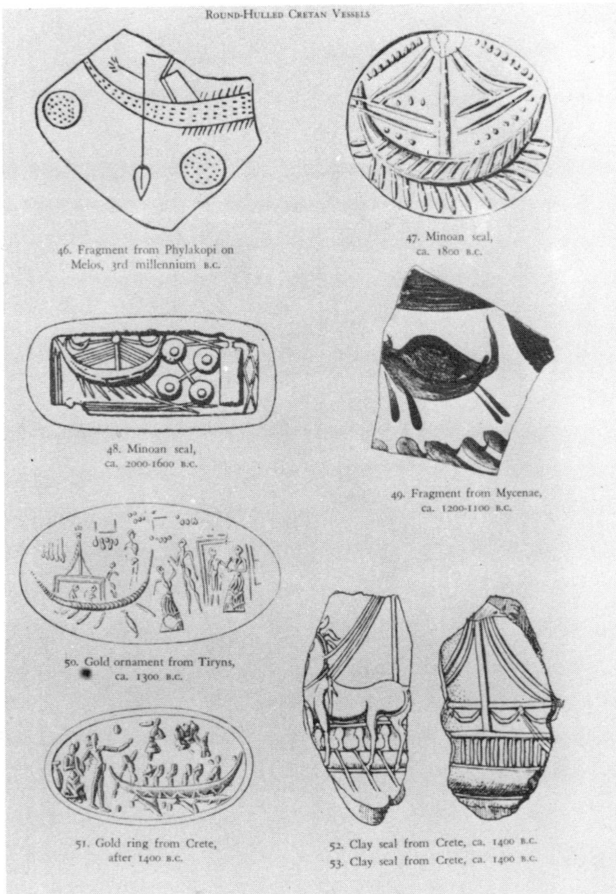

Fig. 9.20 The round vessels appearing on Aegean and Cretan seals around 2200/2000 B.C. (Casson, 1971).

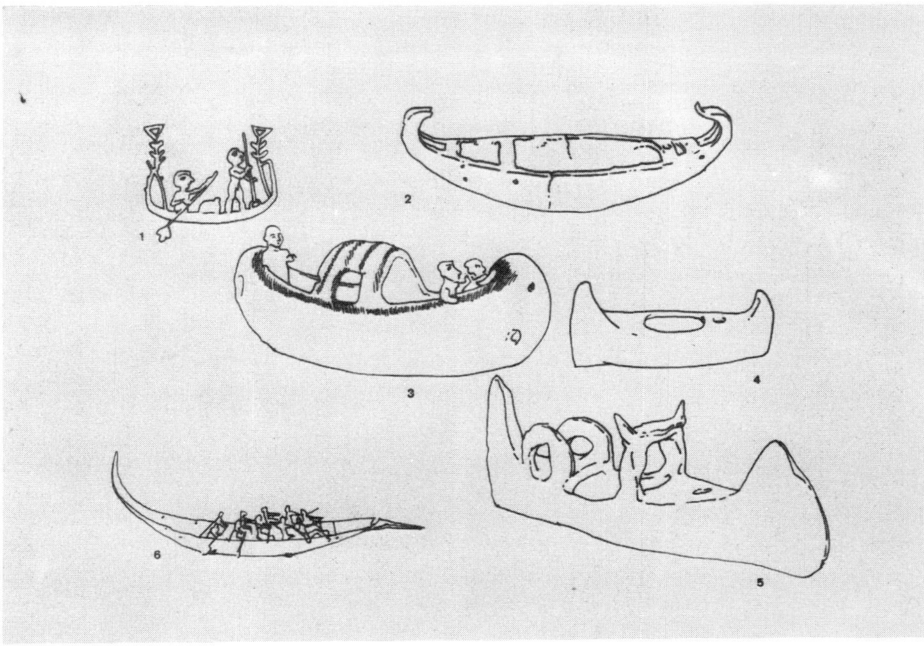

Fig. 9.21 A craft from a Sumerian cylinder-seal (1) and models of proto-dynastic Egyptian craft (2,3,4,5) and a silver model of the II Intermediate Period (from Landström, 1970).

Fig. 9.22 Detail from de Hagia Triada painting (from De argne, 1964).

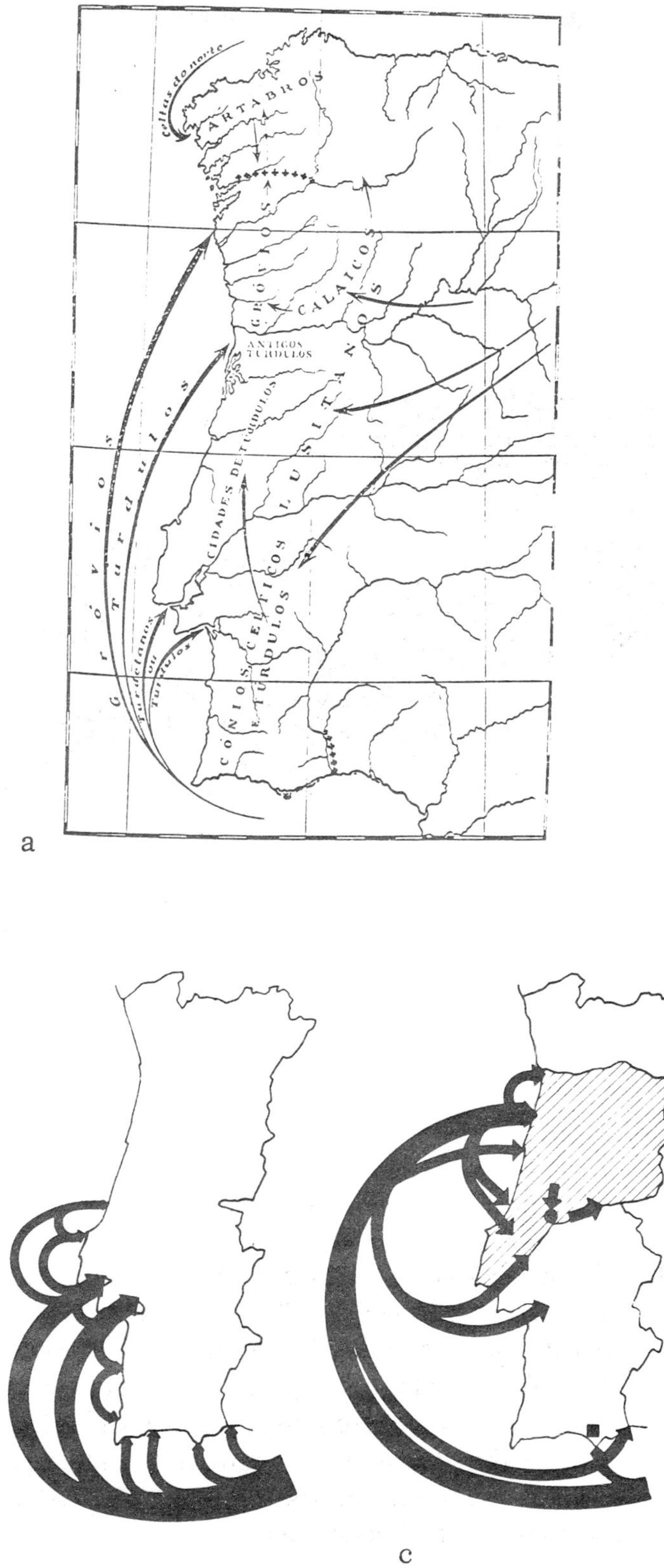

Fig. 9.23 The dispersion of South Peninsular populations after the fall of Tartessus (a) compared with the migration routes of the saveiro and similar craft: original (b), with occupation of the central area, and recent southbound expansion (c).

Fig. 9.24 The main lines of westbound progression: A Ur/Eridu; B Ugarit and/or Al-Alalak/El-Mina; 1 Caravan's route via Aleppo; AE Egypt; 2 Mesopotamian vector; 3 Phoenician vector; C Crete; D Byblos, Sydon, Tyr; 4,5,6 Cretan vectors; a Venice; b Marseille; c Tartessus; 7 control of the tin route; d main area of the Portuguese boats of the Mesopotamian type; e Scilly Is.; f Castro Urdiales (Moorish); 8 Rhône river; 9 Tin route (Garonne); g Archachon.

tainly bring new material to this question, but so far the data at my disposal are not enough to carry any further research.[30]

Jaime Magalhães Lima, a Portuguese scholar writing around 1925, has already suggested that the fishermen of Ilhavo, an important fishing community in the Aveiro lagoon, descended from Tartessian stock.[31] Linked to our own suggestion above, Magalhães Lima's position becomes fairly acceptable, particularly when we notice the coincidence between the area of dispersion of South-Iberian populations (Turduli) after the fall of Tartessus (509 B.C.) and the main areas of distribution of our boats of Mesopotamian breed (Figs. 9.1, 23, 24).

This brief account skips many of the detailed steps that have led me from a purely nautical study of the saveiro, into the broader field of History. Such steps resulted from the impossibility of fully understanding the construction of the saveiro with the sole aid of the classic tools of our trade. It remains nevertheless, that through the boat of the xavega we have found new items to evaluate the current contradictory explanations for the origin of a certain breed of Portuguese fishermen who even today keep their social identity, characterised by their extreme mobility (they are called "sea-gipsies")[32] (Fig. 9.25) and features of bravery and hardiness that not only have once led them to almost monopolise the crews of our cod-fishing fleet in Newfoundland but also gained for them the respect, if not fear, of the other fishermen.

From this stage of the study of the saveiro and her crew, it was only natural to enter the framework of archaeological research now taking place in the Near East and Eastern Mediterranean in the broad attempt to understand the mechanisms of cultural migration and intercourse that by far surpass the natural cautiousness of traditional historians. It is my hope, therefore that a consistent appraisal - alas, far beyond my scope - of archaeological, ethnological and historical data already available at both ends of the Mediterranean may one day result in a comprehensive view of the cultural phenomena that shaped Late Neolithic and Early Metal Europe.

Fig. 9.25 Vila Franca de Xira: boat-houses of the 'sea-gipsies'.

ADDENDUM

It was only at the time of the Symposium that the author became acquainted with Henry Hodges Technology of the Ancient World, Harmondsworth, 1970, Mr. Hodges approach to the moon-shaped vessels of the Eastern Mediterranean and his cautious note on the Portuguese boat shown on Fig. 132 (a saveiro), although not relating them to the Ur silver model, is another valuable contirbution to the more and more frequent convergence of historians and archaeologists studying the progression of Levantine boats across the Mediterranean.

Also during the Symposium Mr. Basil Greenhill published his Archaeology of the Boat where he refers to the research on the Humberside finds, now taking place at the Archaeological Research Centre of the National Maritime Museum, and admits possible similarities between the Humber remains and the saveiro. Independent of the confirmation or not of such similarities by the A.R.C. studies, these will most certainly be fundamental to our better understanding of the technological intercourse between the Mediterranean and Northern Europe from 2500 to 1500 B.C.

ACKNOWLEDGEMENT

The English version of this paper was prepared by Alexandre Vasconcelos.

REFERENCES

Alonso Romero, Fernando. "Relaciones Atlanticas Prehistoricas entre Galicia y las Islas Britanicas y Medios de Navegacion", periodical unidentified, Salamanca, 1974.

Brøgger & Shetelig, A.W. The Vikings Ships, 1950 English Translation.

Casson, Lionel. Ships and seamanship in the ancient world, Princeton, 1971.

Casson, Lionel. "Bronze Age ships. The evidence of the Thera wall paintings", The International Journal of Nautical Archaeology and Underwater Exploration, 4.1, 1975.

Childe, Gordon. The Pre-History of European Civilisation, 1957, Portuguese translation, "A pre-historia da Sociedade Europeia", Lisboa, 1960.

Daniel, Glyn. The Megalithic Builders of Western Europe, London, 1958.

De Castro, D. José. Estudos Etnográficos - Aveiro, vol. II, Lisboa, 1943.

Demargne, Pierre. Naissance de l'Art Grec, Paris, 1964.

Dias, Jorge. Os Arados Portugueses e as suas prováveis origens, Colimbra 1948.

Filgueiras, Octavio Lixa. "Entre Normandos e Arabes, nas margens do Douro", Studium Generale, Vol. X, Porto, 1963.

Filgueiras, Octavio Lixa. "Da Navegação do Douro", Studium Generale, vol. X, Porto, 1962.

Filgueiras, Octavio Lixa. "Sobre as Origens do Barco Poveiro", Boletim Cultural Povoa do Varzim, vol. 1, 1970.

Filgueiras, Octavio Lixa. "Remanescentes de formas de navegar pre-romanas, em uso no Noroeste Peninsular", Actas do III Congresso Nacional de Arqueologia, vol. I, Porto, 1974.

Filgueiras, Octavio Lixa. Commentários Técnicos da Tese do Moçarabismo Nautico, Lisboa, 1975.

Garcia y Bellido, A. de. España y los Españoles hace dos mil años, según la Geografia de Strabon, Buenos Aires, 1945.

Gelling, Peter & Davidson, Hilda, E. The Chariot of the Sun, London, 1972.

Harden, Donald. The Phoenicians, 1962, Portuguese translation Os Fenicios, Lisboa, 1968.

Hornell, James. Water Transport, origins & early evolution, Cambridge, 1946.

Hutchinson, R.W. Prehistoric Crete, London, 1962.

Ibarra Grasso, Dick & Julio. Historia de la Navegacion Primitiva, Buenos Aires, 1949.

Johnstone, Paul & Tilley, A.F. "An unusual Portuguese fishing boat", The Mariner's Mirror, February 1976.

Landström, B. The Ship, London.

Landström, B. Ships of the Pharaohs - 4000 years of Egyptian shipbuilding, London, 1970.

Laviosa, Clelia. "La Marina Micenea", Annuario della Scuola Archaeologica di Atene e delle Missione Italiane in Oriente, vol. s XLVII/XLVIII, 1969/1970.

Madahil, A. Gomes da Rocha. Etnografia e Históŕia - Bases para a organizacao do Museu Municipal de Ilhavo, Ilhavo, 1933.

Magalhães, Luis de. "Os barcos da ria de Aveiro", Portugalia, vol. II, Porto, 1905.

Paris, A. Souvenirs de Marine Conservés, Paris 1882/1908.

Poujade, Jean. La Route des Indes et ses Navire, Paris, 1946.

Rubin de Cervin, G.B. "The Evolution of the Venetian Gondola", The Mariner's Mirror, August, 1956.

Schulten, Adolf. Geografia y Etnografia antiguas de la Peninsula Iberica, Madrid, 1959/1963.

Schulten, Adolf. Tartessos, 1921, Spanish Translation, Madrid, 1972.

Thesiger, Wilfried. "The marshmen of Southern Iraq: and their boats unchanged since the days of ancient Ur 5000 years ago", The Illustrated London News, February 19, 1955 and February 4, 1956.

Thesiger, Wilfried & Maxwell, Gavin. "Marsh dwellers of Souther Iraq - Primitive Ma'dan", The National Geographic Magazine, February, 1958.

Vicente, Eduardo Prescott & Serrão, Eduardo da Cunha. "Contribuição para e conhecimento das relações por via maritima, no Eneolítico e na Proto-Historia, entre as populações da costa centro-ocidental da Peninsula Iberica o os povos do Mediterraneo", Actas do Congresso Internacional da Historia dos Descobrimentos, Lisboa, 1961.

Wooley, Sir Leonard. Ur of the Chaldees, 1929 reviewed edition. London, 1950.

NOTES

1. Strabo, Geographica; Plinius, Historia Natural; quoted by Schulten, 1963, p.78.

2. Strabo, Geographica L.III, Spanish translation by Garcia y Bellido, 1945, pp. 3 & 7; cfr. Filgueiras, 1962.

3. Avienus, Ora Maritima, English translation by Skene, W.F., in Celtic Scotland, quoted by Hornell, 1946, p. 113; cfr. Schulten 1963, p. 302; "tenemos de contar ... con las relaciones estrechas entre Galicia y Bretaña, por un lado, y con Britania, por otro, las cuales se comprueban en numerosas coincidencias arqueologicas y la repeticion del nombre do los Oestrimnios de Galicia (Avieno, 154) en la Bretaña (Avieno, 91), de los Albiones de Galicia (Plinio, 4, 111) en Inglaterra (Avieno, 112)..."

4. The overlapping boards on the sides are laid on successive continuous rows, bridging from stem to stern.

5. Cfr. Filgueiras, 1963.

6. Dias, 1948.

7. Cfr. Filgueiras, 1970.

8. This paper is an abridged revision of the authors work Comentários Técnicos da Tese do Moçarabismo Nautico.

9. Found in the tomb of A-bar-gi, as reported by Sir Leonard Wooley, 1950, p. 41.

10. Thesiger, 1958, p. 212, refers that the Ma'dan do not consider themselves as Arabs, but as direct descendants of older Mesopotamian stock.

11. Ibarra Grasso, 1949, p. 175, show the photograph of a barca serrana (Mondego river, above Coimbra) and in p. 172 they wrote that "... En Portugal también se encuentra este tipo de canoas y su forma es la de verdaderas gondolas aunque las extremidades no se encuentran tan intensamente levantadas como en las de Venecia". Cfr. the Cretan seal reproduced Casson, 1971, fig. 51.

12. Apart from Ibarra Grasso, 1940, Rubin de Cervin, 1956, p. 203, also remarks the resemblance with the old Ur silver model- the same author, referring to the Ma'dan that "... have built up there amongst marshes and lagoons... and shaped their shallow boats much in the same way as the ancient Venetians were compelled to do in order to ensure their means of living...", adds on a footnote "...that these shallow boats, while having raised ends and carvel-built hulls, are coated outside with bitumen, bearing the names of taradas and barkash, which are also to be found in old Venetia: compare tarida and

barcaccia." Cfr. Thesiger, 1958, p. 212; Landström, 196, p. 214 and fig. 518.

13 Cfr. Paris, 1882/1908, plate 86.

14 Reported by Ibarra Grasso, 1949, with no further explanation. Johnstone and Tilley, 1976, p. 20, refer to the "combination of extravagant sheer with horizontal planking" as a feature common to the saveiro, the gondola and the Maltese boats. This is a construction detail characteristic of the Mediterranean, contrasting with the more recent Nordic system of continuous double curved rows of overlapping planks (Filgueiras, 1975, note 7, p. 6; see also note 4 above) but it is irrelevant for any typological identification: none of the Maltese boats as those reproduced in Paris, 1882/1908, belongs to the Mesopotamian plank-canoe family.

15 Gordon Childe, 1960, p. 145.

16 For this purpose I have used photographs from Thesiger's short articles of 1955, 1956, and 1958.

17 Poujade, 1946, p. 265. With this interpretation I propose an answer to Johnstone and Tilley, 1976, p. 15: the rule of the Gabarit de St. Joseph is also evident in the construction of the gondola (Rubin de Cervin, 1956, pp. 211 ff.) Hornell, 1946, pp. 230 f., demonstrated that the Arabs adopted the naval techniques that were predominant in their conquered lands in the somewhat rudimentary Indian Ocean or in the more sophisticated Mediterranean basin. In more recent times, however, shipbuilding, in S.W. Asia went through great developement that even reached such "primitive" groups as the Ma'dan. It is interesting to mention that the late Jaime Cortezão, an authoritative scholar on the Portuguese Age of Discoveries, but lacking any archaeological background, suggested that the saveiro was introduced by the Arabs. His position, however, is based on XIII century sources and other arguments point to the existance of the saveiro before the Muslem occupation.

18 Daniel, 1958, pp. 130 f.

19 Alonso Romero, 1974, pp. 17 ff., relates the boats in the Borna rock engravings to Egyptian and Nubian papyrus models, presumably linked themselves to older Mesopotamian vessels, but even the conventional representation in those engravings is very clear as to non moon-like shape of those boats. Paradoxally, this substantially adds to the archaeological value of the Borna petrogiphs (see note 25 below).

20 Gelling, Peter and Davidson, Hilda E., 1969, p. 156.

21 Garcia y Bellido, A., 1945, p. 220; cfr. Harden, 1968, p. 173. It seems to me that Lethbridge's suggestion, as reported by Johnstone and Tilley, 1976, p. 15, can hardly stand its ground. Regretably I have no first hand knowledge of Lethbridge's work (see note 28 below).

22 Schulten, 1972. Ibarra Grasso, 1949, pp. 171 f. referred to the importance of Crete on the Mediterranean expansion of this type of craft, but without any attempt at explaining the probable mechanisms of expansion.

23 Cfr. Casson, 1971, p. 33. Hutchinson, 1962, p. 96, wrote: "The Tiryns boat may, of course, be a Mycenean variant unknown in Crete, but there are other similar boats....appearing on Minoan seals, usually in religious scenes. These boats resemble an old Mesopotamian type employed by the marsh Arabs and illustrated by a silver model found in the royal graves at Ur". Hutchinson is, of course, an opponent to the thesis of Minoan thalassocracy. Cfr. note 29 below.

24 Laviosa, 1969/1970, p. 19.

25 Hornell, 1946, pp. 49 f.

26 Cfr. Hornell, 1946, p. 230: "The belief held in Ancient Egypt and in India that travel by sea causes defilement difficult to remove except by elaborate purificatory rites. Plutarch (De Isis et Osiris, c. 32) states that the Egyptian priesthood kept themselves religiously aloof from the sea. Porphyry (de Abst. IV, 8) confirms this when he writes - to sail from Egypt....was considered by them (the Egyptians priests) to be one of the most unholy of things".

27 Evans chronology; Hutchinson suggets 2100/1950 B.C.

28 Schulten, 1972, p. 18. (translated for this paper). This statement strengthens my lack of enthusiasm for Lethbridge's suggestion (see note 21 above). In any case many Portuguese boats of this family are fitted with sails.

29 Laveosa, 1969/1970, pp. 21 f. Cfr. note 23 above: It is possible that further research will diminish the role of Minoan sea travelling in Western Mediterranean, therefore confirming Hutchinson's position and bringing to the fore the importance of Anatolian/Syrian navigators. Such development could never favour the thesis of Phoenician vectors - in fact chronological precedure above would refute that thesis.

30 Casson's analysis of the relationships between the boats represented on the Thera wall paintings and those appearing on Mycenean seals and ceramics together with full understanding of "similar" Portuguese vessels will certainly open the way to clarify many of the doubts still existing on some construction techniques.

31 Quoted by Rocha Madahil, 1933, p. 132. Cfr. Vicente and Serrão, 1961: for these authors the Western Central coast of Portugal is only the part between the Tagus and Sagres; hence their association of the Caparica saveiro and the Ur boat which is geneologically correct but of scant historical value as Caparica is a very recent settlement founded by emigrants from the Aveiro lagoon.

32. Luis de Magalhães, 1905, p. 60, refers that: "...estes barcos (bateiras - Aveiro) exploram toda a ria...Fora da sua região propria (a ria) encontram-se om todos os rios onde há colonias piscatorias de Aveiro: o Douro, o Mondego, o Tejo, o Sado, a ria de Lagos e Portimão, o Guadiana. Algumas destas colonias fixaram-se até em portos espanhois, como em Cadiz e Sevilha, mantendo sempre inconfundiveis a sua nacionalidade, os seus hãbitos, costumes e tipos".

DISCUSSION

McGrail: I wonder whether some of the similarities between boats separated widely in time and space may not have something to do with function? Their form may be related to particular tasks or environment, and thus might be seen in different areas without any direct influence or content.

Christensen: In Northern Europe, especially in Scandinavia, we are fortunate to be able to link archaeological finds with modern boats. They look the same, similar boatbuilding tools have been used for the last thousand years, and the terminology is the same, so we are on very safe ground as Hasslöf has demonstrated. However Professor Filgueiras has jumped at length and quickly from ancient times in the Eastern Mediterranean to present day Portugal with very little in between. Is there for example, any linguistic evidence to support his theory?

Wright: Are there similar constructional features, as well as parallels of shape between ancient Mesopotamian boats and modern Portuguese boats?

Filgueiras: Aspects of the construction are the same, for example there is a step by step similarity between Ma'dan building techniques and the construction of the barca serrana. There is no direct evidence of this postulated relationship between boatbuilding in the Eastern Mediterranean and that of the Iberian Peninsular, and there are no historical records. Thus any confirmation must come from parallel sources. The integration of ethnographical, archaeological and historical data I attempted in my paper, may help towards confirmation.

Christensen: The archaeological material will turn up whether we want it or not, but the still living tradition of boat building all over Europe and the Near East is going and it's going fast and as soon as the last boatbuilder puts down his tools his knowledge has gone, totally. But his knowledge may help us to extend ours so it is most important to record modern comparative material.

McGrail: Ethnography can provide us with a series of solutions from which we may get some idea of what our fragmented remains might have looked like as complete boats. Are Ethnographers recording the information archaeologists need?

Crumlin-Pedersen: The important thing is for boat archaeologists to be aware of the actual boatbuilding and boathandling techniques. I wonder whether we are covering the field well enough and what we could do to cover it better. We concentrate very much on the boatbuilding itself, but who can sail a one masted square rigged boat today properly so it gives the right performance? Do we care enough about recording and maybe even continuing, aspects of that living tradition for the benefit of future boat archaeologists?

Hasslöf: Our experience in Scandinavia is that you must go out to the boatbuilders. They will give you the answers.

Naish: There is a very good little book written by R.C. Leslie in 1892 Seaboat, how to build, rig and sail her. The information in it is taken straight from the people who were building the fishing boats that worked off the beaches in Dorset and South Devon, and it is an invaluable record.

Ellmers: In Norway there is demonstrable continuity of tradition from the present day back to the Viking Age boat builders and even beyond. In other countries this is not always so. However there is one nice example from the Lake of Mondsee in Austria where a dugout was built in the same year that an atomic powered ship was built at Kiel. This dugout building was well recorded by Austrian ethnographers, and this was most useful to me especially when I investigated the five or six half finished dugouts which have been excavated in Germany and the surrounding countries. These could only be interpreted in the light of this ethnographic evidence.

Levy: There is another example of living tradition to be seen in Kuwait which has the dhow as its national emblem; they are still building, very largely in the old tradition, with the old skills.

Wilson: I believe that the applications of ethnological parallels to archaeology can be done in a number of cases but I am highly suspicious of carrying a cultural trait from the eastern Mediterranean in 1000 B.C. or thereabouts 2000 miles away and 3000 years later to Portugal. In Scandinavia there is some evidence for continuity: pictorial representation, models, primitive paintings in churches, boatbuilding tools and so on, which do indicate continuity. But I suggest that continuity must be demonstrated for an area and not just assumed to be there. This is especially important when using illustrative evidence for the illustrator may be copying a prototype from a distant region, and he may in any case have little knowledge of boatbuilding.

Filgueiras: As you know I am not an archaeologist, at least not in the traditional meaning of the word, as my research is mainly directed towards the study of archaic survivals of vernacular boats. There are, however, several other reasons for postulating early Eastern influence on the Iberian Peninsular, in addition to the evidence of the boats I have described.

The hypothesis of Eastern influence on the Peninsular was first put forward by land archaeologists and historians namely Schulten's theory of the Cretan origin of Tartessos. These problems were discussed some 4 or 5 years ago at a symposium of land archaeologists dealing with this Levantine question. The paper I have submitted to you aims at bringing additional evidence from boat archaeology and ethnography to help in clarifying this obscure aspect of of Mediterranean history. Only a comprehensive analysis of all available data from archaeology, nautical archaeology, and ethnography, and further research on the points of non-continuity will enable specialists to find the correct interpretation.

McGrail: I take three main points from this discussion. Firstly that there is a tremendous job to be done in recording the dying traditions of wooden boat building and boat handling. Secondly, that this sort of evidence is useful in its own right to widen our experience and to suggest a range of possible solutions to reconstruction problems, so that the archaeologist can ask the right questions from his material. Thirdly, that we must be rigorous in our approach to continuity of cultural tradition - it must be demonstrated and not assumed.

10. SEARCHING FOR PATTERN AMONG THE LOGBOATS OF ENGLAND AND WALES

Sean McGrail

Summary

This paper is divided into four sections: first, the general requirement for quantitative methods is considered; second, the use of quantitative techniques in a recent investigation of the logboats of England and Wales is described; in the third part the application of quantitative methods to the whole field of boat archaeology is discussed in the light of the conclusions drawn from the logboat investigations; and in the final section the creation of an international data bank is proposed.

PART 1 - THE REQUIREMENT

The fragmented, distorted, and ephemeral nature of boat finds produces problems which, whilst not unknown in other branches of Archaeology, are presented to the boat archaeologist in an acute form. The search for pattern - the recognition and interpretation of regularities (Hodson, 1971:30) - amongst this assorted jumble of evidence may thus degenerate into the formulation of hypotheses based on inadequate foundations, and over confident attempts to equate finds with known historical types. Real progress can only be made if we stick to the scientific method, of which quantification is an essential aspect. By quantification I mean something more than the use of numerical methods: I mean the recording of hard fact, and the use of clearly defined, objective methods to analyse and to synthesise this evidence. In this wider sense I hope we shall see references to quantification in all the themes of this Symposium.

The general case for quantification in archaeology has been well summarised by Doran and Hodson (1975: 3-7, 93). It is especially relevant to boat archaeology where we are investigating not only the form but also the function and capabilities of our finds, for these latter qualities can only be fully evaluated by the quantitative methods of the naval architect. Form and function are inextricably linked in our study, and this is to our great advantage I believe, because thereby we can achieve a greater understanding of man the boatbuilder and boatuser. The use of quantification, in its broadest sense, at all stages of research should ensure that terms are defined, that reasoning is made explicit, and that assumptions, abstractions, and methods can be critically appraised by other workers who, if they are convinced by this demonstration can then have full confidence in the published results. Quantification also has the following advantages:

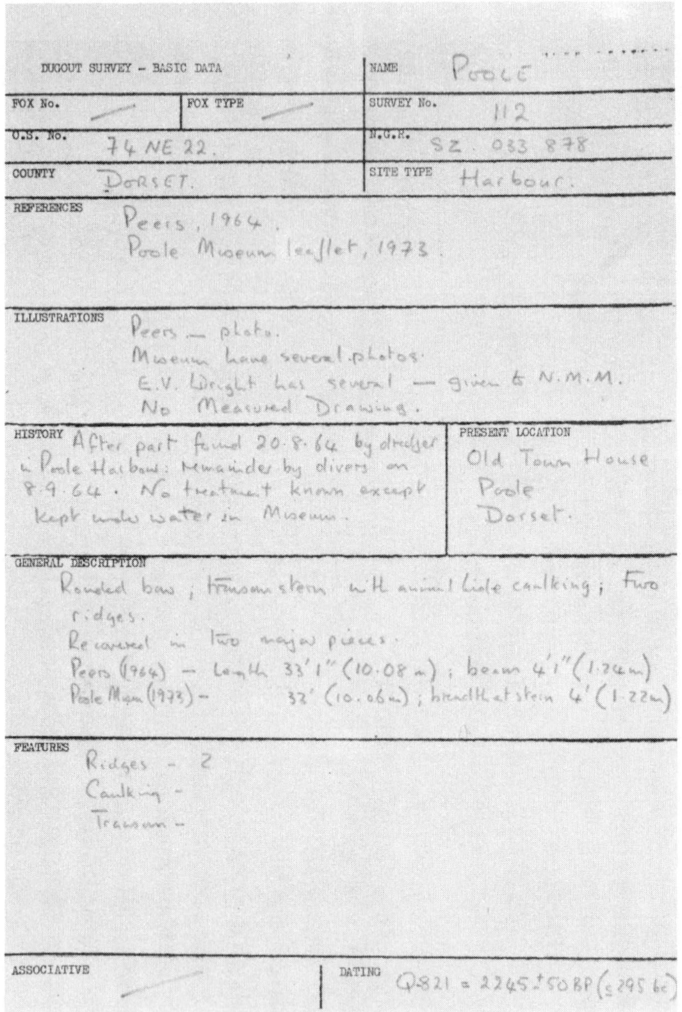
DUGOUT SURVEY - BASIC DATA
NAME Poole
FOX No. —
FOX TYPE —
SURVEY No. 112
O.S. No. 74 NE 22.
N.G.R. SZ 033 878
COUNTY Dorset.
SITE TYPE Harbour.
REFERENCES Peers, 1964.
Poole Museum leaflet, 1973.
ILLUSTRATIONS Peers — photo.
Museum have several photos.
E.V. Wright has several — given to N.M.M.
No Measured Drawing.
HISTORY After part found 20.8.64 by dredger in Poole Harbour: remainder by divers on 8.9.64. No treatment known except kept under water in Museum.
PRESENT LOCATION Old Town House
Poole
Dorset.
GENERAL DESCRIPTION Rounded bow; transom stern with animal hide caulking; Two ridges.
Recovered in two major pieces.
Peers (1964) — Length 33'1" (10.08 m); beam 4'1" (1.24m)
Poole Mus (1973) — 32' (10.06m); breadth at stern 4' (1.22m)
FEATURES Ridges - 2
Caulking -
Transom -
ASSOCIATIVE —
DATING Q821 = 2245 ± 50 BP (c 295 bc)

Fig. 10.1 The basic data for the Poole logboat, compiled from the information available before the logboat was recorded. National Maritime Museum.

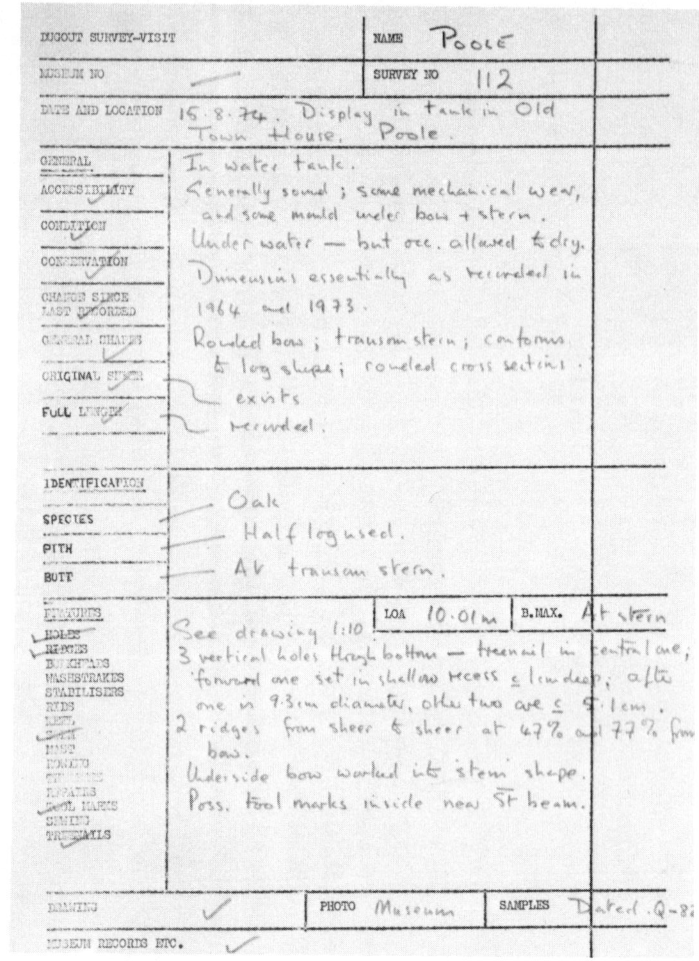
DUGOUT SURVEY-VISIT
NAME Poole
MUSEUM NO —
SURVEY NO 112
DATE AND LOCATION 15.8.74. Display in tank in Old Town House, Poole.
GENERAL In water tank.
ACCESSIBILITY Generally sound; some mechanical wear, and some mould under bow + stern.
CONDITION Under water — but occ. allowed to dry.
CONSERVATION Dimensions essentially as recorded in 1964 and 1973.
CHANGE SINCE LAST RECORDED
GENERAL SHAPES Rounded bow; transom stern; conforms to log shape; rounded cross section.
ORIGINAL SURFACE exists
FULL LENGTH recorded.
IDENTIFICATION
SPECIES Oak
PITH Half log used.
BUTT At transom stern.
FEATURES
HOLES
RIDGES
BULKHEADS
WASHSTRAKES
STABILISERS
RIBS
LEVEL
SEAT
MAST
ROWING
STEERING
REPAIRS
TOOL MARKS
SEWING
TREENAILS
See drawing 1:10
LOA 10.01m
B.MAX. At stern
3 vertical holes through bottom — treenail in central one; forward one set in shallow recess c 1cm deep; aft one is 9.3cm diameter, other two are c 5.1cm.
2 ridges from sheer to sheer at 47% and 77% from bow.
Underside bow worked into 'stem' shape.
Poss. tool marks inside near St beam.
DRAWING ✓
PHOTO Museum
SAMPLES Dated Q-82[illegible]
MUSEUM RECORDS ETC. ✓

Fig. 10.2 A record of the Poole logboat, compiled in August 1974. National Maritime Museum.

(i) A full, objective record of evidence is obtained which can be used by subsequent workers.

(ii) A great quantity of data can be validly summarised.

(iii) Selected information can be displayed in easily comprehended ways - tables, diagrams, curves, etc.

(iv) Numerical methods, and computational aids, can be used on the data to investigate hypotheses rigorously.

By advocating the use of numerical methods I do not mean that all archaeologists must necessarily understand the finer points of mathematical theory, but I do believe that they must become aware of the capabilities and limitations of these procedures, and thus use them wisely. An understanding of the use of confidence limits applied to radiocarbon dates is already commonplace among archaeologists, and indeed, as Spaulding has observed (1971: 15), all archaeologists use implicit mathematical reasoning even though they may be unaware of the fact; what is now required is that this reasoning be extended and made explicit.

Nor does my advocacy of these methods imply that I have no confidence in the work of those who have the apparently intuitive ability to compare and classify artifacts, a facility which is often the fruit of long years of experience. But I believe that the onus is upon them to demonstrate the reasoning behind this ability, and to define their terms, so that others can follow. Authority is not enough; the man with vision must demonstrate by quantification.

PART 2 - AN EXAMPLE OF THE USE OF QUANTITATIVE METHODS

a) The Survey. I will illustrate some methods of quantification by describing aspects of my recent study of the logboats of England and Wales. Here I will not mention the several false starts I made, but I will simply describe the method I finally decided was the best, and ask you to evaluate it against my definition ------ or your own definition ------ of quantitative procedures.

During the three years of this investigation, there were no new logboat finds in England and Wales, and so the research was partly based on other people's data, although in a surprisingly large number of cases I recorded for the first time by measured drawing, logboats which had been excavated up to 50 years ago. My first step was to trace all documentary evidence for the finds, whether published or otherwise, in learned journal, daily newspaper or on museum record cards. Particular emphasis was placed on tracing the earliest drawing or table of measurements, and on establishing evidence from which to judge whether the logboat had been recorded before or after significant shrinkage had taken place. When duplicates had been eliminated I had records of 172 unexpanded logboats. The evidence for each one was summarised on a standard Basic Data record form, (Fig. 10.1), from which I was able to formulate questions to be answered during my survey of those 67 logboats which survive in various stages of degredation.

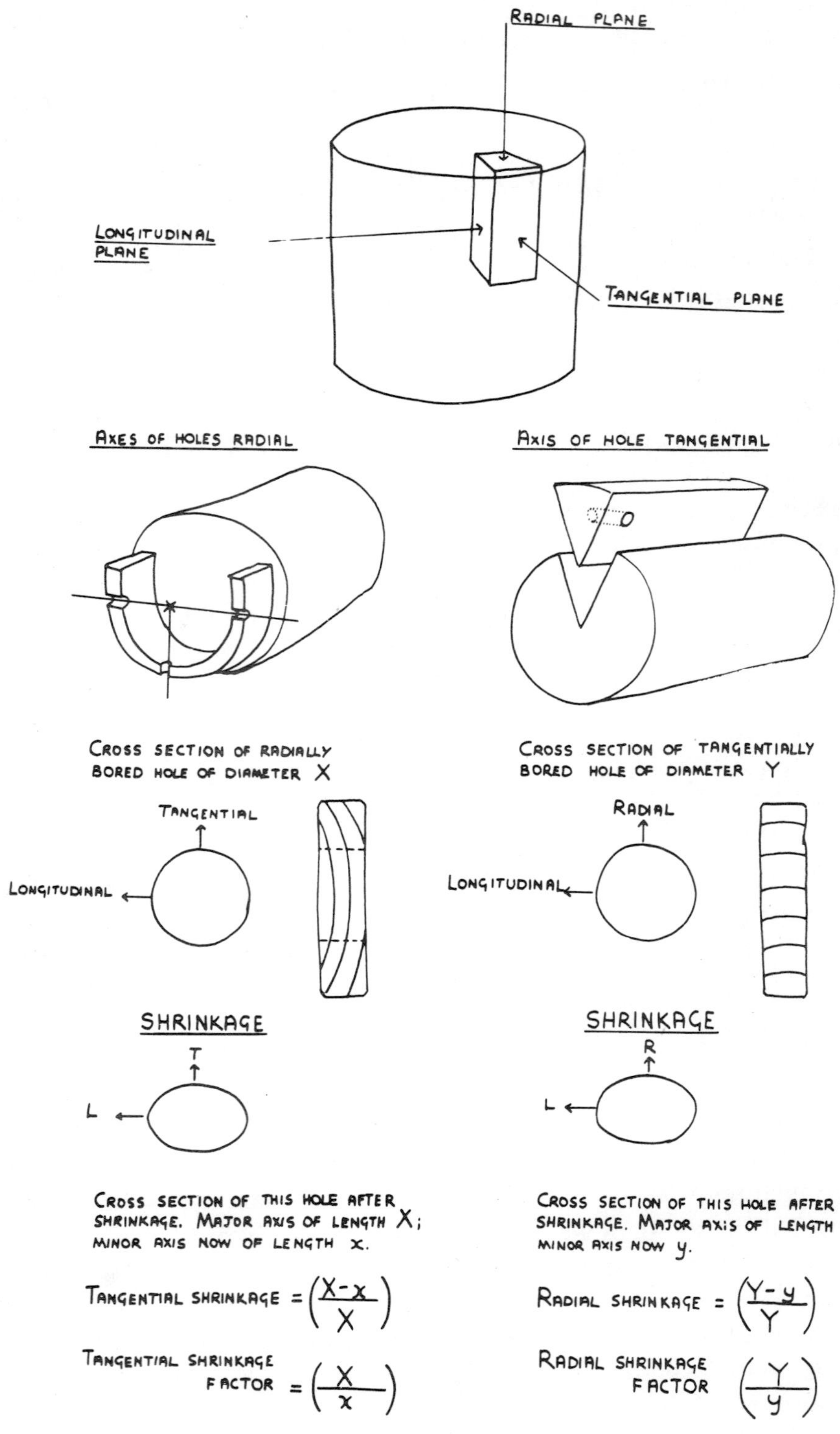

Fig. 10.3 The calculation of tangential and radial shrinkage factors. National Maritime Museum.

Outwitting those few museum curators whose aim seemed to be to keep their logboats hidden, preferably filled with other artifacts, I measured or remeasured 61 of these 67: the other six being inaccessible. The present state of those I measured was compared with their earliest recorded state and these details were noted on a second standard form (Fig. 10.2), on which were also recorded the features and fittings still detectable, and the characteristics of the parent tree - species, position of pith, orientation of butt - details which had seldom been recorded previously. I also attempted to trace any conservation treatment from museum records - but, in general, logboats are badly catalogued, and conservation records are almost non-existent.

A fair drawing at 1:10 scale was then compiled, based on the earliest reliable drawing; to it were added, where applicable, features identified during the survey but which had not previously been recorded. At this stage I considered I had assembled the objective data.

b) Interpretation. An attempt was then made to interpret features such as ridges, holes, and so on, using evidence from the Continent and Ireland where logboats which are more nearly complete have been excavated, and using ethnographic parallels where these were valid. In the event, several features remained unidentified. The next and crucial step was to estimate how much each boat had shrunk between the time it was in use and the time it was first reliably recorded. A mean shrinkage ratio was calculated from two sets of measurements, for those finds where these could validly be made:

(i) The first method was by comparing the lengths of the major and minor axes of the now elliptical cross section of holes in the logboats. It was assumed that the holes were circular when bored, and thus this comparison would be a measure of shrinkage. Wood shrinks differentially, the relative measurements for oak (Quercus Sp.) being approximately 1 in the longitudinal plane, 12 in the radial plane, and 25 in the tangential plane. Longitudinal shrinkage is relatively insignificant, whilst tangential shrinkage is approximately twice radial shrinkage. This differential shrinkage over a period of time thus distorts a round hole into an elliptical one (Fig. 10.3). The vast majority of holes measured had been bored radially into the log and thus had one axis of their cross section in the longitudinal plane of the parent tree and one in the tangential plane. The longitudinal axis of the cross section of the hole experiences negligible shrinkage and thus remains very close to the original diameter of the hole when bored. However, the timber shrinks significantly along the other axis and thus the ratio of major to minor axis is a measure of tangential shrinkage. Two boats in the survey (Kentmere and Llandrindod Wells) have radially split planking associated, and the ratio of major to minor axis of holes in these planks is a measure of radial shrinkage. Oak (Quercus Sp.) dried from the green state to a moisture content of 12%, shrinks c. 7.5% tangential and 4% radially, (Farmer, 1972:147). However when waterlogged wood is allowed to dry, it shrinks significantly more than this, the amount evidently being dependent on the degree of waterlogging, and on the degree of drying allowed to take place. Locally this shrinkage may also be effected by the run of the grain, the orientation of the hole, the presence or absence and the orientation of a treenail in the hole, and the timber species of both treenail and logboat. Thus the ratio major/minor axis of the cross section of bored holes varies not only from logboat to logboat but also between holes in the same boat, (but only in four boats was this range significantly great). The mean tangential shrinkage ratio for 123 holes measured in 31 log boats was 1.27. This is equivalent to (.27/1.27) = 21% shrinkage. The radial shrinkage ratio based on only 2 boats was 1.09 = 8% shrinkage. These values are comparable with measurements made on the timber from the WASA where the tangential shrinkage was calculated, by methods unspecified, to be 16% to 24%, and the radial 8% to 12% (Barkman, 1975:68).

(ii) The second method of estimating shrinkage was by comparing corresponding dimensions recorded at widely separated times, preferably the scantlings near the time of excavation compared with equivalent measurements taken in 1972/5. Åkerlund (1963, 155) used a somewhat similar method during his re-examination of the Nydam boat when he compared Engelhardt's 1863 measurements of the "circumference of the planking amidships" with Johannessen's 1929 measurements, and estimated the shrinkage between these two dates to be 9%. As the Nydam boat had been excavated for 4 months before Engelhardt took his measurements, Åkerlund concluded that 13% - 14% would be a reasonable estimate of Nydam's tangential (?) shrinkage between excavation and 1929. When this comparison method is applied to logboats the principal difficulty is to ensure that the corresponding measurements are taken at exactly the same part of the structure. An additional difficulty is that although many logboats have shrunk uniformly, maintaining a similar shape but of reduced dimensions, some have split along the bottom allowing the sides to open out, whilst others have reacted to the stresses of shrinkage by forcing the sides closer together at the sheer. In these abberant cases comparison of depths and of sheer breadths are invalid; but where the bottom is sound, comparison of breadths across the bottom may be made. A logboat's overall shrinkage pattern can be complex, and it is sometimes difficult to decide whether a particular comparison ratio is a measure of tangential or radial shrinkage or an intermediate figure. Nevertheless, both in individual cases and as an overall mean, these estimates of shrinkage based on scantling comparisons, generally supported the estimates based on elliptical hole measurements. Seventy-five main scantlings comparison gave a mean tangential shrinkage ratio of 1.22.

c) Hypothetical Reconstruction. The logboats selected for hypothetical reconstruction were those for which the evidence reached a minimum standard, defined as:

(i) full length recorded

(ii) original sheer line recorded over a significant length

(iii) representative transverse sections recorded.

These are strict criteria, and only 24 boats reached this standard. The criteria are so rigorous that although there may be variant fittings, reconstruction of the hull is limited to one possibility for each find, whereas in general a unique solution to the problem of deciding the full form of a fragmented find is unattainable, and multiple hypotheses are to be expected

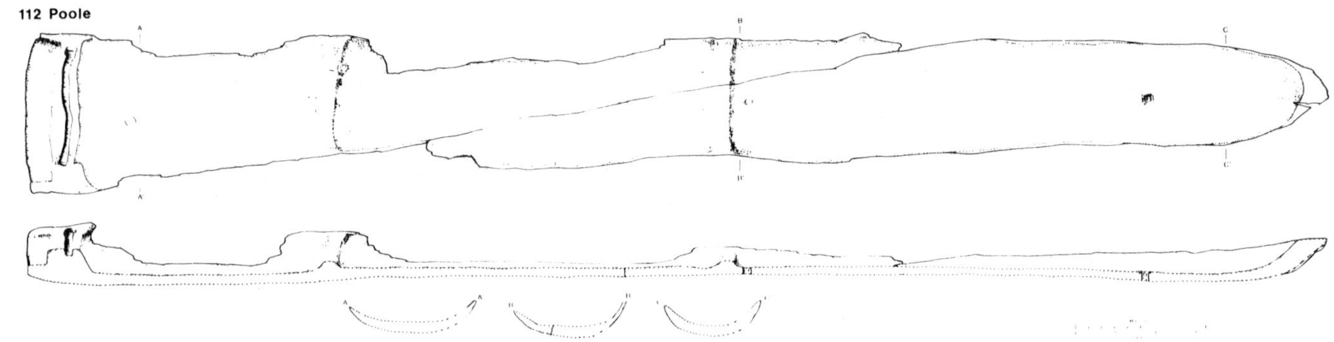

Fig. 10.4 The remains of the Poole logboat recorded in August 1974. National Maritime Museum.

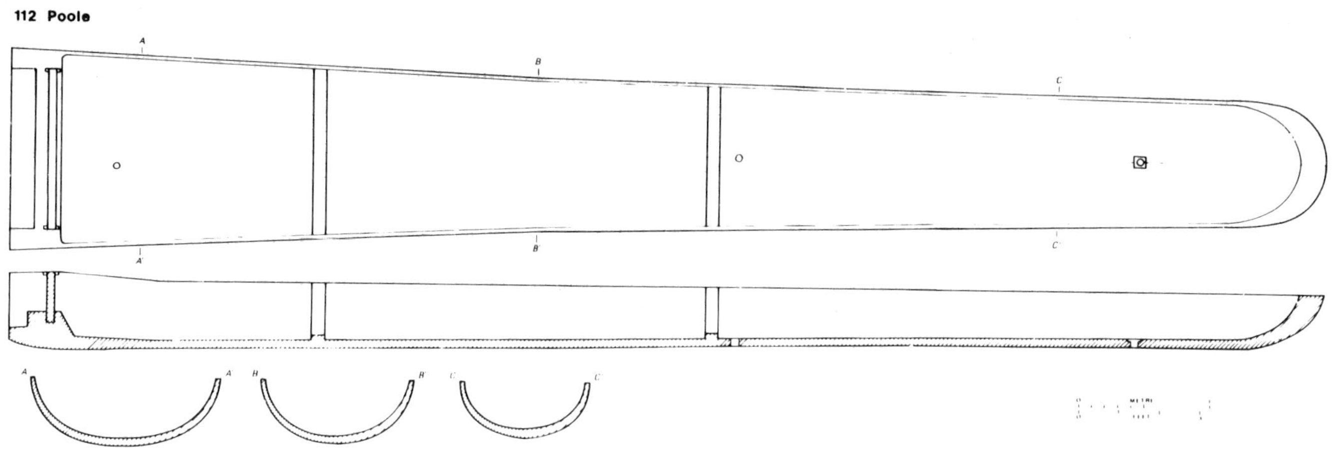

Fig. 10.5 The hypothetical reconstruction drawing of the Poole logboat, after an allowance has been made for shrinkage. National Maritime Museum.

(McGrail, 1976:25). Relaxation of these criteria could result in two or more equally plausible hypothetical reconstructions which could have significantly different hydrodynamic characteristics.

The evidence for each of the 24 logboats chosed for reconstruction was assessed to determine whether the definitive drawing was pre or post shrinkage. The original recorded measurements were used for 13 boats which were deduced to have been recorded before there was any significant shrinkage. But where the drawing was assessed as post shrinkage, a shrinkage factor was used to restore the logboat to its original size: seven boats were reconstructed using their individual shrinkage factor calculated from elliptical hole measurements; shrinkage factors calculated from comparisons of scantlings were used for two, where there were no holes to be measured; and the overall mean shrinkage factor was used to reconstruct two others for which no individual estimate could be made. In these calculations both tangential and radial factors were used as appropriate to the particular form of the logboat and its relationship to its parent tree. The general effect of using these factors to compensate for shrinkage is that, whilst the length of the boat remains substantially the same, the breadth and depth become 10% to 20% greater (Figs. 10.4 and 10.5). This effect has also been noted by Ellmers (1973:33).

d) Parameters. Given that the reconstructed logboat is an accurate representation of the original - and this depends on the sequence of recording and interpretation described above - there are still a number of critical parameters which have to be given a value before estimates of capacity and performance can be attempted. The reasoning behind my choice of specific value will be given in the full publication of the research; in this present paper I list them to make my assumptions explicit and so that they may possibly provoke discussion.

(i) The specific density of wood was taken to weight 60 kg (short and 25% moisture content. Thus oak (Quercus Sp.) weighs 800 kg/m^3 and ash (Fraxinus Sp.) 700 kg/m^3.

(ii) The average crewmember was tken to weight 60 kg (short and lean), and to have his centre of gravity at 1.1 m when standing, at .45 m when kneeling, and at .4 m when sitting. A case can in fact be made for the centre of gravity being at the point of pivot i.e. feet, knees, or backside; the values used in my calculations are therefore a worst case, and calculated metacentric heights may thus be pessimistic ones.

(iii) Cargo was chosen to give a range of density: stone at 2500 kg/m^3, corn at 680 kg/m^3, and turf or peat at 435 kg/m^3. It was assumed that 80% of the hollowed-out volume of the logboat was available for cargo, and that the load could extend above the sheerline where stability considerations allowed this.

(iv) Logboats carrying cargo were generally given a crew of one man, standing so that he could pole and also to give the worst stability condition. But the four longest ones (>8.5 m) were given more crew as they obviously required two or more men to manoeuvre them successfully.

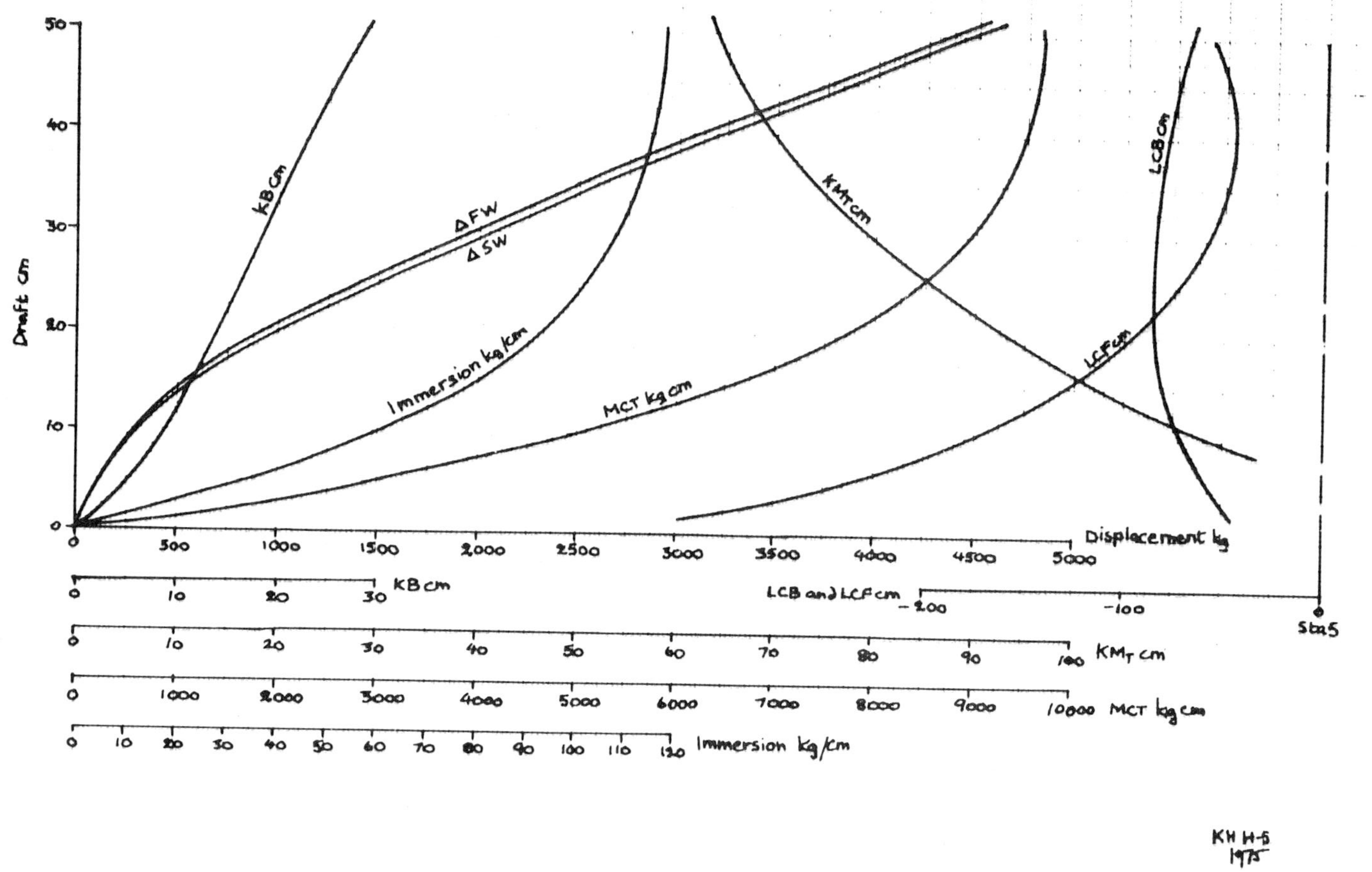

Fig. 10.6 The hydrostatic curves for the Poole logboat reconstruction, compiled by Burness, Corlett and Partners. National Maritime Museum.

(v) Freeboard requirements were taken to be:

Standard = .15 m

Minimum = that at which the sheer would be awash at 10° heel.

(vi) It was assumed that logboats would not be used with a negative transverse metacentric height ($\overline{GM}$), but that the bare minimum of a positive $\overline{GM}$ (.001 m) would be acceptable. ($\overline{GM}$ is a measure of initial stability - see fig. 16.5 in the paper by Coates in this volume).

e) Calculations. After three of these reconstructed boats had been analysed by hand computation and a limited array of hydrostatic curves plotted, this laborious task was kindly undertaken by Dr. Ewan Corlett, and Mr. Jones and Mr. Hamilton-Smith of Burness, Corlett and Partners, produced definitive hydrostatic curves using a programmable desk calculator (Fig. 10.6).

From the survey information and from the hydrostatics it was then possible to analyse each of the 24 reconstructed logboats as a log, as a boat, and as a load carrier. The following information could then be tabulated for each one.

(i) The diameter, girth and the volume of the parent log (minimum size).

(ii) The scantlings, the volume and weight of timber, and the volume of the "hollow" of the reconstructed logboat.

(iii) The results of load/draft/stability calculations at four standard states:

a. carrying the maximum number of men

b. draft restricted to .3 m, and carrying men and some high density cargo.

c. standard draft, equivalent to .15 m freeboard.

d. maximum draft, equivalent to minimum freeboard.

(c. and d.: carrying loads of varying density)

(iv) Coefficients which attempted to evaluate the conversion of the parent log into a boat: log conversion; load space. A coefficient which measured the ability of the logboat to carry loads in the restricted draft state. Five coefficients devised to measure the ability of the logboat to carry a range of cargo at standard draft in a stable condition - that is, with a positive metacentric height. One coefficient to be representative of the logboat's ability to carry cargo at maximum draft. Three coefficients to measure the logboat's ability to carry men. Ten naval architectural coefficients: block (C_M); prismatic (C_P); volumetric (C_∇); waterplane (C_{WP}); slenderness; symmetry; position of maximum beam; draft/length of waterline; maximum beam/draft. These 22 coefficients will be defined, and the reasons for choosing them will be given, when the full research is published. Some in fact, proved to be redundant, and these will not be used in future surveys.

TABLE 1

Coefficients measuring ability to carry various types of load at standard draft, in a stable condition

Logboat	Ability to carry high density cargo measured by: Deadweight / Volume of log	Ability to carry bulky cargo measured by: Volume of Cargo / Volume of Log	Ability to carry men measured by: No. of Crew members / Volume of Log
Baddiley Mere	184	.308	.960
Banks	147	.440	1.198
Barton	119	.183	.542
Blae Tarn	40	1.429	.420
Brigg	282	.356	.738
Clifton 1	75	.251	.933
Clifton 2	89	.331	.955
Ellesmere	113	.169	.273
Giggleswick Tarn	95*	.059*	zero
Glastonbury 1	96	.181	.401
Hardham 1	88	.329	1.228
Hardham 2	103	.293	.976
Hulton Abbey	112	.469	.754
Hylton 2	168	.274	1.395
Irlam	35	.578	.878
Kew	229	.212	.363
Llyn Llangorse	73	.191	.559
Oakmere	42	.978	1.042
Poole	101	.675	1.074
Preston 1	60*	zero	zero
Preston 2	60	.093	.297
Thornaby	111	.238	.351
Walton	163	.320	1.026
Warrington 2	63	.462	1.410

* = Crewman cannot stand, but sits at sheer level so that the logboat may remain stable

The naval architectural coefficients are well known and can be interpreted, in so far as they apply to a logboat form: some permit estimates of propulsion effort to be made. The other coefficients are experimental and they can only be interpreted in relation to the 24 boats analysed. Thus we may say that, on the evidence of one of these coefficients (Table 1, column 2), the Brigg logboat was very good at carrying high density loads when compared with the other 23 boats, and twice as good as the Banks logboat (per unit volume of log); but we do not yet know how Brigg compares with the general capabilities of prehistoric and medieval logboats. A greater number of logboats than the 172 in this survey must be investigated before we can obtain a sample which could be more representative of the original logboat population - a more objective functional classification may then result.

The results I obtained are obviously dependent on the values assigned to the parameters and, although the effect of varying some of them was calculated for two or three logboats, no systematic sensitivity analysis was undertaken due to the relatively great computational effort required. Neither large angle nor dynamical stability were investigated, only initial stability, as without this latter characteristic the other measures of stability, are irrelevant. Transverse initial stability was investigated in detail for each boat, and sufficient longitudinal initial stability calculations were undertaken to show that in general these 24 boats would be trimmed by the stern which in many cases would bring the sheerline approximately horizontal. The initial stability requirement in the transverse plane was sufficiently stringent in fact to modify the apparent carrying ability of the 24 logboats, in one way or another. Thus, when the three ratios in Table 1* were calculated for each boat, significant differences in performance became apparent, and these suggest the provisional functional classification shown in Table 2.

The three unplaced boats in Table 2 had relatively poor performance in all three roles; high density cargo, bulky loads, and men. My hypothetical reconstruction of the Giggleswick Tarn logboat was principally based on measurements recorded at about the time of discovery (O'Callaghan, 1864). The major part of this ash (Fraxinus Sp.) boat has fortunately survived, albeit in 196 separate fragments, which are undergoing conservation treatment here at Greenwich. After conservation it is hoped that it can be reassembled and reliable estimates made of its original size and form. Provisional shrinkage

---

* Notes to Table 1

1. When Displacement was used instead of Volume of Log in the denominator of these three ratios, generally similar sequences of relative ability were obtained.

2. The first and third ratios can be made dimensionless by multiplication by a constant factor.

TABLE 2

Provisional functional classification of 24 Logboats

All-round performance, 1st rate (Relatively good in all three roles: men, bulky loads, high density cargo)	Banks, Hylton 2, Poole, Walton
High density cargo carriers, 1st rate	Baddiley Mere, Brigg, Kew
High density cargo carriers 2nd rate	Barton, Ellesmere, Glastonbury 1 Thornaby
Personnel carriers, 1st rate	Clifton 1, Clifton,2, Hardham 1, Hardham 2
Personnel carriers, 2nd rate	Llyn Llangorse
Bulky cargo carriers, 1st rate	Blae Tarn, Hulton Abbey, Irlam
Bulky cargo carriers and personnel carriers, 1st rate	Oakmere, Warrington 2
Unplaced (Relatively poor in all three roles)	Giggleswick Tarn, Preston 1, Preston 2

Note: None of the 24 logboats analysed had anything reliably associated with it which might have helped to determine its specific use. Nor did the types of site at which they were found suggest any differentiation of function.

calculations indicate that its beam should be c. .7 m rather than the 2 ft. (.61 m) given in the 1864 record and used in my hypothetical reconstruction. A beam of this order would give a more stable logboat, and future work may show Giggleswick Tarn to have been a more useful boat than it appears to be from my calculations based on the 19th century record.

Giggleswick Tarn logboat (Fig. 10.7) is unusual because it is fitted at sheer level with external longitudinal timbers. These were called "auxiliary gunwales" for "heightening the sides" by Sheppard (1910:42), leading subsequent authors to describe them as washstrakes; it is more probable that they are stabilisers. The hydrostatic properties of this boat were investigated both with and without these stabilisers. They improve the transverse initial stability at the deep drafts where they are immersed, and this effect is more marked when the stabilisers are aligned with the longer axis of their cross section horizontal rather than vertical. This is as one would expect, the water plane area, and especially the beam, being increased more quickly with very little change in displacement (see, for example, Rawson and Tupper, 1976:97). It is probable that these fittings would also have a beneficial effect at light draft when subjected to heel, as large angle stability would be increased as the stabilisers entered the water. Constructionally, these timbers must have given longitudinal strength to the boat at sheer level.

It is also possible that the two Preston boats have been hypothetically reconstructed incorrectly. There is not one clear, unambiguous, early account or drawing of either of them, and a composite solution, not using a shrinkage factor, was used as a basis for the hydrostatic curves and my subsequent calculations. The resulting poor theoretical performance may indicate that the composite solutions were unsoundly based, and that the logboats had in fact been recorded post shrinkage. If the recorded scantlings were to be modified by the general mean shrinkage factor of 1.27, the resulting hypothetical reconstructions might be more useful and possibly more authentic. This work remains to be done.

f) The results of the investigation. After this sequence of recording, interpretation and reconstruction, and calculation, it became possible to compile a standard inventory entry for each of the 24 boats. This consisted of details of the logboat's history and its conservation; the earliest reliable measured drawing; a description of the boat, its features, and its parent log; an estimate of the logboat's shrinkage; an interpretation of some of its features; a hypothetical reconstruction drawing, and a coded description of its shape; specification of its carrying ability in four defined states; tabulation of 22 coefficients; and a general description of the logboat's probable use and performance. Inventory entries were also compiled for the remaining 148 boats in the survey in as much detail as the evidence warranted.

g) Conclusions derived from the Logboat Survey.

(i) Quantification is essential is worthwhile results, ones that can be sustained on rational grounds, are to be obtained. The methods, the selection of data, the parameters, the supporting evidence, and the discussion are

Fig. 10.7 The Giggleswick Tarn logboat temporarily reassembled, with one of the longitudinal timbers in the foreground. National Maritime Museum.

then clearly exposed for others to criticise.

(ii) To facilitate co-operation between research workers in this field and to avoid duplication of effort, future logboat finds should be recorded and analysed to agreed standards, and the data made available to all. A standard terminology is an essential prerequisite.

(iii) All feasible attributes/characteristics/features should be recorded even though some may not seem immediately relevant to current problems; subsequent workers will thus be able to use this information should the logboat not survive or should it only survive in a degraded state. It may be thought impossible to record everything, but Doran and Hodson are surely correct when they state (1975:101) that this problem can be solved by a clear definition of the term attribute, by eliminating redundant attributes, and by the use of common sense in not going out of the way to seek obscurity.

(iv) Current research methods including quantification can usefully be applied to old finds, providing the drawings and other documentation are critically assessed and compared with any surviving remains. Certain earlier archaeologists were not rigorous in their methods of recording; there are, for example, drawings of logboats which include elements of reconstruction without this being stated (McGrail, 1976:25).

(v) At present the number of logboats investigated by rigorous methods is small, and therefore the results of these investigations are necessarily relative and tentative.

(vi) Nevertheless even this small sample generates much data, there are many repetitive calculations to be done, and it would also be desirable to investigate the effects of varying the several parameters used in the calculations. There is thus a requirement for aids to computation.

(vii) Classification as attempted by Fox (1926), Graham (1966), and others based solely on morphology is inadequate. In a recent important publication Ellmers (1973), inter alia, attempts to define five logboat types differentiated primarily by the form of their transverse sections, and secondly by the shape of the ends, (Ellmers, 1973:35-40). This is of limited value, especially as it is based on a small sample of finds, and three of the types, (b), (d) and (e) appear to have only one member. Ellmers does discuss the function of some of the boats and gives estimates of their deadweight tonnage, but his methods of calculation are not given, and it is not clear whether he took stability into account although he does refer (1973:24) to some work on this by Timmermann (1956, 1957/8). Quantification could help Ellmers to make his case much more forcibly.

Estimates of performance, such as the ability to carry various loads, can be used to compare and contrast logboats and thus assign probable functions to them. Provenance, date, associated material, the techniques of building and repair, and the additional fittings incorporated in each boat should also be recorded. Only when this range of data is available from a large sample of logboats can comprehensive classification schemes be attempted. A dating programme is an essential complementary investigation, (McGrail and Switsur, 1975).

(viii) Ethnographic accounts of logboats are generally unsatisfactory. There are notable exceptions, but in general they tend to concentrate on form; where building methods are described there is undue prominence given to methods of expansion. Notes and Queries on Anthropology prepared by a Committee of the Royal Anthropological Institute, has a section dealing with travel and transport by water (6th edition, 1971:300-8) which enumerates what should be recorded. There is good advice here, but one would wish to see more emphasis placed on recording the details of construction, on elucidating the function of the features described, and on describing operational techniques such as any specialisation of use, freeboard requirements and loading strategies, apparent stability etc. The aim should be to record the boat ashore and in use in sufficient detail to enable draft, load, and stability calculations to be made for operational states actually used.

PART 3 - MORE GENERAL APPLICATION OF THESE CONCLUSIONS

If we extend these conclusions, especially (vi), from the logboat survey to a consideration of plank boats, we can appreciate more clearly the magnitude of our problems in boat archaeology. Compared with a simple logboat, each plank boat find will, in general, have more attributes (I used 90 attributes in my logboat survey), and will thus generate more data; there will be more calculations to undertake and more parameters to vary; probably there will be several equally plausible hypothetical reconstructions; and the complexities of sail will need to be investigated. Aids to computation and power assisted data storage systems may be desirable when investigating 24 logboats; when investigating plank boats (probably in far greater numbers) these aids are essential, and only electronic computers have the necessary accuracy, capacity and speed. With these aids the following tasks can be achieved:

a) A data bank can be constituted, in which the data can be rapidly updated, and retrieved on demand. Some forms of graphical input and output are possible, in addition to numerical ones.

b) The data can be sorted and analysed systematically; statistical analyses can be carried out, and selected data can be printed out in a variety of formats.

c) Repetitive calculations can be done rapidly and accurately, and the effects of varying the parameters can be investigated.

d) Variant forms of hypothetical reconstructions can be investigated and a series of calculations and simulations undertaken - possibly in conjunction with tank testing of small scale models.

e) Seriation, and classification by objective pattern recognition techniques can be attempted, where there are too many variables for these to be done by other means. Discussion of the application of these techniques to archaeological problems, may be found in Hodson, Kendall and Tautu (1971), Doran and Hodson (1975), and in the proceedings of the annual conference at the Computer Centre, University of Birmingham (Anon, 1974-1976).

It may be appropriate to emphasise at this point that detailed classification of plank boats is as premature as that of logboats, if not more so.

The data base of adequately recorded finds is too small to justify anything but the most tentative and limited classification (McGrail 1976:27). As well as requiring a large sample of finds, future classification studies should be based on an array of attributes and not on a single characteristic. Concepts such as Shell and Skeleton build (Hasslof, 1963), and the recently published Edge-joined and Non-edged joined classifications (Greenhill, 1976:60-88) are invaluable: they clarify some previously muddled ideas, and enable us to recognize distinctive approaches to boatbuilding. But they are two amongst many, and although they may be very significant characteristics, their full diagnostic value will only be revealed when we have comprehensive records.

If we are to make full use of quantitative methods we must aim at recording and analysing boat finds in such a way that we can recognise clusterings of attributes which identify:

(a) functional types e.g. the coastal trader, the river boat, the ferry, the raider, the ocean voyager, the high density cargo carrier (these are not mutually exclusive).

(b) widespread traditions, e.g. "Viking" or "Celtic" - if such existed.

(c) particular "boatyards" - what Crumlin-Pedersen (This volume, p. 164) calls "a certain local practice at a given time and place".

And we should look for those connected series which show progression from one set of attributes to another, orderings which may help to trace the development of the planked boat through time and space. I submit that these aims can only be attained by classification studies based upon many attributes.

PART 4 - A PLAN FOR THE FUTURE

I propose the following action:

a. The compilation of a provisional list of attributes to be tested internationally on a variety of boat finds, and on modern wooden boats.

b. All future boat finds to be quantitatively recorded to an agreed standard, using an agreed list of attributes.

c. Old boat finds to be critically re-examined and, where necessary, re-recorded to the agreed standards.

d. Documentary, iconographic, and ethnographic evidence to be similarly treated, as far as the methods are applicable.

e. The data from a, b, c, and d to be published in a standardised form and also to be stored in a computer data bank. This data bank would be the basis for all future studies, possibly by the methods outlined in Part 3 of this paper.

Doran and Hodson (1975:319-20) have enumerated objections to the idea of a data bank in archaeology; especially significant, and one with much support, is that voiced by those who prefer the idea of an individual scholar collecting and recording his own data, analysing and synthesising this data, and storing the information in his own files, for publication in the manner he considers appropriate, at the time he determines. A data bank, on the other hand, would hold pooled data which would be accessible to all. This may be an innovation but I suggest that therein lies the only way to real progress in boat archaeology; the only way we can handle and investigate in a systematic manner the increasing quantity of data. There are many practical problems (Doran and Hodson, 1975:320-22; Scholtz and Chenhall, 1976) but these are not insoluble. Universal archaeological data banks may never be practicable, but there is general agreement (Scholtz and Chenhall, 1976:94-5; Doran and Hodson, 1975:326-7) that data banks can be successful when they are planned for a specific limited purpose. The specialised and well defined topic of boat archaeology is, I suggest, a suitable subject for a data bank.

Acknowledgement

I wish to thank Professor F. R. Hodson, Professor K. G. Rawson, and Professor D. M. Wilson who kindly read and commented on an earlier draft of this paper.

REFERENCES

Anon., 1971, Notes and Queries on Anthropology, 6th edition, prepared by a Committee of the Royal Anthropological Institute, London.

Anon., 1974-6, Computer Applications in Archaeology, University of Birmingham.

Åkerlund, H., 1963, Nydamskeppen, Göteborg.

Barkman, L., 1975, 'Preservation of the warship Wasa' in Oddy, W. A., (ed), Problems in the Conservation of Waterlogged Wood: 65-105, Greenwich, N.M.M.M.M. + R. No. 16.

Doran, J. E., and Hodson, F. R., 1975, Mathematics and Computers in Archaeology, Edinburgh.

Ellmers, D., 1973, "Kultbarken, Fähren Fischerboote Vorgeschichtliche Einbäume in Niedersachsen". Die Kunde, N.S. 24: 23-62.

Farmer, R. H., 1972, Handbook of Hardwoods, H.M.S.O., London.

Fox, C., 1926, "Dugout canoe from South Wales with notes on chronology, typology, and distribution of monoxylous craft in England and Wales", Antiquaries Journal, 6:121-151.

Graham, J. M., 1966, Dugout Canoes of the British Isles. Unpublished dissertation for the degree B.A., University College, Cardiff.

Greenhill, B., 1976, Archaeology of the Boat, London.

Hasslöf, O., 1963, "Wrecks, Archives and Living Tradition" Mariner's Mirror 49: 162-177.

Hodson, F. R., 1971, "Numerical typology and prehistoric archaeology" in Hodson, F. R., Kendall, D. G., and Tautu, P. (ed) 1971:30-45.

Hodson, F. R., Kendall, D. G., and Tautu, P., 1971 (ed) Mathematics in the Archaeological and Historical Sciences, Edinburgh.

McGrail, S., 1976, "Problems in Irish Nautical Archaeology", Irish. Archaeological Research Forum, 3(i):21-31.

McGrail, S., and Switsur, R., 1975, "Early British boats and their chronology", International Journal of Nautical Archaeology, 4:191-200.

O'Callaghan, X., 1864, "On an ancient canoe discovered at Giggleswick" Journal of the British Archaeological Association, 20:195-6.

Rawson, K. J., and Tupper, E. C., 1976, Basic Ship Theory, 2 volumes, London.

Scholtz, S., and Chenhall, R. G., "Archaeological data banks in theory and practice", American Antiquity, 41:89-96.

Sheppard, T., 1910, 'Prehistoric boat from Brigg', Transactions of the East Riding Antiquarian Society, 17:33-60.

Spaulding, A. C., 1971, "Some elements of quantitative archaeology" in Hodson, F. R., et. al. (ed) 1971:3-16.

Timmermann, G., 1956, "Vom Einbaum zum Wikingerschiff" Schiff und Hafen 8: 130-8.

Timmermann, G., 1957/8, "Zur Typologie der Einbaume", Offa, 16: 109-112.

11. QUANTITATIVE METHODS AND BOAT ARCHAEOLOGY

J. M. Graham

Summary

Information from boat finds could be coded to form a data base for automatic retrieval. Many computer information retrieval systems include simple statistical techniques for summarising and investigating the data. If, in addition to the museum or corpus record the system also held data representing scale drawings of the boats, there would need to be a careful investigation of the rapidity with which the data base would grow. Analysis of the data can be aided by numerical classification techniques such as cluster analysis. At the present time, the wide variety of boat finds (and therefore characters displayed) does not seem to make the application of numerical taxonomy easy. The most promising research area would appear to be in performance prediction - pay-load, manning, speed and perhaps sailing characteristics of ancient craft. These calculations must no doubt go hand in hand with the tank testing of scale model reconstructions of the craft, but should eventually put any discussion of their function and relative merits on a firm basis.

The use of statistics and computers in archaeology is still a matter of research and academic debate. A well balanced survey of the field is given by the book Mathematics and Computers in Archaeology (Doran and Hodson, 1975). Here it is only possible to discuss a few specific problems that face those wishing to apply quantitative methods to the archaeology of boats. For 'historical' boat finds many problems familiar to archaeologists may already be resolved. Records may provide the precise date of the loss, and it may also be possible to say in what year and what shipyard the boat was constructed. The investigators of the wreck of the Dartmouth have no need of quantitative methods to place her in correct perspective. 'The Frigate Dartmouth, 5th rate, sank in a storm in the Sound of Mull on 9th October 1690 while on patrol for William of Orange in his campaign against the Catholic Scots of the Western Isles' (Holman, 1975, 253). Nor indeed could quantitative methods add anything to the place of origin of the Wasa. 'The Royal Warship Wasa, pride of the Swedish Navy, embarked on her maiden voyage on 10th August 1628. After only one nautical mile she met a sudden gust of wind in the middle of Stockholm harbour and capsised' (Bengtsson, 1975, 27). Under these circumstances the date, and the ascription of the boat to a particular class and function, do not rely upon archaeological inference but upon historical scholarship. Nevertheless it will be shown that quantitative methods can contibute to the recording and understanding of the best documented vessels.

The archaeological problems encountered will be discussed under three headings:

I Corpus compilation aided by computer information retrieval.

II Analysis aided by statistics and classification techniques.

III Function aided by calculation of displacement and performance prediction.

Many investigators amass records of sites or artifacts in an attempt to compile a corpus of all the information considered relevant. These frequently include location, some form of description or classification and a list of previously published references. In the case of a boat with good historical information such a record could at the initial collection phase include a definitive statement on the function, date and origins of the find, that would render the analysis phase redundant. However, the normal archaeological problem is to use the data that has been amassed to reach comparable conclusions. At this stage the process of analysis, frequently involves the comparison of scale drawings or measurements taken directly from the object or site, and normally yields a summary classification scheme which together with the measurements or drawings may then be added to the corpus record. The final ascription of function may be a matter of inference and debate in the absence of documentary information. Even though documents make it certain that the Wasa was a warship, there is still room for speculation as to what her handling characteristics in battle would have been had she not sunk on her maiden voyage. It is here especially that quantitative methods could contribute to existing knowledge.

I. CORPUS COMPILATION AIDED BY COMPUTER INFORMATION RETRIEVAL

Computers can be used as though they had infinitely large memories combined with total recall. However all computer services should be judged relative to the facilities that can be provided by other means, in particular by hand. It is impossible to record all the information about a boat, or any other object; it is necessary to select what is considered to be the relevant information. Obviously it is only possible to retrieve the information that has been entered into the storage system. If this includes the name and address of a boat together with its length, then we would expect to be able to retrieve the names of all boats under or over a specified length, or perhaps even to plot out a distribution map of them if an Ordnance Survey grid reference was included with the original information. Usually it is only possible to answer questions that people had in mind when the data was collected. Only in use does the adequacy or otherwise of the data base become apparent. It is therefore difficult to be categorical in stating what information should be entered into a system to help boat research. However a minimum subset would include:

1. Name(s) of boat find.

2. Date - when found.

3. Provenance - country, grid reference, county, parish; river, lake or sea; type of archaeological site if any, e.g. lake dwelling, harbour.

4. Associations - list of objects found in or near boat; and whether stratified with boat.

5. Age - directly by C14 or pollen analysis.

6. Bibliography - all references; or if unpublished, name and address of recorder; details of drawings and photographs and where housed.

7. Museum location - if boat preserved with state/stages of preservation.

8. Physical characteristics - length, breadth, type of wood, type of construction, e.g. dug-out, clinker, carvel. Features present, e.g. for plank-built number of frames, number of strakes, caulking, strake fastening etc. (see Evans and Fenwick, 1971, Fig. 4). Also a special section on repairs made to boats.

9. Propulsion - specific evidence for paddling, rowing or sailing.

10. Type of boat - e.g. dugout class 2 (Graham, 1966), Roman merchant ship, Greek Trireme.

11. ? Drawing of boat - this might include overall plan, longitudinal section, cross section; plus additional drawings of details, e.g. repairs, rowlocks, strake fixings, frame fixings, scarf joints, stem/stern post, mast step etc. For performance prediction the drawings should be held at a maximum reduction of 1:10, and every frame cross-section should be included. 20 cross-sections would not be too many to use in calculation.

It would only take a little thought to add further types of information to this list; but it is already necessary to put a question mark over whether drawings could be included in the information retrieval system, particularly in view of the scale and number of cross-sections required for performance prediction. This is because, while it is highly desirable to have them present, it is vitally necessary to consider the practicalities of the system. If the data base becomes large, then the size and amount of computer time used will be correspondingly large. It is vital to know, before attempting to set up a system, whether there will be sufficient funds available to allow the constant expansion and search of the data base once it gets to a size when it is too cumbersome to manipulate by hand. In other words, while it is possible to think of a computer initially as having an infinitely large memory, it should be always remembered that it is only as big as we can afford to buy or rent. The whole problem of entering drawings into the computer, is not whether it is possible; but rather, because it is necessary to hold them as a series of X-Y co-ordinates, whether we can afford to hold the drawings with sufficient accuracy to make the exercise worthwhile. The representation of a drawing in a computer can only be an approximation to the original, and its accuracy will depend on how many points we are prepared to store and retrieve; for any reasonable approximation this means an exceedingly large number of points, which would very rapidly exceed the amount of data required to cover the information of 1-10 above. Indeed, so rapidly would data from drawings grow that it seems impractical to include them in the data base.

The type of recall that we would expect from a computerised information system would of course include a complete listing of all the information held in the form originally submitted. The information base to cover points 1-10 could be printed out fairly directly on a line-printer or a typewriter (Teletype) connected to the computer; however the X-Y co-ordinates from the drawings would make no sense to the archaeologist if printed directly, but would need to be plotted out to form a drawing again. This can be done approximately using line-printer or typewriter, but for a closer representation a graph-plotter giving a permanent paper record, or a television type display would have to be used. However the real value of automated retrieval is that the archaeologist could select the recall of particular items of information. In this way simple questions about the data base can be answered very rapidly. For example, it should be possible to answer questions such as how many boats were found in this country between 1955 and 1965? in what museums are they? It would then be possible to print out the names of the boats that had not found their way into museums in this period, together with the name and address of their recorder plus references if any. This should make it easy for an investigator, or museum, to follow up these particular finds whether for research or administrative purposes.

In addition the retrieval system might be expected to include facilities for simple statistical summaries of lengths, breadths or other characteristics held in the data base. There remains the problem of what use the boat drawings might be (assuming we could afford to hold them on the computer system) other than printing or drawing them out again on request. The answer, of course, is that, perhaps together with other characters, we might expect to use the drawings to classify the boats in some way. And for this task simple statistics are hardly sufficient; this leads us to consider automatic classification as a separate problem. For a more general account of information retrieval applied to archaeology the reader is referred to James Doran 'Computer-based archaeological data banks' (Doran and Hodson, 1975, Chap. 12).

II. ANALYSIS AIDED BY STATISTICS AND CLASSIFICATION TECHNIQUES

Once more it seems necessary to make it clear that it is impossible to give even a passing summary of the topic of data analysis, for which the reader is referred to part 2 of Doran and Hodson (Doran and Hodson, 1975, Chap. 5-9). Many of the methods described are of relevance to the proposed boat data base and the simplest of these could be incorporated in an information retrieval system. These might include bar charts, histograms, scatter diagrams, as well as tests for similarity and correlation. Needless to say the value of the results would not solely depend on the analysis, but also very heavily on the information collected from the boats in the first place. In particular, if we wish to use automatic classification techniques the characters or summary measurements taken from the objects need to be subjected to scrutiny. They should be chosen to give a good representation of all the variation observed

BOAT NUMBER →

FEATURE NUMBER ↓	1	2	3	4	5	 9
1. L	14.0?	13.76	13.10	17.8	16.5?	
2. B	3.0?	3.35	3.6	4.0	3.5	
3. L/B	4.6?	4.0	3.6	4.5?	?	
4. D	0.9?	0.85?	1.3	0.45	?	
5. #F	13?	12?	20	38	18?	
6. F_{sp}	48.0	85.0	54.0	?	90.0	
7. X-sect	R	R	R	R	R	
8. Nib	✓	✓	✓	X	X	
⋮						
23.						

Fig. 11.1 Table of plank boat features for comparison. Note the small number of boats, the wide variety of features, and the imperfect state of preservation indicated by a question mark. All this means that a quantitative comparison is not very profitable at the moment. (After Evans and Fenwick, 1971 Fig. 4).

in the objects for classification, while at the same time the characters should all be logically distinct and separable. The problem of classification can then be seen as an attempt to allocate objects to classes so that those that have most characters or measures in common are found in the same group.

The discussion of the Graveney boat (Evans, and Fenwick, 1971) included a feature comparison table (Fig. 11.1). They list some 25 logically distinct features that they have observed on the boats under study; however further progress towards any numerical treatment of the data seems to be precluded at the moment, because there are only 9 specimens available for comparison. Nevertheless, this is potentially an ideal data base which we can only hope will be extended in the future with fresh finds and the probable addition of new features. In general there are insufficient numbers of boat finds preserved at the present time to make a numerical taxonomy approach a profitable one. The problem posed by such boats as North Ferriby, Nydam, Sutton Hoo and Graveney is made more acute by their rarity; in this case the typological difficulties are not diminished by automatic classification techniques.

There is however a corpus of boat data which seems to take the right form for a computerised classification scheme, and that is of dug-out canoes of which there are some 400 records from the British Isles (Graham, 1966). This would seem a promising beginning for some form of cluster analysis, though it is only possible to produce scale drawings of about 120. Typical examples are given as figures 11.2 and 3, showing additional stern-board and rowing features respectively. As the numbers of dug-outs are small, it does not seem appropriate to have too many characters or to rely on a great many detailed measurements. Table 1 gives a list of simple shape descriptors that can be applied to the scale drawings - broadly speaking they merely seek to determine whether the boat is square or rounded in plan, cross-section or longitudinal section. These characters were scored for each boat by hand and the similiarity matrix thus produced subjected to analysis. It would be possible, expecially if the drawings themselves were input to a computer system, to formulate numerical rules to adjudge whether a boat was squared or rounded. It would also be possible to use simple measurements like length or breadth; or more complex ratios

e.g. length/breadth
breadth at 1/5th length/breadth at 4/5th length

to add to the anaylsis. This opens the whole question of how adequately complex three-dimensional objects, like boats, can be represented using measures like ratios. For work with hand-axes Derek Roe's 7 or 8 measures seemed to afford one of the few possible approaches for applying numerical techniques to nearly 5000 specimens (Graham, 1970). Roe made it clear that he thought of his objects as being simple and wished to 'devise effective methods of study which are correspondingly simple, and can be used and understood by anyone without special equipment or special mathematical knowledge' (Roe, D.A. 1964: 251). While we concur entirely with his sentiments, we also observe that boats are more complex objects than hand-axes, and it is therefore to be expected that a more complex data base should be maintained and analysed for them.

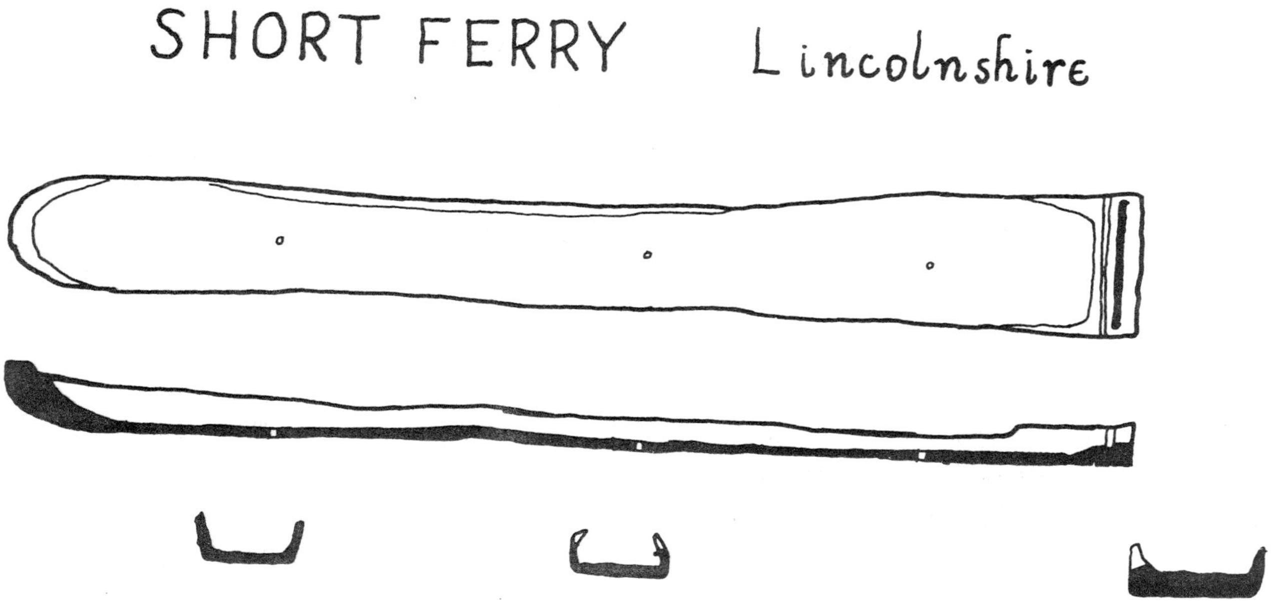

Fig. 11.2 Short Ferry (Lincolnshire). Scale 1:48. Stern-board dug-out. (After Thompson, 1954 Fig. 2).

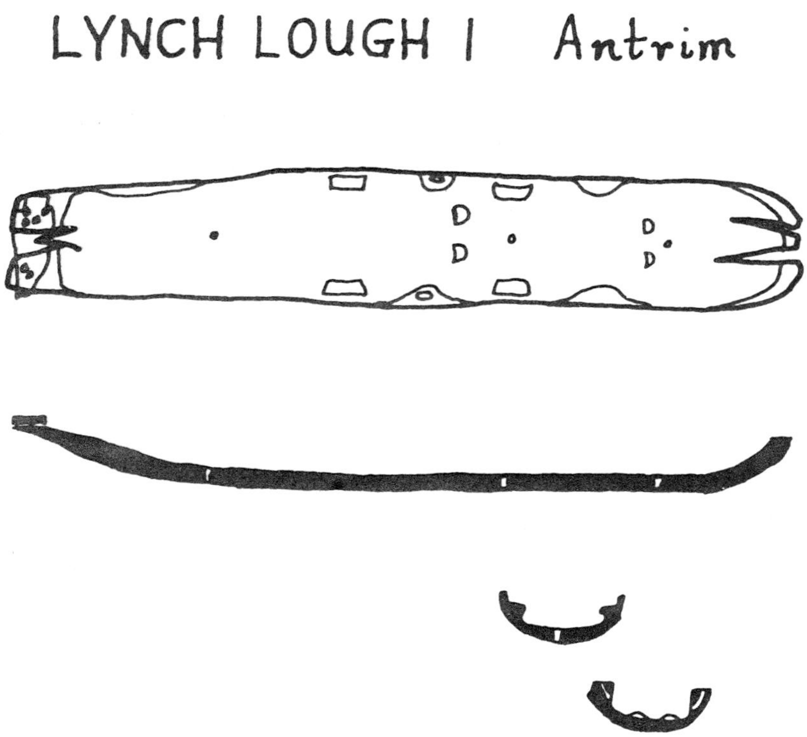

Fig. 11.3 Lynch Lough (Antrim). Scale 1:48. Rowing dug-out.

Table 1

Dug-out descriptive characters

A	Stern in plan - square.
B	Stern in plan - rounded.
C	Stern in plan - pointed.
D	Prow in plan - square.
E	Prow in plan - rounded.
F	Prow in plan - pointed.
G	Overall plan of sides - parallel.
H	Overall plan of sides - converge or diverge.
J	Overall plan of sides - widest amidships.
K	Stern longitudinal section - square.
L	Stern longitudinal section - rounded.
M	Stern longitudinal section - cutwater.
N	Prow longitudinal section - square.
P	Prow longitudinal section - rounded.
Q	Prow longitudinal section - cutwater.
R	Cross-section - flat bottom and square sides.
S	Cross-section - flat bottom and angle or rounded sides.
T	Cross-section - rounded throughout.

In this country there are literary and historical records that show dug-outs in use in the 18th and 19th century. Radiocarbon dates demonstrate their existence from 1500 B.C. to 1000 A.D. (McGrail and Switsur, 1975). It is recorded that in the year 1600, on the Erne Loughs of Ireland alone, there were between 400 and 500 dug-outs (Lucas, 1963 Calendar of the State Papers of Ireland, 1600, 280). Therefore, the boats surviving for analysis may well range in date by some 2000 years, but on the evidence of the Erne Loughs the sample size of 120 is exceedingly small. It is not therefore surprising that preliminary results using the shape characters seem to confirm that typologically similar dug-outs have been used for long periods throughout the British Isles. (Graham, 1966). Using more detailed characters in description would seem to make it even less likely that groups of dug-outs could be identified. Nevertheless every attempt should be made to record as much detail as possible, not just with dug-outs, but with all boat finds in the hope that one day we will have enough examples to make a more complete historical reconstruction.

III FUNCTION AIDED BY CALCULATION OF DISPLACEMENT AND PERFORMANCE PREDICTION

The tenor of the argument so far may have suggested that at the moment quantitative methods have little to contribute to the history of boating. However, when it comes to the consideration of function and performance, calculations can be made from accurate scale drawings of dug-out and plank built boats alike. These are based on the extensive work which has been carried out on hull and sail design applied specifically to the modern yacht. For a general introduction to the topic, the reader is referred to the works of Kay and Marchaj (Kay, 1971 and Marchaj, 1964). In principle, observations and tank test data, usually derived from boat models, are used to refine the accuracy of estimating by calculation yacht performance from hull and sail dimensions. These techniques are well tried and Kay claims that with good tank and tunnel test data from a scale model he can manage to obtain the sailing characteristics at all points within 5%. (Personal communication 27.4.76). This degree of accuracy relies on having data available on the drag characteristics of hull shapes very close to those boats whose performance is to be predicted. What now seems to be called for, is the initiation of a programme of research that applies the knowledge gained from modern yachts to ancient boats; this will inevitably involve some tank testing of scale model reconstructions of some ancient boats. If this programme can be carried out we should be able to ask a whole new series of questions. Simple ones: what is the draught of the craft when empty? What is its load capacity for a given freeboard? More complex questions: how stable is the craft? What speed could it achieve for a given man/sail power? And if a sailing vessel more complicated questions still, about the sail area and possible heave, pitch and yaw figures to try and give some idea of overall handling characteristics. The answers to most of these questions are not yet available; all it is possible to do here is to give some idea of the method by which the more elementary ones might be tackled. Eventually it is hoped that a similar methodology will be applied to the boats from North Ferriby, Nydam, Sutton Hoo and Graveney to help answer both the simple and more complex questions.

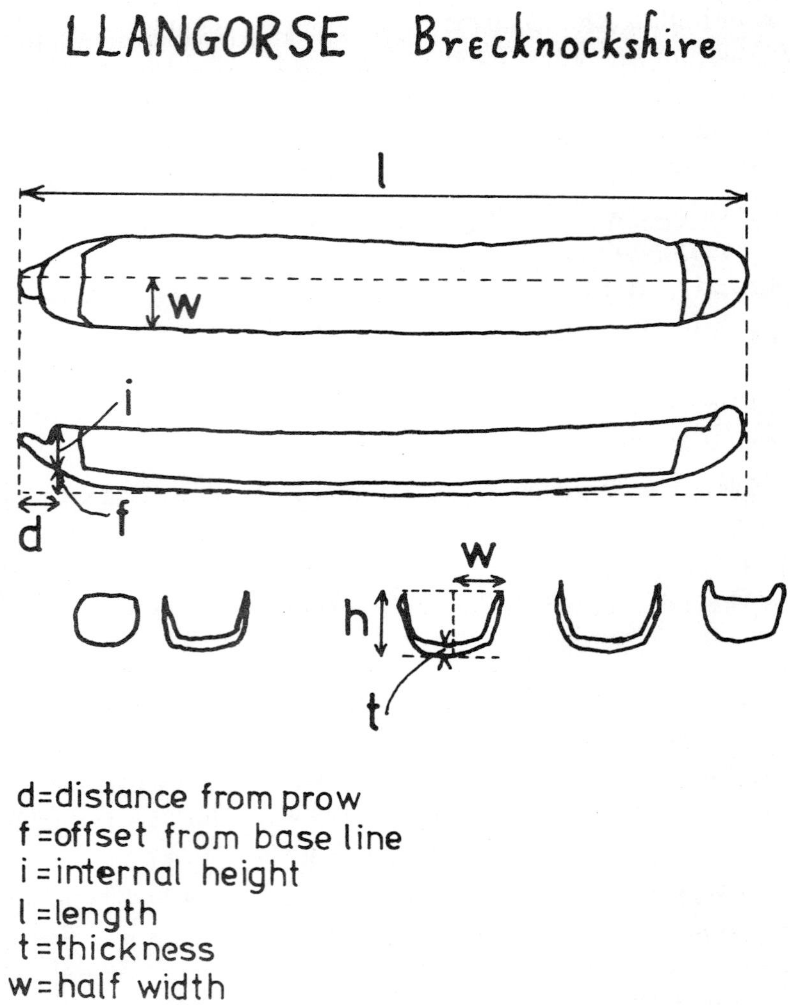

Fig. 11.4 Llangorse (Brecon). Measurements shown are required for calculations. In addition many x, y coordinates need to be taken to represent the outline of the cross-sections both internally and externally. Scale 1:48.

Clearly most of these questions could be answered by making a full-size reconstruction of the relevant boat, and taking measurements; however this is a very expensive and time consuming process and it is suggested that some at least of this information can be provided from calculations carried out with the aid of a computer. In any case it would seem wise to investigate the design by computer before building a model or a full-scale replica.

In order to calculate the draught of an empty vessel, it is first necessary to estimate the volume of wood used in the construction to obtain the weight. This requires some detailed measurements from scale drawings. Figure 4 shows a typical dug-out record. As well as the overall length (L), for each of the five cross-sections the following information needs to be input:

i the distance of the cross-section from the tip of the prow (d)
ii its height above the baseline (f)
iii height of the section (h)
iv its half-width (w)
v its thickness (t)
vi sufficient x, y co-ordinates to define the external and internal shape of the half-cross-section.

The method of calculation uses 19 equally spaced stations along the length of the canoe, and so it is desirable to divide the length by 20 and use this interval to define the position of the 19 on the longitudinal section and for each of these stations supply the required data for points i - v above.

The original data (fig. 11.4) contained 5 cross-sections: this could give representation of the volume of the canoe if taken without reference to the plan and longitudinal cross-section. For example the extreme prow and stern sections are only good descriptors of a very small part of the boat. Hence the input for the 19 stations is used to control which cross-section is used for area estimation. The succeeding steps in the calculation are:

1 Estimate the area of the half-cross-section for each of the 19 stations. Methods of estimating the area under a curve rely on slicing the profile into a large number of short lengths. The area contained within each slice can most easily be approximated by a rectangle. However for little extra computational effort a better approximation to the area of the slice is given by a trapizium. This is equivalent to replacing the profile by a sequence of straight lines, while Simpson's rule is more accurate still since it approximates the curve by a sequence of quadratics. (Kinney, 1962.)

2 The next stage is to find the volume of wood used in the construction and therefore the weight of the empty dug-out. This is given by:

   2 x $\frac{1}{2}$ cross-section area x inter station gap x unit weight of wood

   It is necessary to sum this for the 19 stations to produce the total weight of the boat.

3 The calculation of the freeboard of an empty boat relies upon Archimedes principle. That is that the weight of the canoe is equal to the weight of the volume of water displaced. Re-using the

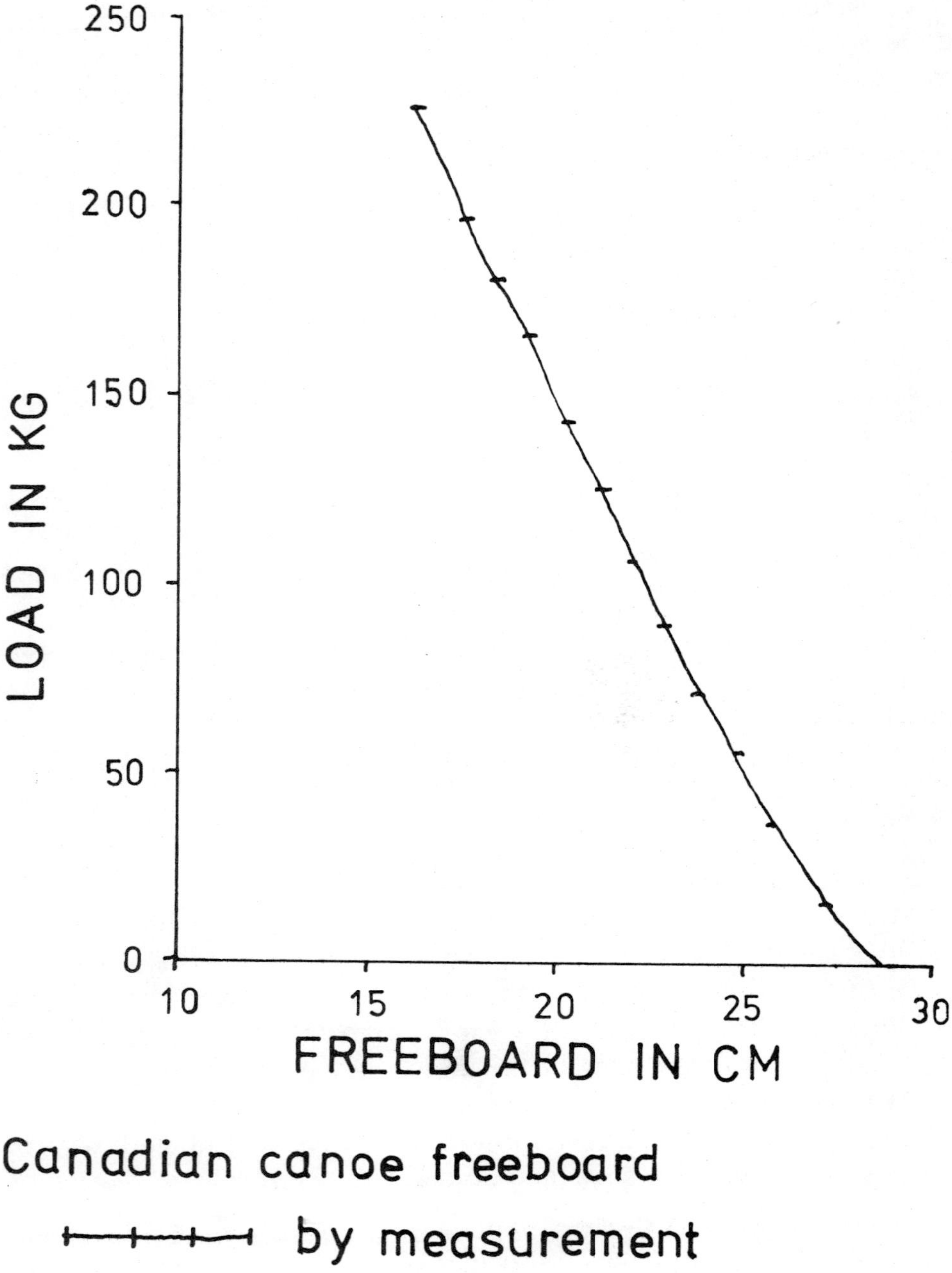

Fig. 11.5 Load x freeboard for a Canadian Canoe. Results from experimental measurement.

profile information referenced in stage 1 above, it is possible to calculate the volume of the boat "underwater" for a particular draught and to iterate to a position where the weight of the water displaced is equal to the estimated weight of the boat. This then gives the draught and freeboard of an unladen boat.

4 In addition the procedure of 3 can be entered with cargo weights, to produce a table of reduced freeboards according to load.

It is of course possible to check the results of such calculations by experiment. For this purpose a fibreglass Canadian canoe of about 4.5 metres was drawn and measured in the same way as the dug-out canoes, and suitably prepared data submitted to the program. The canoe was also floated in water and a whole series of freeboard measurements were taken, both with it empty and as more and more cargo of known weight was added. The results of the actual experiment are plotted on figure 11.5. For an empty boat the observed freeboard differs by 1 cm from the program estimate. However by the time 220 kg have been loaded the observed freeboard was still over 16 cm while the program estimates 11.5 cm. Bearing this cautionary information in mind it is still of interest to look at a program plot of the 4.67 m long Llangorse canoe. The estimated weight of the unladen canoe, assuming a specific gravity of 0.85, is nearly 300 kg; very heavy when compared with a plank built boat such as the Viking Faering replica which although 6.5 m long weighed only 107.7 kg (McGrail & McKee, 1974). The loading line approximates quite closely to a straight line, with a value of 26 cm freeboard for an empty boat. With 220 kg there is still a freeboard of 17 cm, while if the freeboard is reduced to 10 cm some 400 kg can be accommodated. Such a high payload would be inconceivable if dug-outs were thought of as sea-going, but the condition of some rivers and lakes might make a freeboard of 10 cm seem hazardous particularly if the cargo was not stable. If we think of this in terms of two crewmen weighing together perhaps 200 kg, this allows them to carry two passengers or a further 200 kg of payload. Immediately it would seem pertinent to enquire how fast these crewmen might be expected to propel themselves.

The estimation of the speed of a boat, relies on the calculation of the total hydrodynamic resistance of the hull. This can be thought of as two main components: frictional resistance and residual resistance. Marchaj further subdivides the resistance met by a yacht hull as made up of:

i skin friction resistance R_f
ii wave-making resistance R_w
iii induced resistance, or resistance due to leeway R_i
iv heeling resistance R_h
v eddy-making resistance R_e

(Marchaj, 1964, 232)

For a dug-out being paddled or rowed at a maximum of perhaps 5-6 knots, the resistance contributed by R_i, R_h and R_e is either irrelevant or so small that it can be omitted without significantly affecting the results. Thus for the dug-out the frictional resistance can be represented by skin friction R_f and the residual resistance by wave-making R_w.

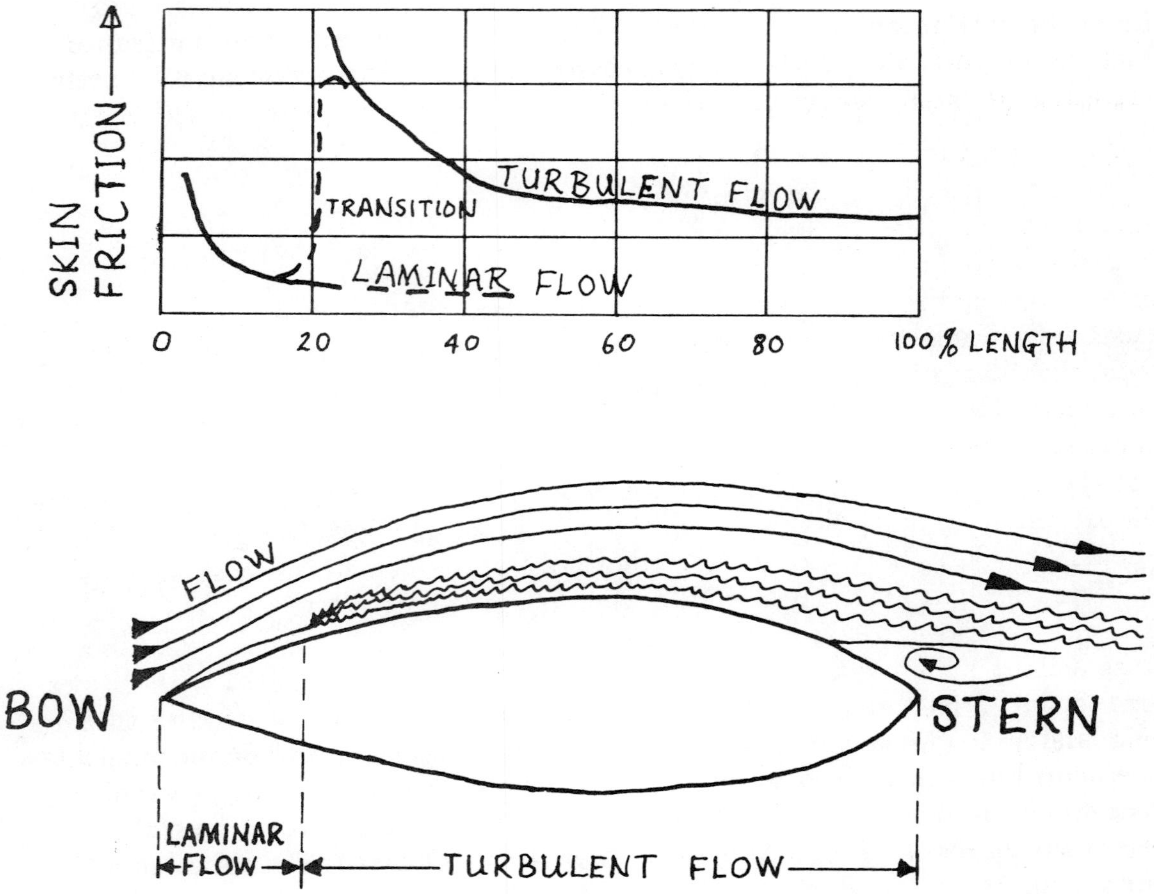

Fig. 11.6 Behaviour of the boundary layer around a typical yacht hull.

Skin friction resistance is to be seen at work in the boundary layer that surrounds a moving boat where water particles are accelerated from stationary to the speed of the hull. Figure 11.6 shows schematically the flow of water round the hull of an International Canoe at about 2 knots. A small portion of the bow experiences laminar flow causing relatively small resistence, while the majority of the hull produces turbulent flow and therefore considerably increased resistence. The point at which laminar flow gives way to turbulance is partly determined by the smoothness of the hull. By comparison with modern yacht finishes it only seems safe to assume that dug-outs would have had a finish somewhere between ordinary wood and average barnacle growth 0.5 mm - 5 mm. Thus laminar flow would normally be restricted to less than 10% of the hull. The actual resistence contributed by the laminar and turbulent flow is a function of

i skin friction coefficient C_f
ii the mass density of water
iii the velocity of the boat
iv the wetted area of the hull.

Unfortunately the skin friction coefficient is not a constant associated with the hull surface alone but varies according to the nature of the flow in the boundary layer, the length of the hull and the velocity as well as the roughness of the hull surface. Reynolds number can be used as a guide to the associated frictional coefficient C_f for a particular case.

Residual resistence also needs to be considered for its contribution to the total resistence offered to the hull. The full-scale tank tests of the International Canoe show this very clearly (Fig. 11.7). At a speed of 5-6 knots, residual resistance represents approximately 50% of the total resistance. The portion of the residual resistence relevant to a dug-out is caused by the movement of water particles that dissipate the kinetic energy imparted to them by the passage of the hull. In general terms the resultant drag is a function of the density of the water, and the area, shape and velocity of the hull. However, it is necessary to consider that wave-making is caused along the water-line length of the hull. Bernoulli's equation gives a theroretical explanation of this (Marchaj, 1964, 245). The waves transverse to the direction of travel, as opposed to the divergent bow and stern waves, require special attention. It is the height and number of these waves along the waterline that is vital for resistence estimates. As the waterline length of the hull is known and the distance between wave crests is a function of velocity, the number of transverse waves generated can be determined. This is known as Frounde's number or the speed-length ratio, and is used for non-planing hulls to give a forecast of the maximum speed of a yacht. At low speeds the wave length and amplitude will both be small. However the height of the transverse waves, a critical parameter of residual reisitence, is effected by hull shape and for this reason towing tank measures are frequently used in place of analytic calculation in the determination of residual resistence.

Figure 11.8 shows a resistance curve produced experimentally for the Gokstad Faering (McGrail & McKee, 1974). Two people were able to propel her at over 7 knots while one person could achieve somewhere between 4-5 knots. A good test for a computer program written to estimate speed would be to check

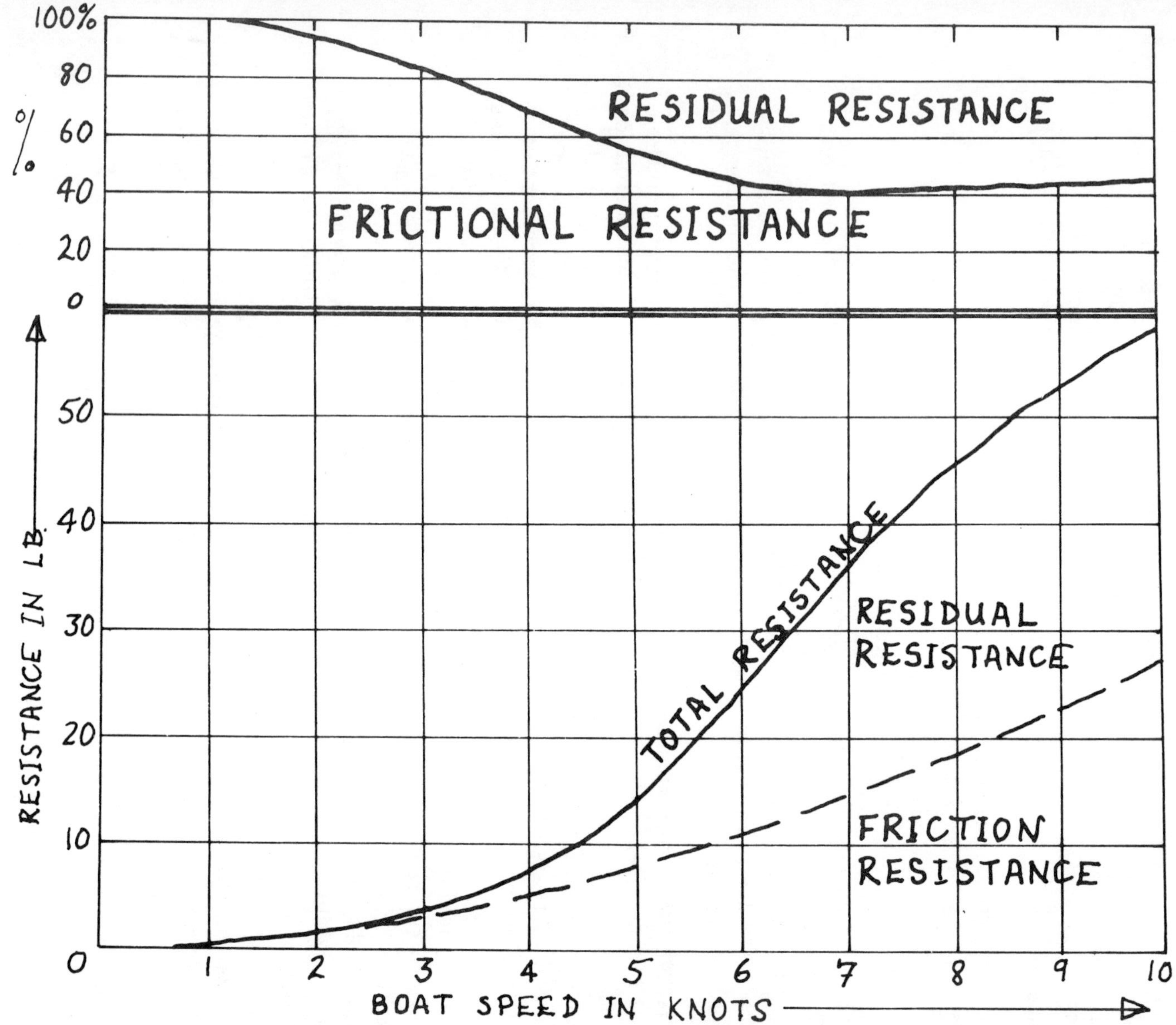

Fig. 11.7 Typical yacht hull showing contribution of frictional and residual resistance from full scale tank test data.

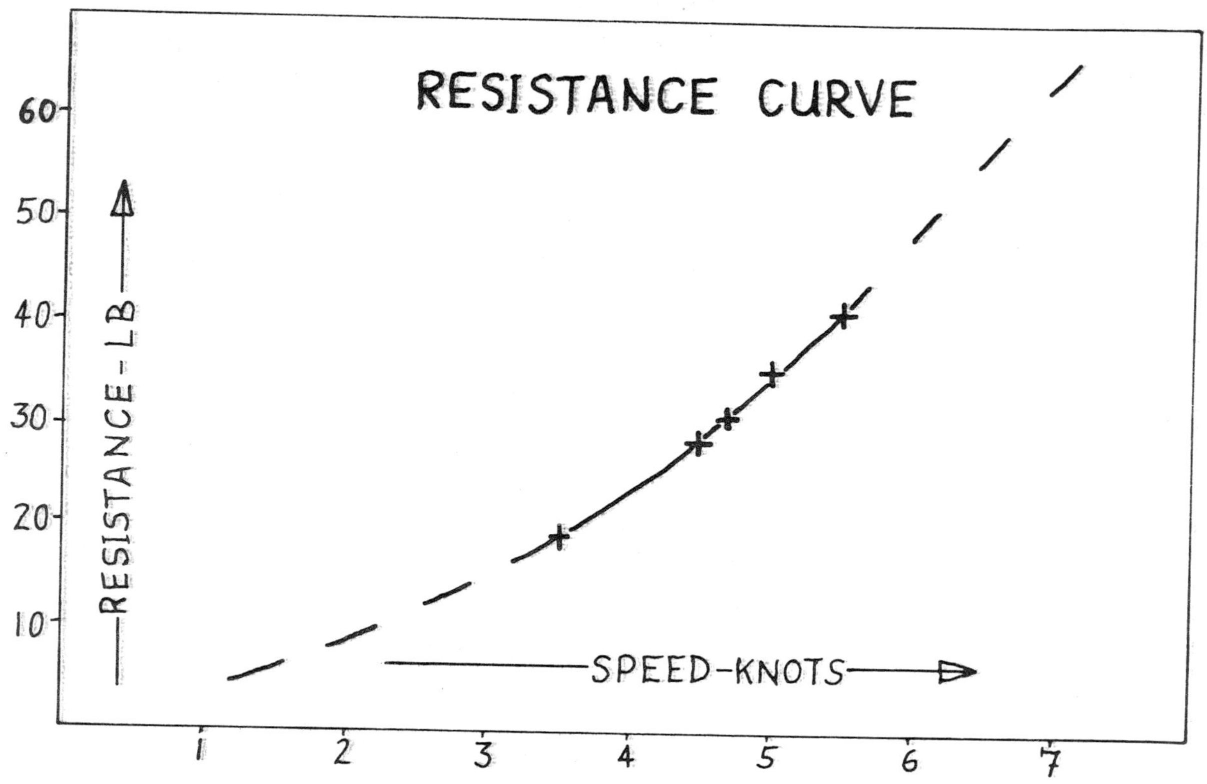

Fig. 11.8 Gokstad Faering replica resistance curve produced experimentally. (McGrail and McKee, 1974. Fig. 19).

for agreement with these results, before being applied to the Llangorse dug-out in answer to the question how fast one or two people with paddles might go. Dug-outs vary in length from 2-16 m, and so in discussing their function and performance it would be desirable to have comparative figures for a number of specimens. At the same time data from plank-built boats would facilitate more precise comparison of those examples previously mentioned from North Ferriby, Nydam, Sutton Hoo and Graveney.

REFERENCES

Bengtsson, S., 1975. The sails of the Wasa. Unfolding, identification and preservation. Int. J. Naut. Archaeol. 4: 27-41.

Doran, J.E. and Hodson, F.R., 1975. Mathematics and computers in archaeology.

Evans, A.C. and Fenwick, V.H. 1971. The Graveney boat. Antiquity, 45: 89-96.

Graham, J. M., 1966. The dug-out canoes of the British Isles. Unpublished dissertation for the degree of B.A., University College, Cardiff.

Graham, J.M., 1970. Discrimination of British Lower and Middle Palaeolithic hand-axe groups using canonical variates. World Archaeol. 1: 321-342.

Holman, R.G., 1975. The Dartmouth, a British frigate wrecked off Mull, 1690. Int. J. Naut. Archaeol., 4: 253-265.

Kay, H.F., 1971. The science of yachts, wind and water.

Kinney, F.S., 1962. Skene's elements of yacht design.

Lucas, A.T., 1963. The dug-out canoe in Ireland. The literary evidence Särtryck ur Varbergs Museum Årsbok, 68: 57-68 Sweden.

Marchaj, C. A., 1964. Sailing theory and practice.

McGrail, S. and McKee, E., 1974. The building and trials of the replica of an ancient boat; the Gokstad Faering. National Maritime Museum, London.

McGrail, S. and Switsur, R., 1975. Early British boats and their chronology. Int. J. Naut. Archaeol., 4: 191-200.

Roe, D. A., 1964. The British Lower and Middle Palaeolithic: some problems, methods of study and preliminary results. Proc. Prehistoric Soc., 30: 245-67.

Thompson, F.H. 1954., Archaeological notes for 1952 and 1953. Lincolnshire Archit. Archaeol. Soc., New. Ser. 5: 75-82.

DISCUSSION

Bruce Mitford: The Nydam ship as a result of Akerlund's work appeared as something which was essentially a good hull form and a very seaworthy vessel. This kind of conclusion as to seaworthiness and performance based on calculation of shrinkage, is of great interest to archaeologists like myself concerned with the Anglo-Saxon invasions.

Are there different shrinkage factors for different species of wood?

Dating is absolutely essential; any programme such as we have heard advocated should include properly controlled carbon 14 tests on what is after all material available in some bulk. At the recent International Congress in Nice, Ole Klindt Jensen described some unpublished dugouts and other boats found on the island of Bornholm. There were 42 boats or fragments of boats excavated by him on this island which were in almost all cases shadows, stains in the sand. They date from about the birth of Christ to the 4th or 5th centuries A.D. Some of the boats had one or two strakes a side, but most of them were dugouts with a single gunwhale strake laced on. It was said that these boats were identical in all respects with some well dated 9th century A.D. examples which had been excavated near Stockholm and the lecturer then went on to say that the same type of boat was either in current use in Finland or had been in use up to 1930. He was able to show photographs and films of these dugouts being made. So here you have a time span of nearly two thousand years with a virtually identical primitive type of boat continuing to be made. I would have thought that the absolutely basic thing is to try and get datings in this country on types of boat which can last for a very long time without very much technical change. If there proves to be no change this would suggest that there is a continuity of the economic need which gave rise to that particular design of boat.

McGrail: We do have a dating programme for the logboats I have investigated and there are some 40 samples being dated at Cambridge. There are problems in dating some of them because of difficulty in tracing the conservation treatment they have had. Similarly one has to know exactly where in the log the sample is from. There have been other logboats dated in the past and we are not clear now where in the log those samples were taken, and so there may be errors of three or four hundred years. There is a paper by McGrail and Switsur in the International Journal for Nautical Archaeology, 4 (1975) discussing aspects of this work.

Shrinkage is a measure of the change in dimension relative to the original dimension, and this is what Dr. de Jong calculated in his paper for the various types of waterlogged wood, and I have calculated it in my paper. I also use another term, shrinkage factor which is a rearrangement of the numbers so that I can multiply the existing scantlings by this factor to get back to the original pre-shrunk dimension.

Akerlund published shrinkage data for the Nydam boat, and so did Barkman for the *Wasa*; these are percentage shrinkage compared with the original dimension. All these figures - de Jong, Akerlund, Barkman, and McGrail - are for the shrinkage of waterlogged wood when allowed to dry. The shrinkage of most species of fresh wood is well known and the figures are published for example by the Forest Products Research Laboratory in their Handbooks. But the data for waterlogged wood is not well known, and we are only now collecting it. The shrinkage of waterlogged wood is generally much greater than that of fresh wood. I have not yet worked out whether there is a correlation between time of deposition and shrinkage, but that may come out one day.

McKee: How can you be sure that the original builder used a tool that drilled a round hole? One must examine the tools as well as the boats.

McGrail: Until I get firm dates for these logboats I cannot correlate the shape and size of holes in them with the corresponding contemporary tool kits. So for the present this is an assumption, but I think a reasonable one. The shrinkage factors I used were mean values calculated from several holes in a logboat, and the individual values generally had a small range. There are some anomalies which I will discuss when my full results are published.

Wright: In Britian we have a relatively small sample of dugouts compared with our friends on the Continent. Is any progress being made on the Continent towards agreeing on a common system of recording data? We cannot deal with any one country in isolation in this field.

Ellmers: In Western Germany I am the only one working in this field. I have only looked into a few museums in Germany and I have published the dugouts from only one museum (*Die Kunde*, NS24 (1973): 23-62). There are many other dugouts awaiting investigation.

de Weerd: There is so much material in the Netherlands that it is hardly possible to put all the data in a computer. I believe, that at this stage of boat archaeology the most important thing is to make files and computers will help us to be as perfect as possible in making these files.

McGrail: I doubt we will get full agreement here today, but what I would like to see come out of this symposium would be a small working group which would look further into the matter. As I see it the problem is that I do not have access to de Weerd's or Crumlin-Pedersen's information and they do not have access to mine. I have given you some facts today, but even when I publish my full report it will not contain all the data. By having an international data bank we would all have access to each others data. But before we can set up a data bank we have to agree on a standard recording system.

Each archaeologist can operate on the international data in the way he thinks best. He can investigate some of the possibilities described by John Graham, and he can do this by hand computation if he wishes or he can use a computer or a desk calculator. We can each do our own investigation of this pooled data, but obviously there may be advantages in working together and benefitting from each others experience.

Crumlin-Pedersen: When Sean McGrail talks about a data bank everyone thinks of a computerised system but let me mention that the International Journal of Nautical Archaeology is an excellent data bank, the files of Museums are very fine data banks, although they may not be efficient for this purpose. I think it is very important to remember that John Graham mentioned that the numerical system is a very bad system for the storage of drawings. The Danish Research Institute for Modern Shipbuilding has developed a computerised system for calculating all the hydrostatic data for a given ship and they take any of our construction drawings, and give us a complete set of curves for that ship. If we give them the maximum and minimum solutions as used by Eric McKee for the Graveney boat reconstruction (Fenwick V.H. (ed) The Graveney Boat) they can do it all in one machine and give us the full set of information for the whole range of variations. I think we will basically have to rely in the future on visual presentation of the finds in drawings, supplemented by numerical analysis.

Evans: One of the first things that one does want to establish is a fixed range of measurements that should be taken. John Graham pointed out that when Valerie Fenwick and I were working on the Graveney boat we took a selection of ships that had been published and we did our best to compare and contrast features to see whether there were any points of similarity. We found that in each case there was one vital measurement left out that we particularly needed. I think, as Sean McGrail said, that perhaps a working party could be set up to explore possibilities of, I hate to use the word, parameters within which boat archaeology could work, I think this is the way for the future recording of ships. When we are next faced with a boat excavation, we may then have an immediate idea of what measurements to take.

McGrail: I think we should attempt to get a standard list of features or characteristics, so that we don't miss anything when recording. I think there were 15 features listed in Angela Evan's table of comparisons with the Graveney Boat, but I used 90 on my logboats. For a plank boat there could be an incredible number. I think however that we must look ahead and I feel that unless we do record boats in a standard manner and unless we get the data on to a computer, we won't be able to cope with the information.

Cutbill: You cannot run an international data bank through the good will of an archaeologist in one Museum who happens to be enthusiastic; that does not give you the continuity. Secondly you have to have data to put in it in this case computer data. We have heard some of the difficulties of collecting this data and it seems to me we are some way off having the kind of data that we can usefully bank here. Thirdly data banks are expensive because of the organisation. I do not want to be pessimistic but, while it is very reasonable to use computers for doing calculation for individual researchers, and for researchers with access to computers to use them for information retrieval, their use for international data banks either by Museums or archaeologists, is a very different problem because of the organisation and the money involved. These are the sort of questions you should be asking if you are looking towards international co-operation. To get computers the techniques are easy, now think about the organisation that you are going to have to set up, what it will cost and how you will run it.

Taylor: The International Journal for Nautical Archaeology has begun a series of articles comparing the details of published Mediterranean wrecks. This should draw readers attention to wrecks which may be similar to their own and then they can get full information from the excavator. Possibly we could do something similar to launch Sean McGrail's scheme.

de Weerd: We have had an archaeological data bank at the Centre for Archaeological Investigation in the Netherlands for some years and it is working fairly well, but also fairly slowly because of the expense. There are only 5 or 6 classes of monuments and finds being put into the computer. The time you have to spend on it is quite considerable. Possibly we can arrange to exchange information between such a data bank working on general archaeological material and a future bank working for boat archaeology. Maybe we can exchange experience on the setting up of such banks.

Sutcliffe: The English Channel was recently crossed by a three piece suite. This means that the successful factor in what might be called primitive craft is the man on board and this is a highly uncomputable factor. If prediction of performance is presented in a scientific form, do you think there is a danger of formalising interpretations for the future, simply because they have a numerical and scientific basis? This does in fact frighten somebody working intuitively and may also alter interpretation quite erroneously.

Graham: Yes there is a danger of this, however any published calculation of predicted performance will be checked by others working in the same research area. The calculations should give some guidance of how a boat is going to respond under various conditions.

Wright: Would representatives of the continental countries welcome an initiative by the National Maritime Museum here in setting up a basic, simple, recording scheme which had international acceptance?

Ellmers: I think we should have a sort of standard table of what is to be recorded on an excavated dugout or an excavated ship. Of course there will be some individual things which must be added to this table for particular finds. In this way we will help each other to improve methods and other research workers will not have to start right at the beginning and unknowingly repeat work in compiling a standard table which has already been done.

Marstrander: We have many dugouts in the Norwegian Museums, mostly a burden on the Directors, and I think it would be most fruitful for us to get international co-operation. I also emphasise that we have to agree on a common list of which points we have to record.

Bruce Mitford: I am not going to attempt to summarise the discussion but there does seem to be one clear point on which a consensus has emerged and that is there ought to be a getting together to agree on exactly what features and details should be recorded in boat finds. The National Maritime Museum would seem to be a very good place to have such an international get together as a real basis for advance in this field.

Editorial Note: This topic was further considered in the discussion after Basil Greenhill's Summary.

12. SOME PRINCIPLES FOR THE RECORDING AND PRESENTATION OF ANCIENT BOAT STRUCTURES

Ole Crumlin-Pedersen

Summary

The paper stresses the need for a detailed first hand recording of all parts of ancient boat structures to be undertaken after the excavation stage and before reassembly. Three methods: scale drawing, full scale tracing and stereo-photogrammetry are described and discussed in relation to the problems of reconstruction and publication. Fold-out plans of all planks etc. are recommended as the link between the excavation plan and the naval architect's plans of the reconstructed ship. The publication of a boat find should provide sufficient information for a modelbuilder to produce a model of the vessel correct in every detail and with a clear indication of the limits between facts and fiction.

The basic problem for boat archaeologists is to bridge the gap between some mute fragments of dirty, old wood, claimed by him to be parts of an ancient boat - and a proper presentation of the vessel involved, telling the public as well as the colleagues as much as possible about the boat or ship - the technical, economical and social aspects of this structure from the past - but also clearly stating the limits between guesswork and facts.

We have a few very good examples of vessels, which need very little interpretation to be understood - a classical example being the Oseberg Royal vessel excavated in 1904 and now to be seen restored at Oslo with less than a few percent new wood in her construction.

A few other examples of this kind of extremely well preserved vessels are found: The Cheops Boat, The Bremer Cog, The Wasa.

But the general situation for the boat archaeologist is to find more or less the bottom part of the vessel, leaving him or her in a bad situation, as a fragmentary boat-find require even more care (and thus money) to be properly recorded and presented than a complete boat does, though it is far less spectacular.

One of my points is, however, that the only justification for spending considerable time and money on the excavation and conservation of ancient ships rests with the increase of knowledge gained - and that is basically dependent on the principles of recording and presenting the find. So we have to find ways and means to get the most out of our recording and presentation, - and to do that, we must know precisely what is needed for a proper presentation of the boat. What are out present and future standard requirements?

First of all we must realize that the facts of date, type, size and lines cannot be observed directly - they will have to be deduced from observations of the details and general pattern of the actual remains. So what is actually left for us of the boat must be examined and recorded carefully by methods giving us a maximum of relevant information.

Up till now the standard documentation in publishing a boat find has been based on the traditions of archaeologists and naval architects working by two methods:

1. The standard archaeological excavation plan prepared by various methods to record the condition of the wreck site, below or above water, by a projection on a horizontal level and some cross-sections.

The excavation plan (Fig. 12.1) is needed to show the overall situation of the site, including the general appearance of the distorted structure of the hull, and the relation between this and parts of the cargo and the surrounding features, Some constructional features of the hull will also be visible on the excavation plan, e.g. the general pattern of frames and beams, but quite evidently the details of shape, nailing etc. could not be amply reproduced here.

2. The naval architect's standard plan of the complete ship, showing the lines as well as longitudinal and transverse sections of the hull, sometimes even showing the arrangement of decks and rigging.

The naval architect's plans (Fig. 12.2) are needed as the generally accepted way of presenting the curved shape of a vessel by showing the lines, which appear when the (reconstructed) hull is sectioned according to certain rules. The midship section and a longitudinal section will show the general arrangement and the scantlings of all principal elements of keel, framing and planking. The naval architect's plan is sufficiently detailed for him to describe the hull-shape and to allow an exact calculation of displacement, stability etc. of the vessel - because these are the requirements to which naval architects prepare their plans. Generally the minor details of the construction, including the nailing and fitting together of the elements are only shown in a very general form or not shown at all in a drawing prepared for the boatbuilder, as these details follow a certain local practice at a given time and place.

Both archaeologist and naval architect use these standard methods to ensure easy reference between sites as well as between ships, and the "ship archaeologist" will need both these standard methods in the future as well as in the past.

But for the study of the history of boatbuilding the details defining "a certain local practice at a given time and place" are of primary importance - and often these are in fact the only constructional features to be established beyond doubt. The presentation of the general appearance of the original vessel may be rather hypothetical as it will include parts of the hull, which have not been preserved, and usually it is beyond control, as there is nothing or very little in between the excavation plan and the naval architect's plan to document the transformation.

Thus what we need is a set of methods for a proper recording of the individual elements of the construction and a good visual presentation of these and their detailed interrelationship. This will enable anyone interested in the ship to study the details of the construction, to compare these in detail with other finds, and to try out other possible reconstructions.

RECORDING STRUCTURAL DETAILS

The relevant features to record on the structural elements of an ancient boat - planks, frames, beams etc. - are not always easy to define and separate from the non-relevant ones. The position and nature of nailholes, original edges etc. of side-planks of the shell of the hull are part of the relevant information, which may even enable us to reconstruct the true shape of a ship better than could be done from fragments of the frames alone. In some cases changes in colour or texture of the wood will indicate the position of ribs etc. and thus these features may also be of interest to record. The wet surface may show toolmarks which are of importance to the study of the boatbuilder's trade, and the cross-sections of the planks and beams may tell us about the utilization of available timber resources, just as samples of the wood, caulking material etc. may provide good clues to the origin and date of the ship. This all indicates that the methods of recording should supply us with a very detailed presentation of each piece of wood from the boat, and that no simple method is likely to cover the complete range of observations alone.

Recording of the details should always be carried out just after the excavation when the structural members have been taken apart - but before preservation treatment starts.

The archaeologist in charge of a boat-find of this nature will have to select the appropriate combination of the methods of recording. Sample taking, and close-up photography in normal and oblique light will be needed in all cases. Casting is a good way to safeguard the preservation of delicate tool marks etc. and casts should be prepared of parts of the ship which are removed to serve as samples for analysis or conservation experiments. But the major part of the work will have to be done in one of the following three different ways of recording the elements:

1. by scale drawing
2. by full scale tracing
3. by stereophotogrammetry

1. Scale Drawing:

This Universal method for recording archaeological finds has been applied for some parts of the Skuldelev Ships as well as for the Graveney Boat at scale 1:5 or 1:10 (Fig. 12.3).

In principle the manual scale drawing technique is discontinuous, as it is based on the measuring of the coordinates in three dimensions of a number of single points of an object and the plotting of these points on the paper. As this is a time-consuming process one is forced to select a limited number of points on the object to plot and then complete the drawing of the remaining features by eye. This is a good way of preparing a presentation of a well defined object, but for fragments of an ancient boat with tool- and wearmarks,

Fig. 12.1 Example of an excavation plan. The Seventh-Century Byzantine wreck found at Yassi Ada and investigated 1961-64 by a team from University of Pennsylvania. The very fragile remains of the ship represent mainly parts of port side amidships and aft.

Fig. 12.2 Example of a reconstruction plan. The Byzantine Yassi Ada wreck, showing lines and some constructional features as developed after a series of scale models had been built based on surviving hull remains and their relation to the topography of the site.

OUTSIDE

Fig. 12.3 Example of a 1:10 scale drawing of hull plank from the Graveney Boat, measured by Eric McKee 1971-72.

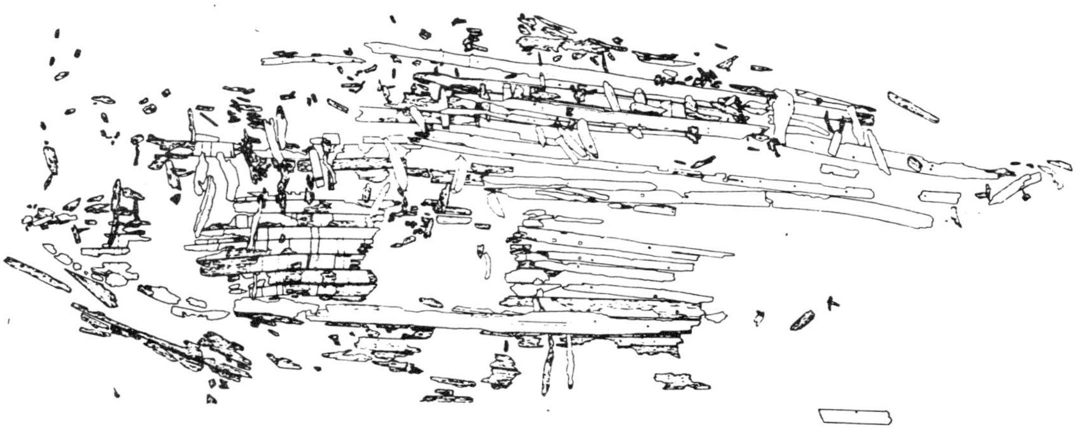

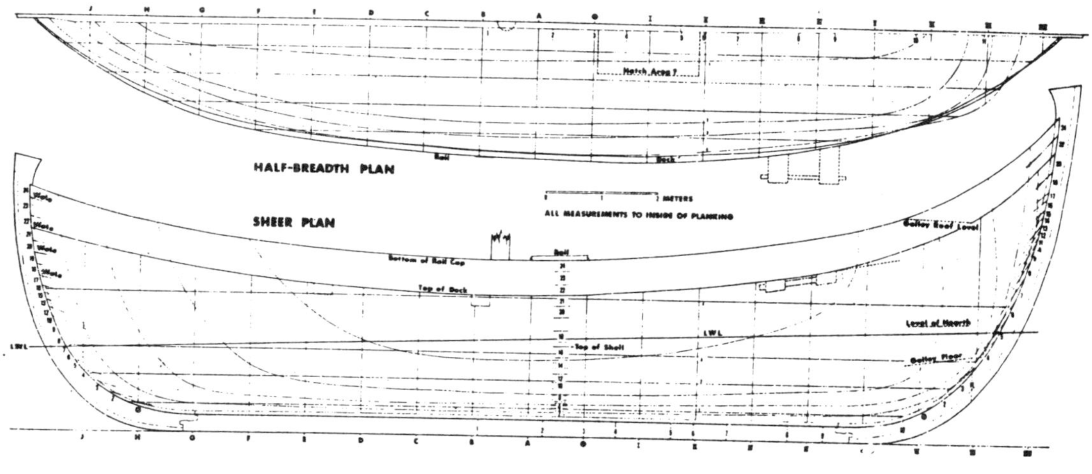

Hatch Area ?
Rail
Deck
HALF-BREADTH PLAN
METERS
ALL MEASUREMENTS TO INSIDE OF PLANKING
SHEER PLAN
Galley Roof Level
Rail
Bottom of Rail Cap
Top of Deck
Level of Hearth
LWL
LWL
Top of Shelf
Galley Floor

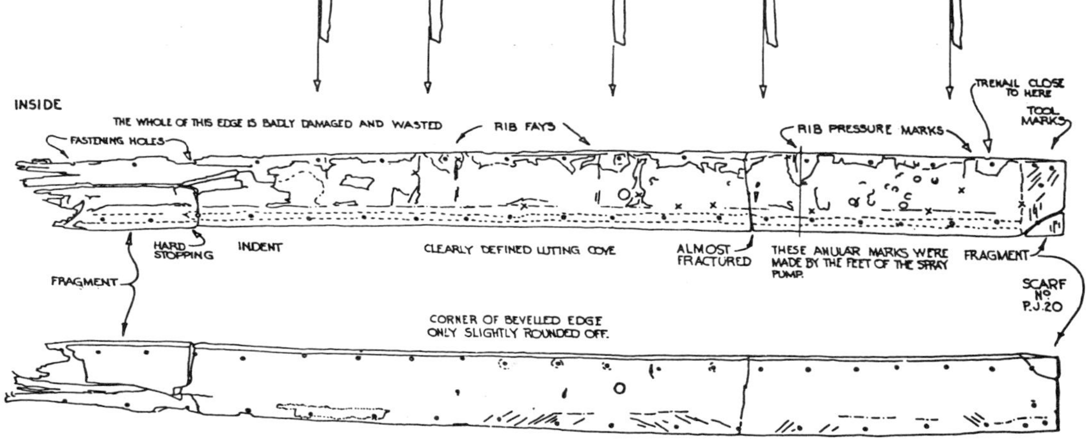

OUTSIDE
INSIDE
TREHAIL CLOSE TO HERE
INSIDE
THE WHOLE OF THIS EDGE IS BADLY DAMAGED AND WASTED
RIB FAYS
RIB PRESSURE MARKS
TOOL MARKS
FASTENING HOLES
HARD STOPPING
INDENT
CLEARLY DEFINED LUTING COVE
ALMOST FRACTURED
THESE ANULAR MARKS WERE MADE BY THE FEET OF THE SPRAY PUMP.
FRAGMENT
FRAGMENT
SCARF No P.J.20
CORNER OF BEVELLED EDGE ONLY SLIGHTLY ROUNDED OFF.

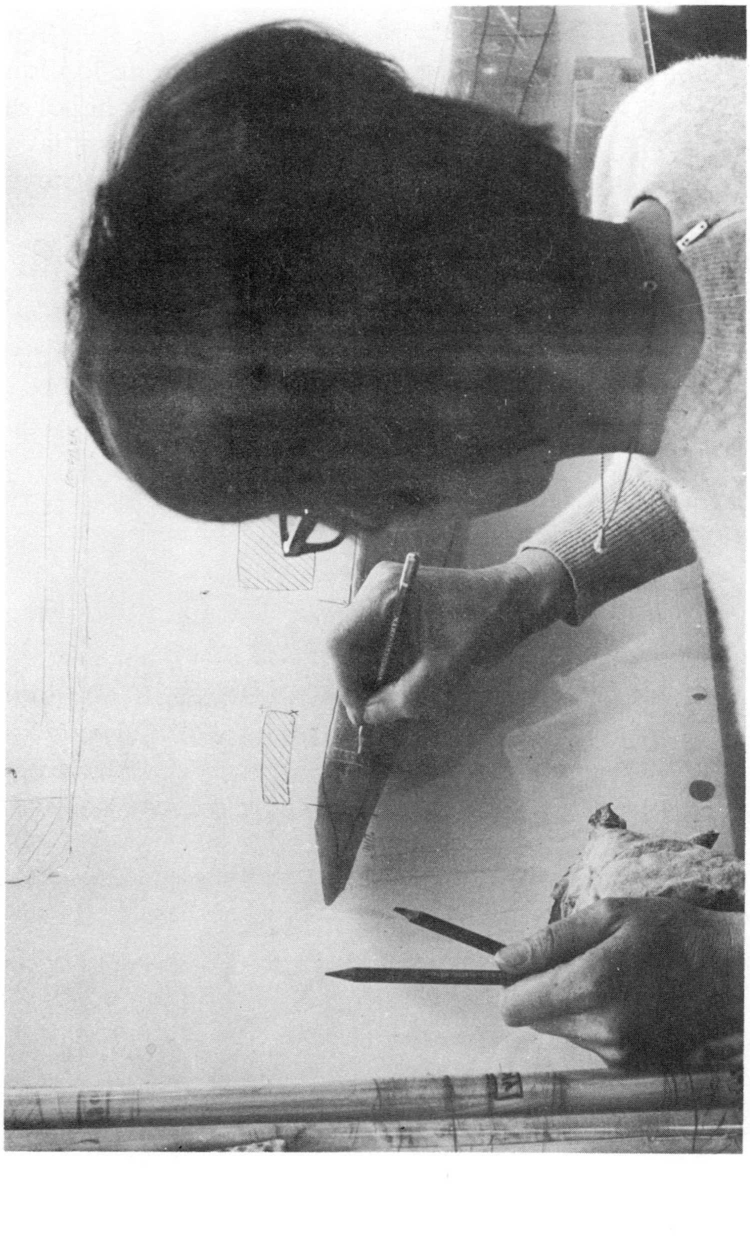

Fig. 12.4 Tracing in full scale of a Skuldelev Viking Ship frame on to a clear drafting folio, supported on a glassplate over the timber. Colour codes are used to indicate original edges, broken or damaged parts, and hidden lines (e.g. direction of nail holes in the wood).

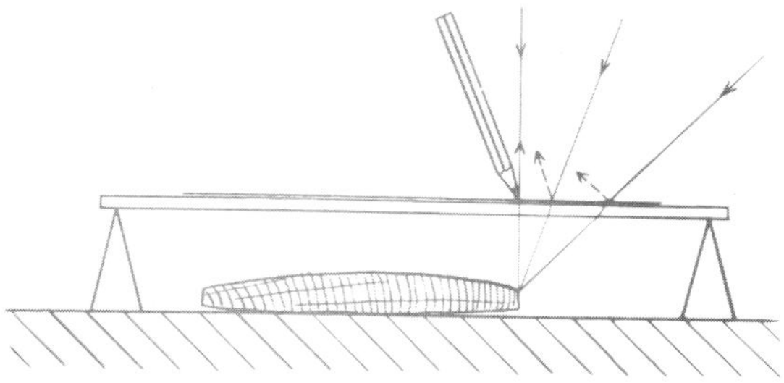

Fig. 12.5 Principles of full scale tracing on transparent folio. The folio is supported by a glassplate over the object. When the eye is held over the plate so that the reflected image of the eye is seen, the pencil will trace the true size of the object following the features to be drawn.

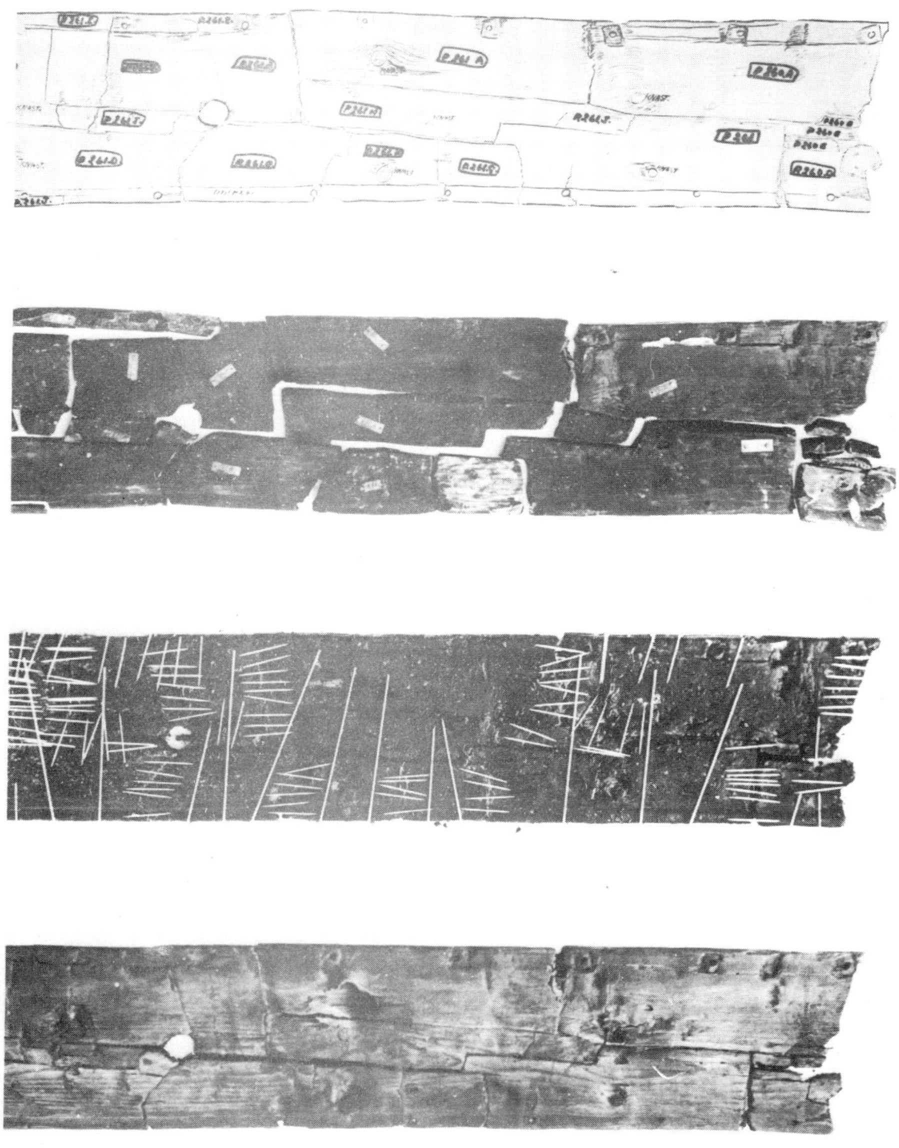

Fig. 12.6 Tracing of a plank from Skuldelev 1 (top) compared with the fragments after conservation, during reassembly (with toothsticks etc. on top of plank to indicate internal strengthening system), and after cleaning of surface (bottom).

Fig. 12.7 Arrangement for reproducing full scale tracings in scale 1:10. The camera, a 13 x 18 cm Monorail MPP with cassettes, is hoisted in a calibrated position ab. 3 meters above the tracings to give negatives in scale 1:10.

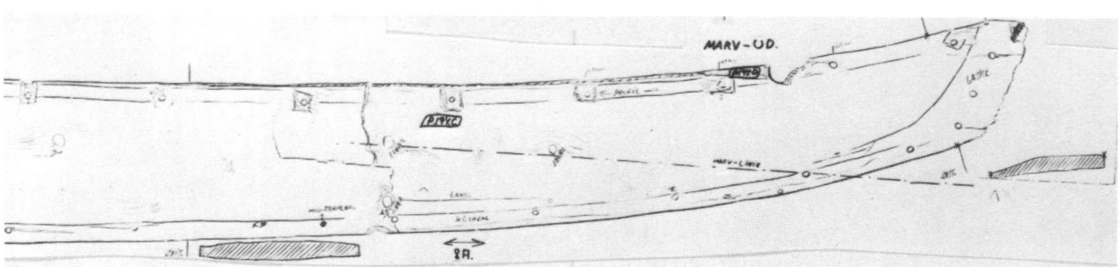

Fig. 12.8 Example of a full scale tracing of hull plank from the Skuldelev 1 Ship, reduced photographically to scale 1:10 (see Fig. 4 and Fig. 7).

original and broken edges, changes in texture and colour, this presupposes that the work is done by an extremely careful and competent draftsman, e.g. the Graveney Boat drawings prepared by E. McKee for the National Maritime Museum.

In general an outside observer is not able to check the accuracy of a scale drawing, as it is not customary to mark the plotted points distinctly, and therefore it may be difficult to distinguish between the work of a careful draftsman plotting many points and that of a skilled artist working primarily by eye. In many cases the minor differences resulting from their different approach may not mean very much, but when the drawings are to be used directly or indirectly for future reconstruction of the vessel in model or full scale, the accuracy within the last few milimetres may be crucial.

2. Full Scale Tracing:

This method was developed after the excavation in 1962 of the Skuldelev Viking Ships, to provide a continuous and accurate full-scale tracing of all elements of the ships before preservation (Fig. 12.4). It was carried out on long lengths of transparent drafting folio, suspended on top of a glass plate over the actual plank (Fig. 12.5). Each length of folio recorded the detailed features of one element or one series of interrelated elements (e.g. the planks forming one strake from forward to aft) and was fitted with a round stick at each end as a double roll to ensure easy handling.

These tracings serve as an accurate and detailed firsthand documentation of all elements and fragments of the five ships (Fig. 12.6) and although they comprise a total length of more than 2000 metres of folio, the rolls of drawings take up very little storage-room. Besides, the full-scale tracings have been indispensable for the practical work of keeping a check of all fragments, and of controlling the effect of the conservation, and they have served directly as patterns for the moulds prepared for the final reconstruction of the actual ships.

Special photographic equipment has been built to reproduce the full-scale tracings to scale 1:10 (Fig. 12.7). The photoprints in one-tenth scale of the drawings are prepared from the negatives as contact prints on transparent film (Fig. 12.8). This makes the fitting together of several prints into a complete one-tenth plan of a long plank or other element very easy.

This technique is ideal for flat or slightly curved planks, beams etc. and may even be used for frames which are curved in one plane only. The draftsman must be instructed to trace carefully all constructional features such as edges, nailholes etc. and to record (to the specified degree of detail) the broken edges, wear marks, changes in colour etc. There is no need for the draftsman to be an expert in boatbuilding, as he is just supposed to trace continuously the visible features without interpreting these or selecting "the most important" ones.

As a simplification of the method the folio may be placed directly on the top side of the plank without having the glass-plate in between, as done on some of the Graveney planks with strips of polythene sheet.[1] In this way the equivalent flat shape of a strongly curved plank may be obtained. This may be a quicker procedure, but it tends to provide less detailed and accurate tracings.

The reason to hunt accuracy so strongly is that the drawings should provide the ship archaeologist with a well defined two dimensional geometric representation of the plane shape of the elements of the boat, mainly the planking. On this basis the original lines of the hull can be established more accurately and detailed than from the cross-sections provided by the frames and often subject to considerable uncertainty, expecially in the upper parts. To do this properly requires a set-up that accurately records the overall shape of all planks, so that a good scale model of each plank can be made and assembled to make a model of the hull.

3. Stereophotogrammetry:

The special techniques developed for surveying by photogrammetry have been taken into use by archaeologists on land as well as under water.

In Sweden, Carl Olof Cederlund has been working on the problems of recording the size and shape of individual elements of a wreck by photogrammetry.[2] He had hundreds of loose pieces of a 18th century wreck brought to the surface with a need for a quick yet detailed recording of these before they were brought back to be stored in the sea bed beside the wreck. A simple arrangement provided a base for the stereo camera equipment as well as a background for the objects. In the first instance no drawings were prepared from the photos, as these themselves were taken as the medium of reference, realizing that at any time they could provide information on the exact measurements and curvature of the elements.

Special markers were fitted to the timber to indicate nailholes etc., as these were often difficult to distinguish on the photos. This is a serious drawback, as many of the important structural details will not show up on a photo taken from a position some metres away from the element. Besides, small fragments and long lengths of planking, keel, beams etc. could not be recorded properly in the same set-up of the equipment.

It may, however, be possible to get the work of recording done properly by a combination of photogrammetry to give accurate size and shape, and sketches/notes to give details of structural features etc.

PRESENTATION

Ship archaeology has given us several examples of ancient wrecks which have been interpreted very differently by different scholars, e.g. the Bruges Boat[3] and the Galtaback Ship.[4] Often very far reaching conclusions have been drawn from reconstructions of a rather hypothetical nature. A critical analytical approach to the conclusions drawn from the source material has been difficult because of the lack of traditions for a proper presentation of the primary matieral, and I think the crucial point for us is to establish standards for this presentation.

To meet this requirement we must sort out the deductions and speculations, and simply in the first instance present the primary observations related to the excavation situation (traditional excavation plan plus descriptive notes) as well as to the recording process of the individual elements of the boat. It may seem to be overdoing the things to present every single element of a boatfind as a scale drawing in the publication, but with a systematical approach to the problem of arranging the elements visually in relation to one another it may well be possible to show the plankings all folded out as well as the timbers grouped logically on a few composite drawings (Fig. 12.9).

These drawings will have served the archaeologist as a basis for the structural arrangement and lines of the vessel to be shown in the reconstruction and only by publishing these drawings, can a link be established between the excavation plan and the final reconstruction. Ideally the presentation of the elements should allow anyone to check the reconstruction in all details.

In the publication of the Cheops Boat finds,[5] a plan is included showing all the planks of the hull drawn up before assembly. On the basis of this a complete model of the hull may be constructed. The length, shape and arrangement of the planks come out very clearly, and may serve as a source of reference for other similar boatfinds, for the study of the timber problem of ancient Egypt etc.

In fact the fold-out plans of the planking of the Skuldelev ships have turned out to be a very useful and handy set of reference for most of the details defining "a certain local practice at a given time and place" - as well as for repairs etc. I am sure that a more widespread use of drawings of the planking (to a certain degree also the more schematical strake diagrams as published for the Graveney boat) will provide us with a reference material, which may be best be described as "fingerprints of the boatbuilding traditions" - maybe even with similar possibities for analysis as real fingerprints.[6]

After having laid the foundation by publishing the primary material fully, the next stage is to analyse and describe the various characteristic constructions of the vessel. For the visual presentation of complicated details of the construction the practise of showing sections projected on horizontal and vertical plans in not too good. Personally I prefer perspective or isometric drawings, which when shown from the most instructive angle - or "exploded" - may transmit more directly an understanding of the interrelationship between the elements. (Fig. 12.10).

This type of presentation can be used even if the shape of the hull and the principal dimensions have not been established in detail, whereas a standard projection will show e.g. the cross-sections in a very authoritative way.

Finally the plans showing the reconstructed ship are needed, following the naval architect's standards, thus providing a basis for calculations of cargo capacity, stability etc, as far as this is possible on the basis of the available evidence (Fig. 12.11). As a supplement to the drawings one may use photos of a model for the presentation, and indeed a working model is an ideal tool for investigating the possible solutions to the reconstruction of missing parts, provided the model is build carefully from the detailed records of the individual parts.[7] In this way the main dimensions, cargo capacity etc. should be defined within certain limits.

A proper record of the principles of construction as well as of the individual elements of a certain boat-find will have been given in the publication, if anyone who is interested and capable, can build an exact model of the boat as reconstructed by the archaeologist, with a full knowledge of the limits between facts and fiction, and thus with the possibility of investigating other solutions to the reconstruction problem.

Fig. 12.9 Example of scale drawings of all elements of a fragmentary boat find. The Egernsund barge, from southern Denmark, dated c. A.D. 1100 and dredged up in 1966 (compare Fig. 10 and 11).

Fig. 12.10 Primitive 'exploded drawing' illustrating the structural lay-out of the ends of the Egernsund barge.

Fig. 12.11 The Egernsund barge, reconstruction plan drawn to correspond with the presentation in Fig. 12.9 of parts of the vessel actually found. Even at this small size of reproduction of the plan (original in 1:10) the evidence for all detail of the reconstruction can be checked.

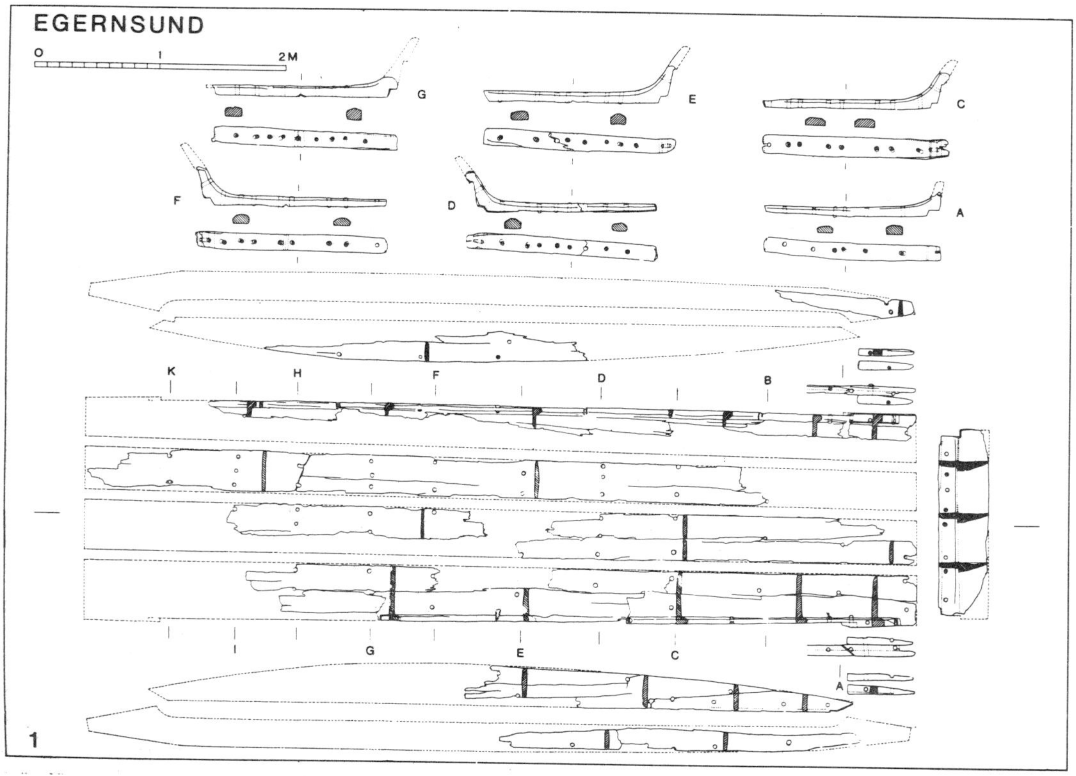
EGERNSUND
0
1
2M
G
E
C
F
D
A
K
H
F
D
B
I
G
E
C
A
1

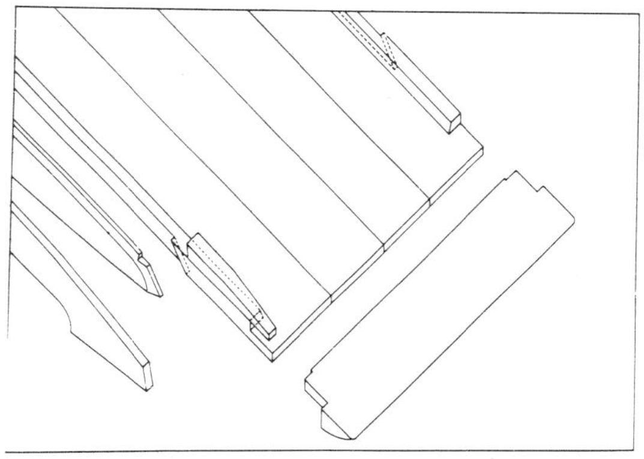

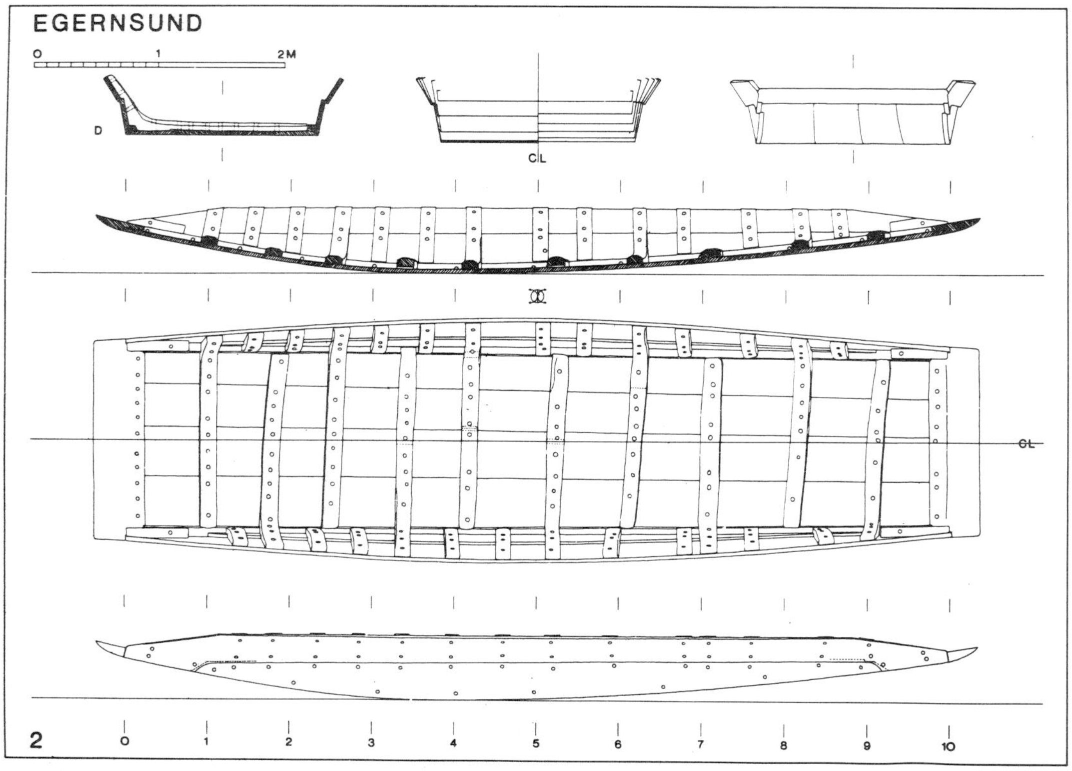
EGERNSUND
0
1
2M
D
CL
CL
2
0
1
2
3
4
5
6
7
8
9
10

These requirements for documentation put a heavy burden of work on the archaeologist and his/her assistants. In the words of Valerie Fenwick in the first volume of the International Journal of Nautical Archaeology, where she discusses these problems: "The marine archaeologist may, on reading this, be downcast at the prospect of such huge and detailed records".[8] Well, he will get the satisfaction that there will always be full documentation behind the statements in the publications, and besides, he may start revising the standards for the excavation plan, which may be relieved of many details.

On this basis the presentation of a boat find will follow four stages of observations and conclusions drawn from the material:

1 Primary observations, being direct observations of size and shape of elements of the boat, general lay-out of the construction, species of wood, wood-working technique.

2 Deductions from 1, being the lines and detailed construction as developed when working directly from the primary material.

3 Supplementation, being the additions which the archaeologist find necessary to add to 1 and 2 to come to a complete boat or ship.

4 Analysis of functions, being the statements given about the capacity, behaviour etc. of the vessel (as defined by 1 + 2 + 3), based on practical experiments and/or relevant calculations.

If we keep this pattern clear in our minds - and our publications - we will have a sound base for analysis and comparative studies.

NOTES

1. Cdr. J.E.G. McKee: 'The Graveney Boat Find - a reconstruction.' The Naval Architect, July, 1973.

2. Personal communication from Carl Olof Cederlund, Stockholm.

3. Peter Marsden: 'A Boat of the Roman period found at Bruges, Belgium, in 1899, and related types. The International Journal of Nautical Archaeology and Underwater Exploration (1976), 5.1: 23-55.

4. See the summary in Detlev Ellmers: Fruhmittelalterliche Handelsschiffahrt in Mittel- und Nordeuropa. Neumunster 1972, p. 317-18.

5. Abdel Moneim Abukakr and Ahmed Youssef Mustafa: The Funerary Boat of Khufu. Beitrage zur Agyptischen Bauforschung und Altertumskunde. Heft 12. Wiesbaden 1971.

6. See McKee's drawings in Clenched Lap or Clinker, Greenwich 1972, and Basil Greenhill: Archaeology of the Boat, London 1976, fig. 155.

7. Peter & Joan Throckmorton: 'The Roman wreck at Pantano Longarini.' The International Journal of Nautical Archaeology and Underwater Exploration (1973), 2.2: 243-266.

8. Valerie H. Fenwick: 'Thoughts on the recording of old ships.' The International Journal of Nautical Archaeology and Underwater Exploration (1972), 1: 180.

13. THE RECORDING AND PRESENTATION OF FINDS FROM NAUTICAL CONTEXTS

Angela Care Evans

Summary

Although archaeology seeks to be constructive it is destructive in that it disturbs and destroys the context of the site or object that it is examining. Because of this, accurate recording is essential. On land such recording is an integral part of any work, but underwater this is not always the case.

For wrecks underwater or on land there is an urgent need for the establishment of an accepted framework of measurements which can be applied to all wreck sites for comparative purposes. Apart from this the underlying objective of detailed recording should be the provision of enough information to reconstruct what remains of an archaeological wreck should its timbers not be recoverable.

INTRODUCTION

It is a well known maxim that although archaeology seeks to be constructive, it is nonetheless destructive, in that it disturbs and ultimately destroys the context of the site or object it is seeking to explain. This is especially true of, for example, a multiperiod land-site, but it is equally true in the case of a boat buried on land, or in the sands and muds of the seabed. It is therefore vital for the fullest recovery of evidence, that anyone who practises archaeology, be they professional archaeologists, week-end amateurs, divers or students, should fully recognise not only the responsibility of fully recording what they excavate as they are excavating it, but the equal importance of communicating their find to their colleagues.

The accurate and detailed recording of sites and their finds has long been accepted practice on land, but underwater around our shores, the application of archaeological methods to a site as a whole is still not always regarded as necessary. Fortunately, through the untiring work of a few crusaders, the awareness of the value of archaeology underwater as opposed to the random recovery of finds is becoming more widespread, and there is growing recognition of the responsibilities that the diving archaeologist must bear if historic wreck sites are to be exploited to the full in an academic sense. This responsibility is reflected in the Protection of Wrecks Act of 1973, and with the implementation of this Act archaeology underwater has, if not come of age, at least reached incipient adulthood - although it is still hampered by the occasional delinquent activity.

The division between underwater archaeology and land archaeology is to a great extent an artificial one. The same principles should and can apply to any site or object that is being examined as part of an archaeological exercise. The techniques of excavation are essentially the same but obviously the principle inhibiting factor to the application of archaeological techniques underwater as far as the recovery of information from wreck sites is concerned, is that they are, quite simply, underwater. The corollary of this is the urgent need for divers with archaeological training and, even more so, for archaeologists who can dive. Until such a corps exists around our shores the amount of excavation and recording of ancient wrecks that can be undertaken necessarily will - or should - be limited.

But apart from the immediate problem of applying archaeological principles to excavating and recording, what of the problems of presentation? For the land archaeologist working on a major site, although faced with a potential delay in publication, the opportunities and choice of outlets are many. Apart from the specialist national, and international archaeological journals, which welcome material relevant to their temporal range, there is in this country a considerable number of regional journals which publish locally-found material. In addition to these journals, there are now increasing numbers of magazine-type publications which publish short accounts of up-to-date material faster than the principal archaeological journals. There is then a wide choice for archaeological papers and notes; the principal criteria being the dissemination of knowledge and a high standard of presentation. At the moment those involved in nautical archaeology, who are not necessarily professional archaeologists, are hampered in certain ways: there is a lack of specialist journal space, and uncertainty in some cases of the potential of locally based journals, and there is also, occasionally, a lack of appreciation of the level of presentation that the publication of any archaeological material demands. Archaeologists have, after all, only gradually learned that high standards of presentation are vital to enable those who want to use published material for comparative purposes to draw sensible and accurate conclusions. But the publication of archaeological finds spreads backwards in one form or another over a couple of hundred years, while the publication of finds from nautical contexts is still in its infancy. Nautical archaeology's main journal is only a few years old and is already expected by some not only to absorb material from all over the world but to achieve immediately the same standards of presentation that have taken other archaeological journals years to achieve. Protagonists are warring quite rightly about standards of excavation, recording and presentation, and sometimes lose sight of the fundamental purpose of all three - communication. It is not enough to excavate to a high standard, just as it is not enough to record accurately in three dimensions; to take samples, photographs etc. The material must be published and not necessarily in great detail, but in sufficient detail to enable anyone working in related fields to know a) that an object exists and b) where to find it if more information is needed for his own specific purpose. When publishing his finds, no archaeologist can ever hope to satisfy everybody unless he publishes on the grand scale in the sense of a single volume devoted to a specific find. Even then the specialist can ask why a particular object has not been subjected to a more detailed treatment, simply because out of a wide range of material that one object is of vital importance to him, and yet is of only marginal relevance to the find as a whole. It may seem banal to even write in such simplistic terms, but such is the concern amongst those involved with all aspects of nautical archaeology, that it seems not unreasonable or indulgent in this paper to attempt to pare away the assumptions that have accumulated over many years and lay bare the principles that must underpin all archaeological work.

RECORDING

When a student, the would-be archaeologist learns that one of the first principles of controlled excavation is to answer specific questions. No archaeologist goes blind into his chosen field. Background material and relevant comparative material is sought and studied, and a plan of campaign is evolved to enable the director of an excavation to recover information that will complement known fact and possibly open up new avenues of thought. However, while many excavations begin within this hopeful framework, few fail to burst out and lead the excavator away from his original direction - this is after all how fact building on knowledge helps recreate the rather tenuous past. But these facts - in the case of a boat, details say of construction - must be recorded in such a way that not only the excavator but other specialists may benefit. It does not matter how the facts are recorded, there are many systems, two of which we shall look at in some detail below, what is vital is that accurate recording takes place. If one considers, say, a wreck site, the materi al is of potential value to students working in several different fields - not everyone is necessarily interested in the structure of a wreck, some are interested in the way that a wreck will break up in different conditions and to them the spread of the cargo and the disintegration of the timbers are of more value than the distance between frames or plank-joins. Others, specialising in pottery, find that the recovery of a certain range of, say, bellarmine flagons, can help in the construction of a chronological sequence for differing forms often found only in isolation on land sites. Yet others will specialise in guns or even in the use of different woods for specific purposes in the construction of a ship or boat. Thus it is clear that the excavator of a wreck in any context must record with a degree of accuracy that allows for maximum information retrieval in a number of different ways: the surviving structure and finds within the wreck site must, even after excavation, be capable of being restored academically to their original relationship. For while specific questions are in the excavator's mind at the outset of an excavation, the quality - and quantity - of the finds will impose their own problems of interpretation, often only after the excavation has ended, when the finds are being written up and related to other similar finds. Thus it is vital that the excavator can go back to his notebooks, plans and photographs and be able to re-establish the exact context of a find and study it in the light of any new information that has emerged.

The principles of accurate recording are clearly then of as great importance to the archaeologist involved in nautical material as they are for archaeologists working in other fields. But, although the recording of a multitude of facts is almost inevitable in any context, the blanket recording of everything visible is not necessarily the most efficient means of recovering specific information, and within the overall discipline of recording the need to establish parameters within which certain specialised objects may fall, is becoming increasingly urgent. Ships and boats fall well into such a specialist group. Some overall language of recording is surely well within the grasp of those working in what should be a close-knit field, but even now, too often opportunities are lost for bringing some fixed pattern to the recording of ship structures. This is becoming increasingly necessary with the huge potential not only of computer-based information retrieval, but of the use of the computer for comparative purposes and even more so for experimental projections. It is becoming vital that each ship or boat, no matter what its context or state of preservation, should be treated within a similar framework so that directly comparable results may be achieved. Within this overall framework, details may then be recorded with confidence in the knowledge that on other crafts, similar details will be recorded in the same way. This does not necessarily mean an inflexible system, for within and without the chosen parameters, additional facts may and should be freely recorded. In essence the first objective of recording an ancient boat in the field should be the retrieval of a given set of measurements which would be common to all such structures, followed by the recording of as much ancilliary material as is possible within the limitations of the excavation. The emphasis on field recording may be justly stressed when one considers that, if a boat is lifted, the timbers may vanish for conservation purposes for a considerable length of time, and if the timbers are not lifted then field records are the only source of direct facts short of impractical re-excavation.

However, apart from the problems caused by long term conservation, there is also the very real danger that details, crystal clear when they are first uncovered, will be damaged, even lost, during the process of excavation, conservation and reconstruction. Edges, once sharp, become blurred; carpenters marks are found to be ephemeral; friable wood collapses; all these things are more or less common-place and the necessity for full recording in the field barely needs underlining. What of metal finds? Unless immediately treated corrosion can quickly obscure the fine surface detail of, say, an inscription in bronze; iron corrosion can 'explode' and destroy the object it once represented, silver can quickly dissolve into a heap of silver chloride. Thus it is essential not only for objects to be related to their context by three-dimensional measurement, but that they should be photographed both in situ and after lifting, and, if possible, drawn. This kind of in-depth field recording throws a great burden of responsibility onto the director of any excavation, but especially on the director working on a wreck site underwater where the conditions of preservation might be excellent while the object remains undisturbed, but who is faced, once the immediate environment is altered, with the sometimes extraordinarily rapid onset of destructive processes that can only be retarded by prompt conservation.

But while objects can be lifted and conserved without overwhelming difficulty, what of the timbers of a ship? There can be no doubt that lifting any boat from any context can overstrain the conservation resources of even a national museum and is virtually an impossibility for a local museum's conservation team. Unless a special unit can be set up to deal with conserving specific craft, as has been the case with the Roskilde boats, the Graveney boat and the Punic ship found in Sicily, it is clearly impractical to haul hundreds of feet of waterlogged timber out of their protecting environment into an atmosphere where they will quickly twist, warp and ultimately disintegrate, without the possibility of prompt treatment and ultimate reconstruction. And while this is a possibility with craft like those found at Graveney and Roskilde, what are excavators to do with future finds of larger vessels? The answer in the short term, must be to record in as great a detail as possible and to leave the vessel secure in the environment that has preserved it for so long. But what if it is threatened with destruction? Does such a threat presuppose a measurable loss of knowledge? If a vessel is unique, as so frequently is the case in early contexts, then lifting, if the timbers are sufficiently robust, and preserving for reconstruction becomes a goal to be strived for. But if this course is impossible can recording alone suffice?

The classic example of recording simply having to suffice is of course the Sutton Hoo ship – a ghost-ship with no timbers, surviving only as an impression in the sand, with only lines of iron rivets preserving the run of her planking, concreted shapes in place of her frames and discoloured sand representing the outline of her keel and stem- and stern-posts. When the ship was first excavated in 1939, the hull was surveyed, and a series of detailed measurements as well as record photographs were taken by a team from the Science Museum in London, led by Lieutenant-Commander Hutchison. Shortly after the excavation, the survey and measurements were translated into two plans, one of the lines of the ship and one containing the details of the ship's construction. The former is still the basis of the accepted hull shape, but the latter, apart from being inaccurate in several small details, has always posed serious problems of structural interpretation when studied in depth. Unfortunately, the field records relating to the ship were lost during the war, and the only official record apart from the two plans to survive was the limited number of photographs taken by the Science Museum and those of Charles Phillips who had directed the excavation of the burial chamber which lay in the centre of the ship. It was by the greatest good luck that over five hundred photographs covering every stage of the ship's excavation and illustrating all her structural details were taken during August 1939 by two schoolmistresses, Miss Mercie Lack and Miss Barbara Wagstaff, keen amateur photographers who were holidaying in Woodbridge. When work began again on a reappraisal of the Sutton Hoo ship, it was largely due to the photographic record of Miss Lack and Miss Wagstaff that details of her structure, destroyed in the years that intervened between the first excavations in 1939 and those of 1965-7, could be recovered.

During the re-excavation of the ship which began in 1965, every visible detail was recorded on a series of field plans drawn at a scale of one inch to the foot (feet and inches being used in this instance to enable simple correlation between the 1939 plans and those of 1965/7). The ship was divided on a grid system with squares and baulks and excavated as any other land monument, with sections of the infill being recorded from the ground surface to the surviving skin of the ship. When the rivets, still surviving in their original strake lines amidships though jumbled in the prow and stern, were encountered, they were labelled numerically with copper tags and recorded in three dimensions - tied in by two lateral measurements to their grid square and in depth to the ship's datum line, itself an integral part of the

Fig. 13.1 The Sutton Hoo ship looking towards the bow. The shape of the ship survives only as a fragile impression in the sand with corroded iron rivets showing the run of the strakes. C. W. Phillips.

site datum. Various systems of recording the rivets of the ship were used in an experimental way - in the first season, cross-beams, levelled into the site datum were used as fixed points to plum down to the surface of the ship. Later a more flexible system of portable measuring frames made of 'dexion' and mounted on four legs was adopted. These frames had spirit levels attached to each side of the square and could thus be made absolutely level. By tying each frame into the overall datum of the ship, they could be used as local datums to measure more rapidly the individual rivets in three dimensions. When the measurements of each rivet had been plotted on the relevant field plan, a 'portrait' of it was drawn, showing the direction in which it lay and its relationship to its neighbours. On the plans the lines of the ship's frames were also plotted in a similar manner as were areas of disturbance along the keel, where cross-sections had been cut in 1939, and surface deposits on the skin of the ship. Apart from recording all the rivets that survived in situ, stray rivets found in the infill of the ship were recorded during the excavation for consideration in the final analysis.

In both excavations the ship was treated in almost exactly the same way as an archaeological monument. In 1939 cross-sections were cut on the lines of certain ribs and outside the prow and stern in an attempt to discover the shape of the keel and the stem and stern-posts. A series of sections was cut across the gunwale strake amidships to starboard and an area was carefully excavated to starboard outside the hull alongside the frames of the steering system in an endeavour to recover evidence of a steering 'button'. All these efforts at recovering the detail of the Sutton Hoo ship in 1939 were fortunately recorded photographically and enough survives on the photographs for them to be used as the only evidence for certain aspects of the ship's structure where physical remains have since been destroyed. In the re-excavations, the cross-sections cut in 1939 were re-cut, drawn and photographed, and every change in the colour of the sand was recorded; it was even possible, for example, to trace the dark stains of disintegrated wood which had formerly been the stem-post. Apart from basic recording of the physical remains of the Sutton Hoo ship, a plaster cast of the remains was taken and this was used to make a permanent record in fibre-glass of the hull shape, whose condition was too bad to preserve and as it was essential to dig through the impression and look beneath the ship in case of other burials. Samples were taken with specific problems in mind - both in relation to the ship itself and to its burial deposit. For example, in the burial chamber alone, over 200 samples of sand were taken on a tight grid in an endeavour to plot the contours of any residual phosphates that might have been derived from the burial, other samples were taken as each rivet was lifted so that the density of phosphate within and without the burial chamber could be compared. Finally samples were taken which had a direct relevance to the building of the ship - fine 'hair-like' fragments were lifted for laboratory examination, sand from the overlap of the strakes was packed and labelled for analysis and samples of a whitish deposit that occurred here and there on the surface of the compressed sand were taken to see whether this was the remains of a surface tarring of the long-vanished timbers. The results were both positive and negative. The fine hair-like strands turned out to be roots and not caulking material but to compensate for this, other samples from the strake overlaps showed a trace of animal protein - perhaps the only surviving evidence for

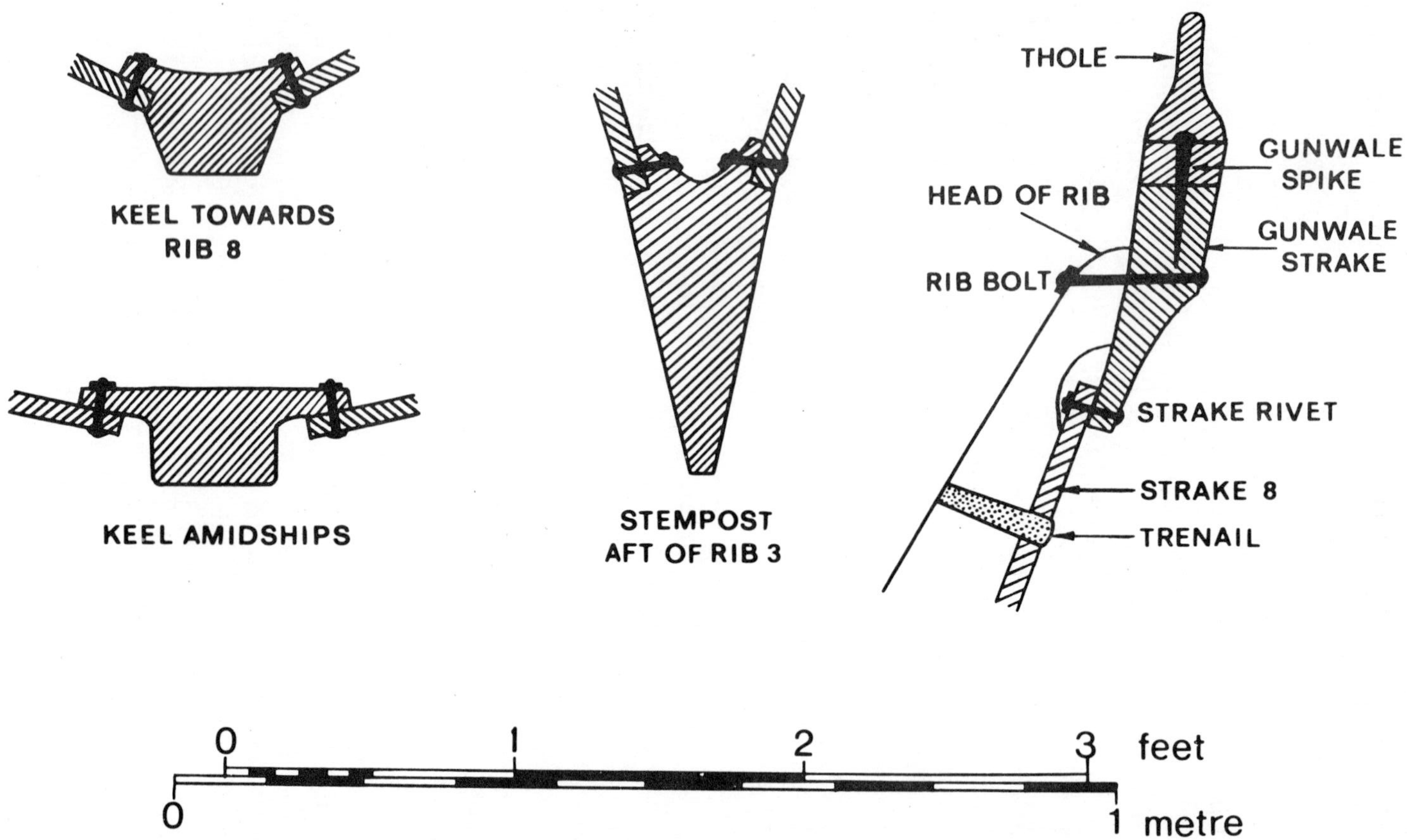

Fig. 13.2 Reconstructed cross sections of the keel, stem, and gunwale strake of the Sutton Hoo ship, based on the evidence of wood grain that survived on the rivet shanks and on cross sections through the ship. British Museum.

animal derived caulking material. Analysis of the white deposits were ambiguous - they showed clear traces of resin, but with no artifical addition surviving it is impossible to say whether these traces are of an applied resinous tar.

From this kind of attention to detail and recording in the field it has been possible in the case of the Sutton Hoo ship to build up an accurate record both of her original dimensions and her basic structure. Furthermore by examining the lifted rivets and recording details of their wood grain preserved in iron oxides, and especially by sifting through the 1939 excavation photographs looking for specific details, it has been possible to discover smaller details of construction - for example the thickness of the planking, the development of the stem-post, even the use of trenails to fasten the frames to the planking. Evidence of repair to the hull has been recovered in several places and from the photographs of cross-sections cut in 1939, the shadow of a rectangular keel has been recovered. All these small details gradually add up to a remarkably complete picture of the Sutton Hoo ship, and although there are questions that can never be properly answered - did the ship have a sail? was the hull built of oak; how was the steering paddle fastened to the hull? it is not unreasonable to regard the recorded detail of the Sutton Hoo ship as sufficient in itself, although recent work in various fields, particularly photogrammetry, make one wish that the ship could have been first excavated in this decade rather than nearly forty years ago.

In both 1939 and 1965/7 the Sutton Hoo ship was excavated using purely archaeological principles and the problems of her reconstruction were approached from an archaeological rather than a nautical stand-point. The Graveney boat, too was tackled as an archaeological project, but in this case, as a rescue operation. In 1939, a month was allowed to the excavators of the Sutton Hoo ship; in the succeeding excavations, time was ample to undertake the kind of detailed recording that was thought necessary. In the case of the Graveney boat only ten days were available to record and lift the timbers of what was clearly a unique vessel, and in that ten days an enormous effort was made to record the ship in every detail: she was closely measured and photographed, a plaster cast of the surviving hull shape was made and her timbers were lifted, washed and rephotographed before packing. The underlying objective was that if the timbers could not be preserved, the recorded detail alone would enable her reconstruction to be made. The Graveney boat thus presented its excavators not only with the challenge of fully recording her structure in a couple of days, but with the problem of lifting the planking for future reconstruction. That such an operation, which included the making of a plaster cast, could be completed in only ten days seemed a colossal task, particularly as, with uncertainty in the very early stages of the excavation about the future of the waterlogged timbers, it seemed that the basic recording might have to stand alone and it therefore had to be consistent and accurate.

In the first instance, the boat was measured from random offsets set up across the hull using nylon string tagged at 10 cm intervals to simplify the recording process. Plank runs were recorded in three dimensions at specific intervals as were the floor-frames, with details of depth, the position of trenails and rebates for side-frames; and, on the midship frames only, the shallow insets cut into their upper surfaces. The surface width of

Fig. 13.3 The Graveney boat - much of the planking, several ribs, and parts of the plank-keel and one stem, survived. National Maritime Museum.

the stern-post and keel were taken at specific intervals. Where the frames did not survive, trenail holes were recorded and the position of plank-overlaps - with the direction in which they faced - were noted and drawn in on the preliminary plan. Sections were taken across the planking at specific points along the hull and cross-sections of the stern-post and keel were added as they were obtained. Other details, the position of the scarf of the keel and stern-post, a hole hacked in the sixth and seventh strakes, axe marks, fragments of rope, etc., were all planned on this provisional drawing. Apart from this basic record of the boat's measurements, which was augmented by a detailed photographic coverage, all the timbers and associated fragments were labelled with tags of brightly coloured "Dymo" tape nailed to the wood with silicon bronze pins, so that when the wood was lifted there could be no confusion as to identity or where, in any possible reconstruction, an individual timber should belong. The overlapping plank ends were also numbered sequentially in this way. The frames were individually numbered and tagged and, before lifting, silicon bronze nails were driven into the planking along their edges recording the exact lie of the timbers against the shell of the boat.

Apart from the necessity of accurately recording as much detail as possible as quickly as possible, samples were taken of both the surrounding mud which had preserved the boat so well and also of the debris trapped beneath each frame, in the hope that some light might be thrown on the boat's environment and, perhaps, the contents of her cargo. Finally as the frames were removed from the hull, and before lifting the planking, a cast in plaster of Paris was made to provide a permanent record of the lines of the hull. After the cast was lifted, the strake timbers themselves were dug out of the mud and, before packing for transport to the National Maritime Museum, were washed and detailed photographs were taken. This however was not the final record of the timber - although the final phase of recording in the field - for while in storage in water tanks at the National Maritime Museum, all the timbers were drawn, at a scale of one to one, onto polythene sheeting using different colours so that the various features on the surface of the planking could be quickly identified. Thus, should some unimaginable disaster overtake the Graveney boat, the mass of detail collected both in the field and in the Museum will serve to produce an accurate reconstruction, as will indeed similar field-records made of the Tudor fishing boat found in the Thames so soon after Graveney excavation - a boat which sadly had to be left at the bottom of a forty-foot coffer-dam, to be entombed in the concrete of the new embankment.

PRESENTATION AND CONCLUSION

Assuming that an archaeological site which includes a ship or boat with associated finds has been adequately recorded, what is the most constructive way of presenting the material so that other people working in similar fields can use the material to its fullest potential? There are two ways of looking at this problem - one is to regard the published material as a useful guide to the range of the primary material which can be examined at first hand if necessary; the other, is to use the evidence of published material as a basis for comparison with one's own primary material. The latter must presuppose the accuracy of the published material, and while this is quite acceptable if one is working with photographs and scale drawings of small finds, from a wreck site — coins, pottery, clay pipes, buttons and all the generalia that may well be discovered — is it a realistic way to approach published plans of excavated ships? Much can be said about the accuracy — and indeed the inherent inaccuracy — of published plans and their value for comparative or reconstructive purposes, and it must be the responsibility of anyone who publishes the plan of a ship or boat to ensure that it is as accurate and as clearly drawn and lettered as possible before it is reduced for publication. It goes without saying that any plan must be augmented with a detailed text and with scale drawings of structural details supported by photographs which should themselves include a scale.

Even if all these objects are achieved, can small-scale plans of boats and ships be used other than as a broad guide to comparative details? Despite their limitations, they are the only realistic source for comparative purposes. Much must be taken on trust as it is sadly impractical to hope to examine personally the original plans and drawings from every excavation relevant to one's own work. But are the published plans sufficiently consistent in their detail to be used as a basis for drawing structural conclusions on projected hull lines? A quick study of a range of plans will reveal that, while some record exactly the kind of information that one would need for a detailed project, others leave out perhaps one measurement or one detail that is essential for the completion of the particular survey that is being undertaken. Because of this the achievement of an established framework within which nautical archaeologists can work would surely benefit future studies. It might be that the answer lies with those who translate ships and boats into computer terms. Working within a strict framework of their own, they could perhaps suggest specific guidelines for a consistent pattern of recording that could be used on any hull structure (regardless of its condition) found on land or underwater. If such a system of recording a range of specific details were evolved and applied consistently to ships and boats would not the usefulness of published material increase?

REFERENCES

Evans, A. C., in Bruce-Mitford, R. L. S., (ed) 1975. Sutton Hoo Ship Burial, Volume 1, Chapter 5.

Fenwick, V. H., (ed) forthcoming. The Graveney Boat.

Olsen, O. and Crumlin-Pedersen, O., 1967. 'The Skuldelev Ships', Acta Archaeologica (Copenhagen) 38:73-174.

14. RECORDING THE REMAINS OF A SEWN BOAT FOUND AT SKEPPARGATAN 4 IN STOCKHOLM

Carl Olof Cederlund

SUMMARY

The fragmented remains of a sewn boat, excavated in 1896, were recently recorded by the University of Stockholm and the Swedish National Maritime Museum, as part of a course in nautical archaeology. Several recording techniques were used: written description, photography, 1: 5 projection plans, full scale documentation as used for the Skuldelev finds, and strake diagrams. These methods were considered to be supplementary to one another.

The hypothesis is advanced that characteristics of the construction indicate that the remains are those of a boat used on Russian rivers or in the coastal waters of the Eastern Baltic, and that they are representative of an East European boatbuilding tradition.

In 1975 a course in marine archaeology was held at the University of Stockholm, attended by about 60 students. The course included exercises in recording ships and boats, two finds belonging to the National Historical Museum in Stockholm being used. One of them had been described in a report to the museum when it was recovered,the other had not been recorded. Neither has been published.

One wreck was found during excavations for a new house at Skeppargatan 4 in 1896, near one of the bays in the water system in the central part of Stockholm. According to the notations made in the registration at the museum in 1896 it was the size of a "skuta", that is to say at least a length of about ten metres, Four parts of the wreck were salvaged and brought to the museum and have since been stored there.

The other find consists of ribs, bites, two planks, and one bite-knee of an approximately 7-8m long, clinker-built boat which was salvaged from the lake of Tingstäde träsk at Gotland, when a lakedwelling ("Bulverket"), dated to late Viking age, in the same lake was investigated in 1930. The salvaged pieces of the boat are similar to the corresponding pieces of boats from late Viking age and early medieval times in Scandinavia and in the areas on the southern shore of the Baltic.

About 40 students took part in the recording of the boat remains, and several different kinds of recording were used. The usefulness of these different methods for the documentation of the find was then discussed.

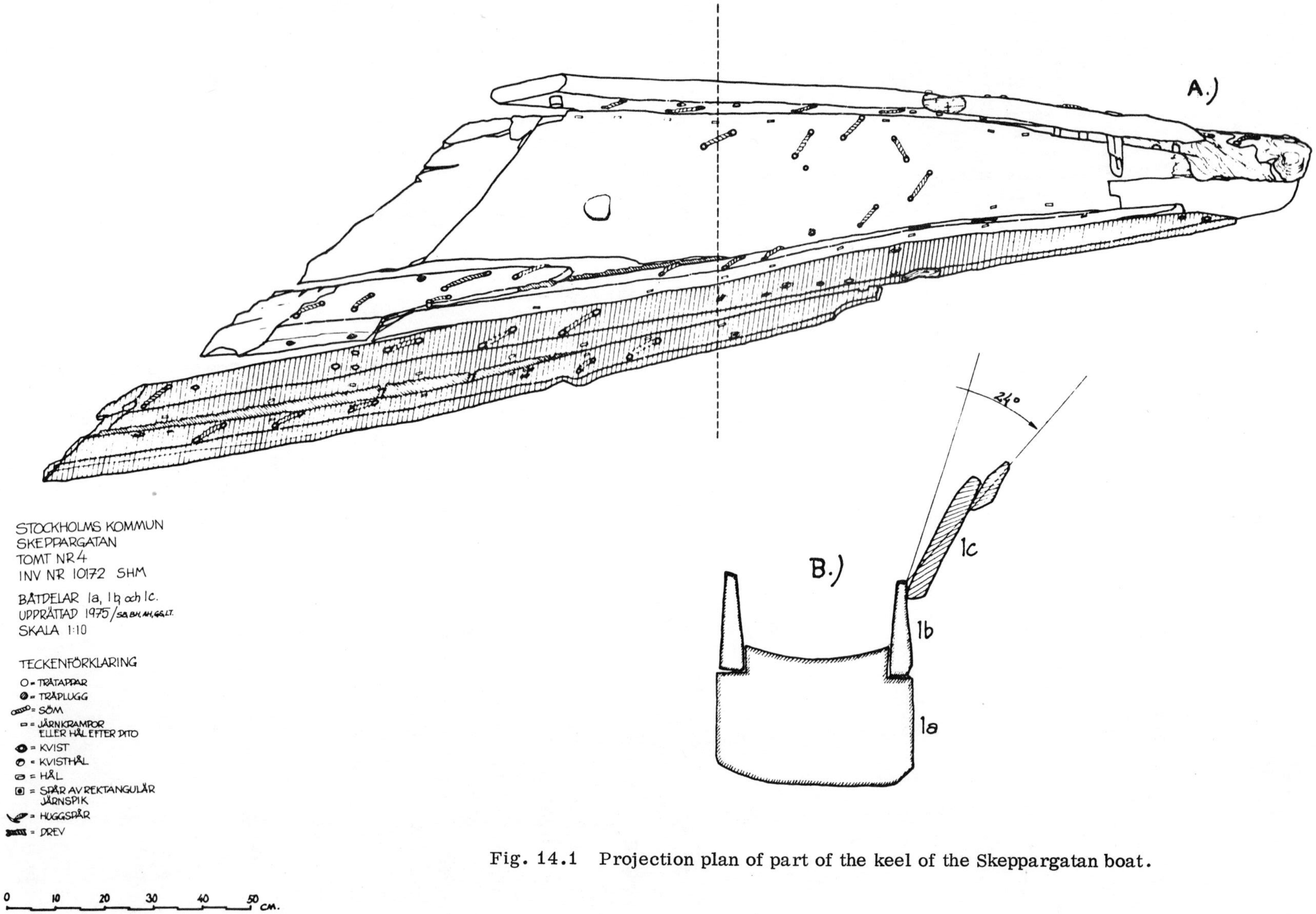

Fig. 14.1 Projection plan of part of the keel of the Skeppargatan boat.

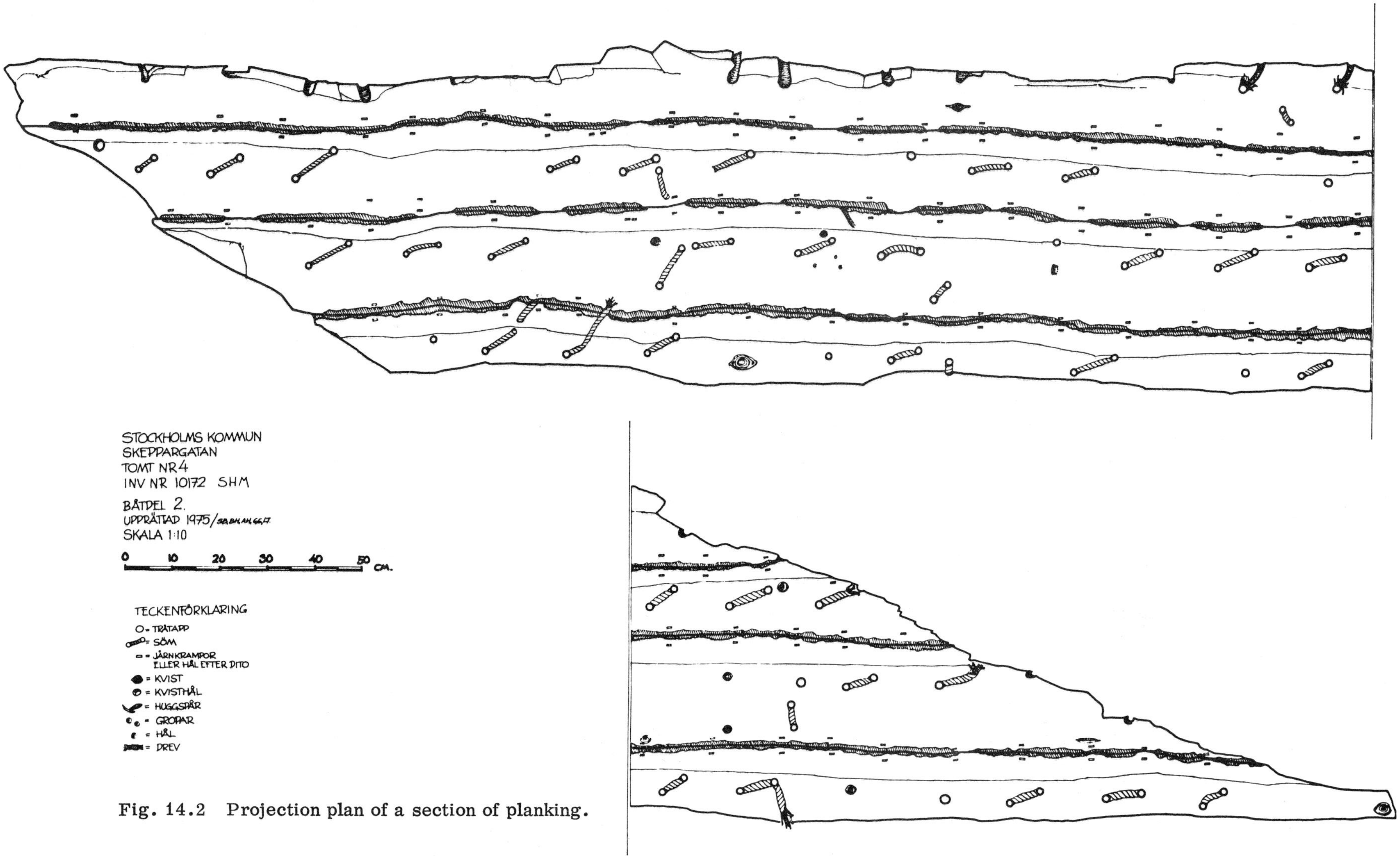

Fig. 14.2 Projection plan of a section of planking.

Fig. 14.3 The surviving part of the keel with the keel-stem scarf. Three fragments of side planking have been placed in their original position relative to the keel. At the far end of the plank nearest to the keel a fragment of a bottom-plank fastened to the keel can be seen.

Here some of the results of the recording of the boat remains from Skeppargatan 4 will be presented.

The salvaged pieces of this boat had special characteristics which are not common in Scandinavian local craft. She was clinker-built, had a very heavy structure, and had been double-ended, according to the observations recorded during the excavation. The surviving pieces consist of part of the keel near one of the stems; two sections of planking; and one loose plank.

The planks had been sewn together with withies of spruce. The same type of fastening had been used to join the stem to the keel and planking to keel. Even if the ribs were missing one could discern their fastenings in the planking - in the form of spruce withies.

Pine had been used for the keel and the planking. The caulking consisted of moss, which was kept in place by battens of pine both on the inner and outer side of the planking. These battens were fastened to the planking by iron-clamps.

The age of the boat is now being investigated with the help of C14-dating, According to the location of the wreck, in an area of the water system in Stockholm which was filled out during the 17th century, it cannot be later than that century.

The recording of the boat remains comprised a thorough written description of the construction; a photographic recording of the same characteristics; projection plans in the scale of 1:5; full-scale documentation according to the method used on the Skuldelev ships; and strake diagrams. Also different analyses are now being performed on different kinds of material from the ship. A botanical analysis and pollen analyses will also be made of the moss used for caulking.

The five different kinds of recording complement each other. Projection drawings in the scale 1:5 were made of the three largest surviving parts; of the boat. For a find of this kind, with only a few parts of the construction preserved, this documentation is primary. These projection plans give a direct and easily interpreted picture of the construction of the pieces as a whole and in detail (figs. 14.1-2). This recording was supplemented by the photographs which also give a clear picture of the various elements in the find and record the details of the construction (fig. 14.3). The photographs enhance the picture that the projection plans give, as they can be exposed from freely chosen angles.

The third kind of recording which can be said to be essential for finds of this type is the written description. This defines the different items that the archaeologist has observed and recorded in drawings and photographs.

Documentation in full scale could, because of the limited time available for the exercise, only be performed on one object of the find, the loose plank. This documentation shows how much information one can collect from one piece of timber by very thorough observation and recording. The method gave a much better record of the details of the construction of the planking than, for example, the drawing of projection plans in the scale of 1:5 (Fig. 14.4). The method is however time-consuming. For the full-scale documentation of one plank much more time was used than for the drawing of several projection plans of the three other surviving parts of the boat.

A)

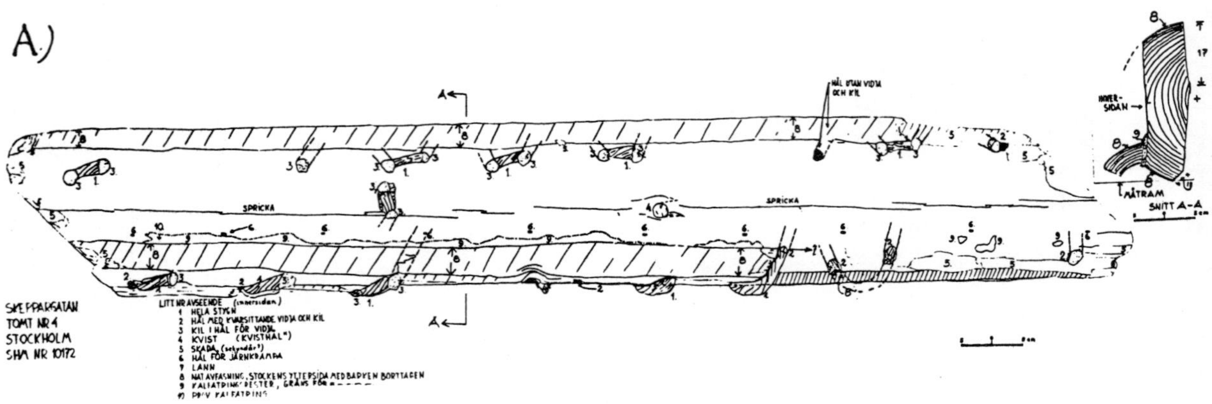

B)

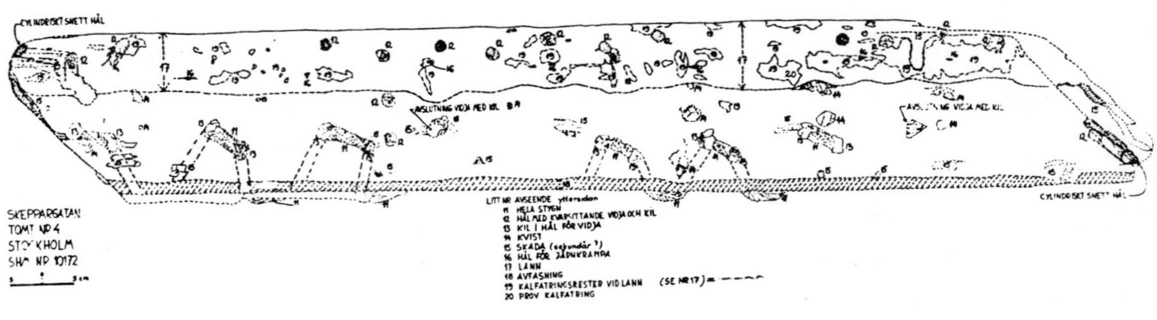

C)

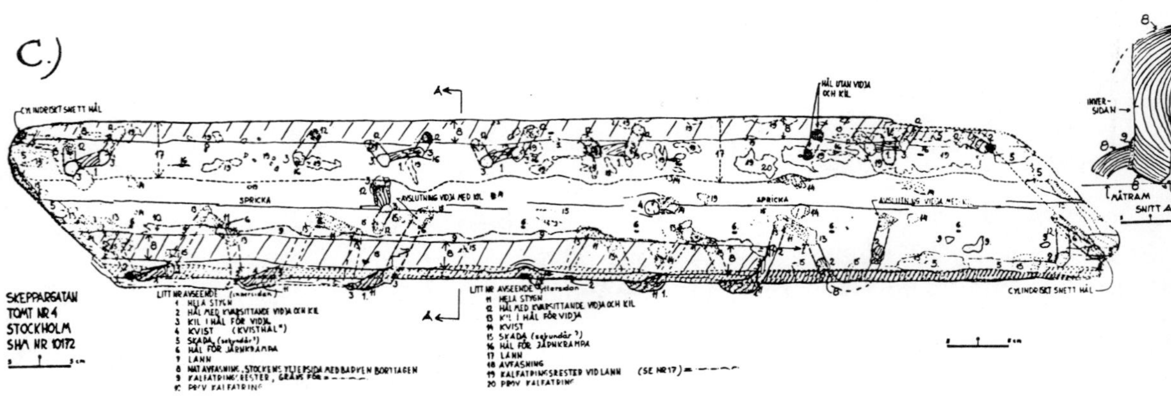

Fig. 14.4 Full-scale documentation of both sides of a loose side plank.

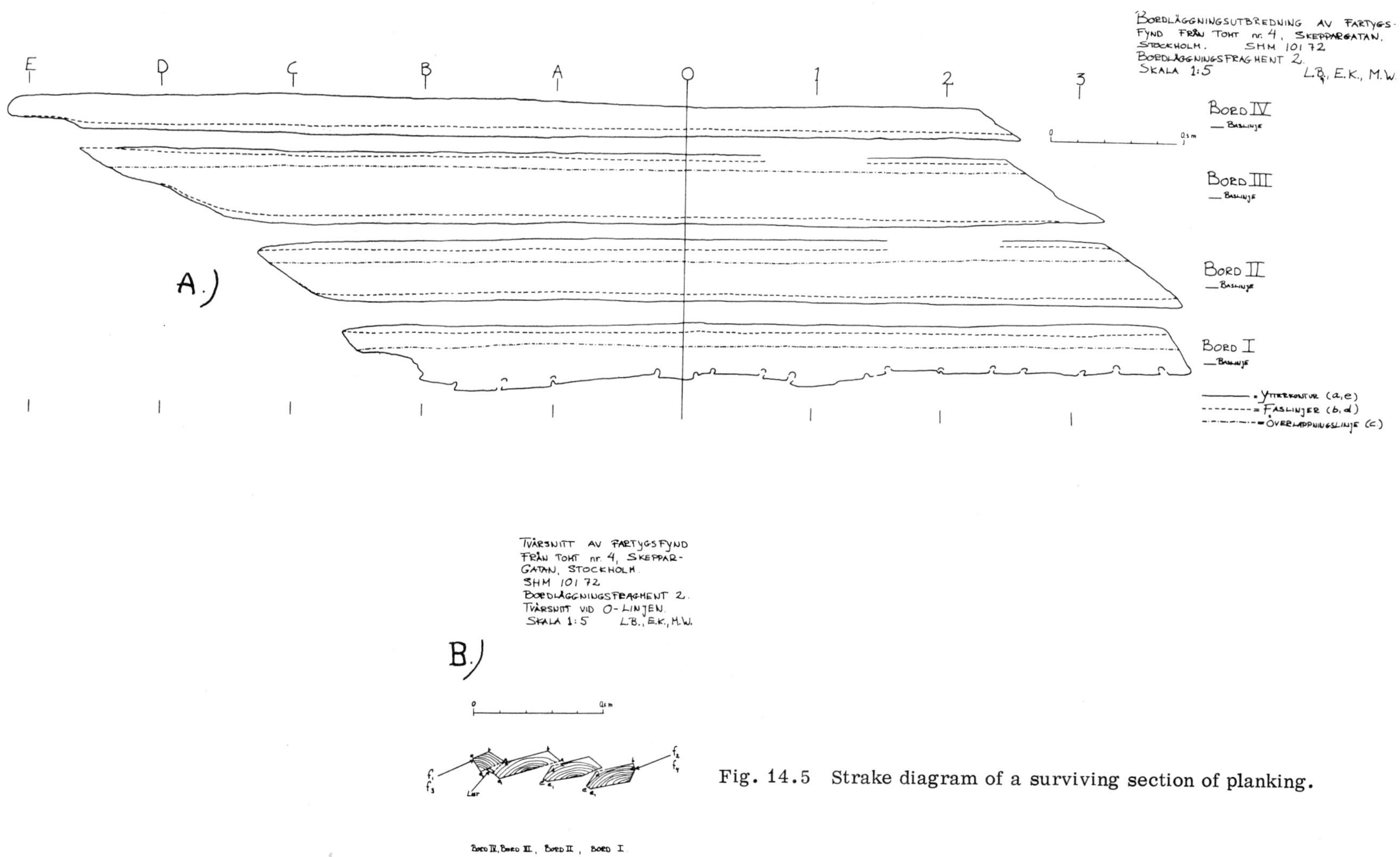

Fig. 14.5 Strake diagram of a surviving section of planking.

A.)

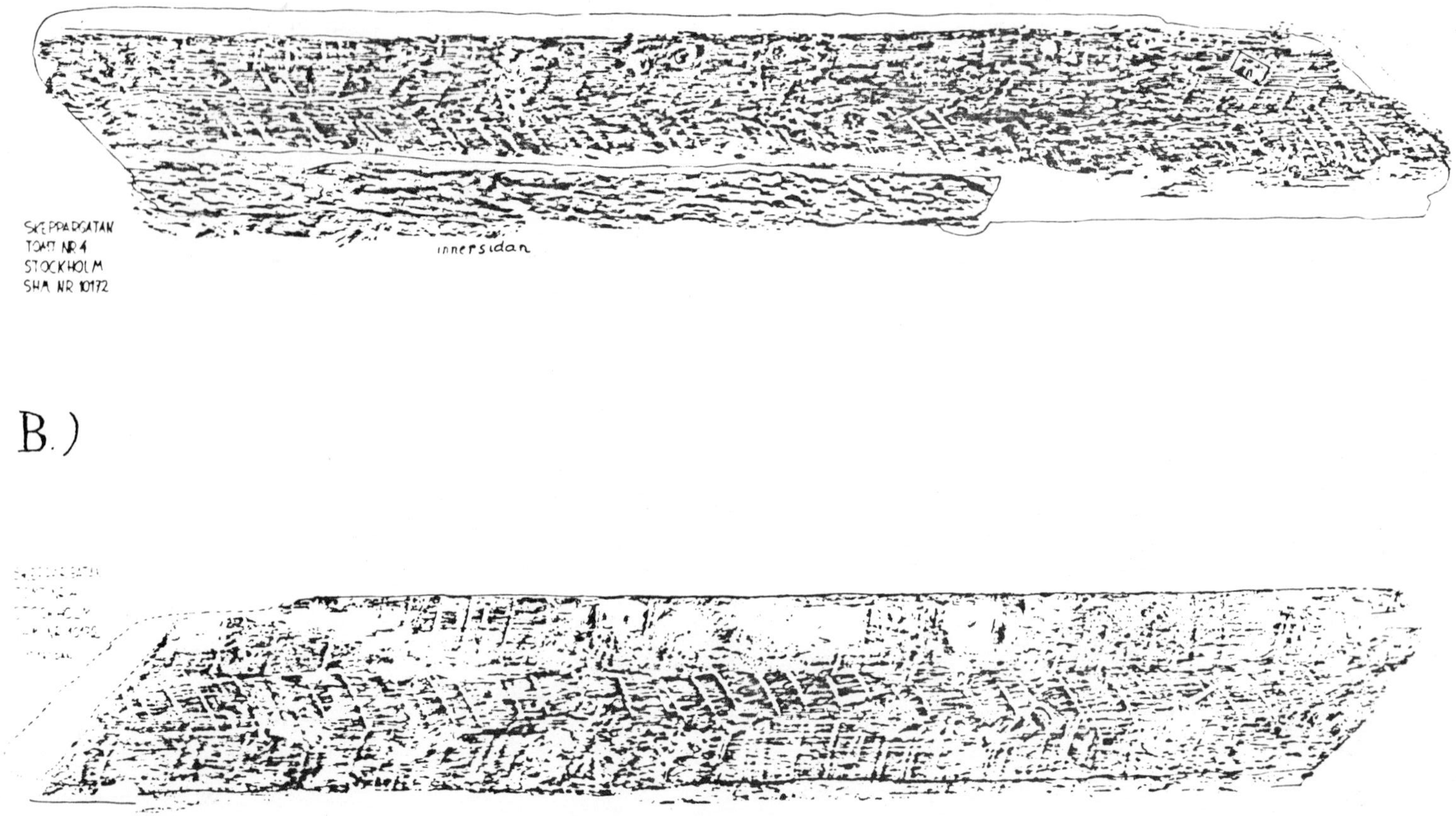

Fig. 14.6 To record tool marks on the surface of the plank in Fig. 14.4 a thin paper was placed directly on the plank and then rubbed with a piece of chalk.

To obtain strake diagrams of the two parts of the planking, their shape was measured from a lath fastened ot the planks. This kind of documentation, together with the cross-sections of the planking, permits us to study the shape of the planks and also to discern how these have been manufactured. Evidently the planks in this case had been produced in the following way: logs of pine had been split in two halves. From each half log a plank had been cut with an axe. One can see on the cross-section that the edges of the planks are the original surface of the logs (Fig. 14.5).

A special kind of recording was tested on the loose plank. It was difficult to record the marks of the axe on the surface of the plank on the full-scale drawing. To be able to give a picture of how these marks spread over the surface of the plank, a thin paper was placed directly on the plank and the structure of the surface was then registered on the paper by rubbing it with a piece of chalk (Fig. 14.6).

The analyses described above are primarily aimed at obtaining evidence of the age and the building-place of the ship. The identification of the different species of wood used in the building of the ship indicated areas in which such trees grow and thus where the ship might have been built.

It is my opinion that recording this kind of boat find consisting of too few parts to allow us to establish of calculate the dimensions and shape of the hull to a fuller extent, should concentrate primarily on the following items:

the recording of details giving information of
1. the origin and building-place of the boat
2. the building technique used
3. the shape and dimensions of the hull - as far as is possible
4. the original function(s) of the boat.

After this short description of the documentation which we performed, I will now discuss the type and the origin of the boat. This is a preliminary report as the comparison of the boat's construction with recent boatbuilding methods in the area and with evidence from historical sources such as written evidence, pictures etc,, has not yet been made. Such an analysis will be presented in an archaeological dissertation on ship documentation which the author is working on. The find itself will be published in a separate publication by the the National Maritime Museum in Stockholm.

The boat was clinker-built. Elsewhere in this symposium it has been said that we have a clear definition of the clinker-built-tradition as it is known in the northern part of Europe in prehistoric times and later. I think that this find deviates in certain important aspects from this clinker building-tradition as we know it, and that it implies further studies are required concerning the building-traditions and shiptypes in those parts of Europe in which clinker-building has been used.

My hypothesis is that this roughly built, clinkered boat with fastenings in sewing technique, with a flat bottom and a stempost at a rather sharp angle to the keel, represents an East-European ship-building-tradition with its origin in the rivers in that area. The assumption of an eastern origin I am primarily basing on certain details in the construction.

1. the sewing technique which is known in Russia - in different times and areas
2. the use of moss as caulking which also is known from different finds in the same area
3. the use of battens covering the caulking and kept in place by ironclamps which is known from finds in Novgorod and from the Bremer-kogge.

As a further hypothesis, which I will subsequently have to prove, I assume that this wreck is the remains of one of the heavy, flat-bottomed boats used mostly by the Russians, which, according to Swedish written sources, were common as transport craft and small merchant-vessels in the rivers in the Baltic states and west Russia and in the eastern part of the Baltic in the post-medieval period, They were adapted and used by the Swedish navy as transport vessels in Swedish waters for military operations in the Baltic states at the beginning of the 17th century. Two types of boat mentioned in written sources which were used by the Swedish navy and with an eastern origin, were "lodja" and "struss".

I think that you could also say that this find points towards new subjects of research within European boat archaeology. Important results have been reached during the last decades concerning the origin and development of types of ships in North-Western and Central Europe which have played important roles in European economic and cultural life, whereas very little is known about the shipbuilding traditions in eastern Europe and it's possible influence on European society through the lively seafaring connections between the different areas since prehistoric times.

DISCUSSION

Martin: It is vital that excavators, recorders of boat information or indeed of any sort of archaeological information should state their methods of recording and the standards of accuracy they set out to achieve, and be honest about the degree of accuracy which they consider they have achieved. The standard archaeological method of looking at something and drawing it from a limited number of measurements does have inherent defects but it also forces the recorder to observe and to try to understand the nature of what he is recording. Drawing the individual elements of a boat seems to me to be a particularly good way of recording. This presents the maximum information with the minimum of potential confusion.

Marsden: Having to work with a model maker who has to translate your drawings and photographs into a three dimensional model is a very valuable experience because you find the ommissions in your recording.

The people who are going to deal with some boat finds may not know one end of a boat from another. Perhaps there should be some general education, possibly through the Journal for Nautical Archaeology, on the first-aid recording that can be carried out by these people who can then store the timber for perhaps five or ten years until it can be properly dealt with by nautical archaeologists. Perhaps doing a total tracing is not really necessary for everything, a combination of measured drawings and full photographic coverage, and where necessary polythene tracings of some of the more technical details might be adequate.

Martin: Important and relevant things only come out of ships which have been thoroughly and accurately recorded and I wonder where our standards should lie, where is the dividing line, what is the acceptable accuracy beyond which we would really be better not having information at all?

Crumlin-Pedersen: What we might achieve in this Symposium is to indicate some standards which were desirable, and then to find out on the basis of these standards which way to go to find economical ways of maximising our information up to that standard or as close to that standard as possible.

McGrail: We recorded the Brigg 'raft' photogrammetrically on site, we took photographs, and we did the normal on site drawing. The post excavation recording is being done but is not yet complete. One problem in recording the Ferriby (Wright, E.V., 1976, The

North Ferriby Boats) and Brigg (McGrail, S., 1975, 'The Brigg Raft re-excavated' Lincs. History and Archaeology, 10:5-13) boats is that they have cleats integral with and projecting from the planks, therefore we cannot do contact tracing and there would be parallax errors if we used Crumlin-Pedersen's stand-off tracing method. We don't do scale drawings for the reason that Crumlin-Pedersen gave in his talk, and we cannot afford to record by photogrammetry. So what do we do? In fact we record by photography, and Brian Tremain, the Chief Photographer has devised a simple method of photographing each element, plan and side elevation and end-on, at a standard distance so that the contact print is one-tenth scale. From these we can build up a continuous picture of the 'raft'. The wood is very fragile, so before we record the under side of the planks we make a plaster cast of the upper surface with a cleat inside, and level off the top of the cast. We then turn the whole thing over so that the underside of the plank is now on top, and we can draw, take notes, and record by our photographic method.

Wright: A good standard line drawing with a caption is generally speaking worth pages of description, but photographs should be included so that the reader can judge the extent to which the line drawing is precise. I didn't have a cofferdam round my Ferriby finds and they were therefore exposed only at low water, the maximum time which we had available on any particular access was $4\frac{1}{2}$ to 5 hours, during which one had to clean the work, decide what to do, measure and then back fill all on a tide. Therefore one is forced to compromise unless one has resources which are generally not available.

Howe: I find that attempting to make a model brings out points that are not obvious on the line drawings. There may be amateur makers of models who could help archaeologists with this task.

Martin: I found that my very amateurish attempts at making scale models and components from my own drawings enabled me to appreciate the meaning of what we had excavated in a way that I could never have done simply from looking at the drawings alone.

Varrall: Working from photogrammetric drawings or photographs, leaves the model-maker with a certain amount of interpretation to do. So I don't think that you can honestly ask an outside model-maker to come in.

Crumlin-Pedersen: May we distinguish between models for display in a Museum and working models to serve the archaeologist as a tool in investigating the range of possibilities in reconstruction. A boat find with shell construction gives a range of different shapes: if you curve the bottom more, you get the stem and stern closer to one another and the breadth will increase; the more transverse members you have the more stable the model will be. These features could not be properly worked out on paper alone, it requires a working model.

Coles: Marsden described the need for some standardised simplified recording procedure and conservation procedure that people who discover boats could adopt. It would be very appropriate if such a set of rules could be published in a popular journal such as Current Archaeology.

McGrail: May I ask Crumlin-Pedersen about page 2 of his text where he says, "as there is nothing or very little between an excavation plan and a naval architects plan to document the transformation". The naval architect's plan is the basis for the reconstruction, and for deductions about the vessel's performance. Should there not be records, published or in a data bank, where you describe how you get from your excavation plans to your reconstruction?

Crumlin-Pedersen: In the final report on the Skuldelev ships there will be an excavation plan and excavation notes about each ship, and there will be a full set of plans showing all the elements in detail. These will serve as a basis for the model which will determine the shape selected to represent the reconstructed ship, probably given as maximum and minimum solutions, or as a mean value with plus and minus so many per cent in cargo capacity, and so on. In this way I think I will be able to give you the information you can use the day you find one of our ships over here, better than if I gave you pages of numerals.

Christensen: In support of Cederlund's east European attribution of his finds, there are unpublished fragments of White Sea Russian boats found at Spitzbergen in Norwegian Museums not as coarse in the building technique but sewn in exactly the same way with spruce roots, interrupted stitches, and moss caulking.

Editorial Note: This topic was further considered in the discussion after Basil Greenhill's Summary.

15. HYPOTHETICAL RECONSTRUCTIONS IN BOAT ARCHAEOLOGY

J.E.G. McKee

Summary

Unless one goes on to work out what the missing parts of an ancient vessel might have been like, much of the effort of recovering, preserving and recording her remains is sterile. Comparative study of similar finds is an added impetus to observation when it is directed towards reconstruction. This also calls for a careful choice of techniques and assumptions; until these have been proved adequate the resulting reconstruction is only hypothetical. Therefore, special care is needed to present solutions in a way that they will not be taken as being factual.

There are many ways of arriving at and presenting a reconstruction. The methods chosen should reflect the degree of reliance placed on the evidence and the confidence the public is to have in the solution. While it is wrong to represent a weak solution in an elaborate or conclusive manner, it is as bad to fail to draw proper attention to sound or alternative answers.

The paper discusses some of the ways of developing and presenting a reconstruction suitable for various levels of approach. If interest and criticisms are stimulated, some over-presentation of a solution may be justifiable.

Close to where the road running south out of Weston-super-Mare crosses the river Axe, there is a public house. It is called the Hobb's Boat and has been there for a long time. Field work showed that the present one was of the "Coaches Welcome" sort with an inn sign with a Viking long ship on it. It took a pint to find out why a Viking ship had been chosen for this lonely Somerset inn. There is a local legend that Danes under a pirate called Hubba had once sailed up the river, but the natives were ready and had brought the raiders to action on Bleadon Hill nearby. During the action an old lady cut the ship's moorings so that they drifted off, denying the defeated Danes their means of escape. One is left to assume that the inn site is close to where Hubba's ship stranded. This was too good a yarn to be left. From local history it seemed that the Axe had been a waterway, navigable up to Axebridge, 6 miles further up the river from the ferry at Hobb's Boat. When the bridge came a wharf was built and this like the ferry needed an inn. Later, in 1841, the Great Western Railway was built across the river a mile downstream, so the wharf was moved leaving the inn with only a road to serve. It looks as if the inn sign could equally well have shown a simple river crossing ferry.

This opening is not altogether idle, as it illustrates just about all I have to say about Hypothetical Reconstructions, or, as I rather think of them, Reasoned Cock Shies, (Concise Oxford Dictionary. "Object set up to be thrown at with sticks, stones, etc..).

When the remains of an old ship are found, in theory it should only be necessary to record them exactly. This may well involve recovery and preservation so that if better methods of observation are devised the original material will be still available. Otherwise the remains have little use beyond the diversion of museum visitors, as the busy scholar will generally refer to the facts listed in the official report.

However, the exercise of creative thought drives one to wonder what the rest of the vessel might have looked like. In this century the inspired approach to reconstruction has been replaced by more disciplined methods. These, though more pedestrian, bring added significance to similar features in other finds as well as an improved perception of one's own. Recording a find as a catalogue of facts and recording it in the knowledge that one is going to have to develop a reconstruction from one's observation are two different occupations.

The development of a reconstruction can take a long time, during which much of the contributory information may be modified, both in its nature and relative importance. It is not easy to keep track of how one reached all ones conclusions or indeed if some of them ought not still to be guesses. A properly conducted reconstruction needs an ordered frame work, in which the results of completed work together with tasks still to be done are apparent. Above all the final findings remains hypothetical until every assumption can be shown to be valid.

NATURE OF A RECONSTRUCTION

The opening story shows that even an inn sign can be used to present a reconstruction. There are better ways; from the full size replica of a named vessel or type, through mock-ups and models, conventional draughts, artist's impressions, written or spoken descriptions, to mental pictures. These are all as honest as the work put into them. At the most elaborate end of this list, full scale reconstruction is extravagent of resources, and once built may prove hard to alter when trying alternatives. At the lower end, nothing costs less or is more adjustable than the images of the mind. Unfortunately, they convince nobody else and are too easily forgotten.

Seeing is believing, it is said; and the inn sign shows how convincing a visual presentation can become. The Victorians were ruthless reconstructors. When they had run out of genuine old churches, castles and houses to improve, they built brand new examples of near Gothic and Renaissance architecture, even having specially written text books to help them. These have confused ideas of what these styles were really like and produced some odd tastes for what is good or bad in old houses.

If this country had had the fortune to have uncovered more of her ancient ships, I fear we would have treated them like the old buildings. The Viking ship on the inn sign reminds us what can happen after a total reconstruction of an early find. Anything as visually captivating as one of these manifestations of scholarship is only justifiable if it is perfect. There can be few adults who do not think they know just what a Viking ship looks like; when, I dare say, there are a few in this room who claim to. But, in our every day life, stamps, match box covers, beer bottle labels and book jackets plug the same popular image; while history books, encyclopaedias and museum models confirm it. If anyone tried to alter this, it might take a century.

The story also infers that more than one reasonable answer is possible. You might have accepted either of the two given, had you not been told of the other. Even then, those that feel that there is much truth in folk-lore might favour the Viking ship, while others would depend upon old newspapers and county records and prefer the ferry. Maybe the ferry had been cannibalised from the raider, or turn out to be a market boat that called there regularly.

Actually, there was only one inn sign. The landlord here needs to catch the eyes of his customers as they speed past. The Viking ship, being instantly recognisable would do this well. Had the inn been a cosy local half hidden down a side lane with a circle of regulars to cosset, the less arresting picture of a rural ferry might be better. It wasn't only the old lady who might have had an axe to grind.

STARTING GUIDELINES

It is as well to draw up a set of guidelines before starting a reconstruction; and you may wish to add your own comments on the following list:-

The find must have been adequately recorded.

The find will not be altered in any way.

The progress of work will be written up daily.

Assumptions will be recorded as they are made.

Validity of these assumptions will be tested regularly.

The extent of the reconstruction to be specified.

This reconstruction policy to be reviewed in the light of subsequent progress.

ORDER OF WORK

There is no set procedure for tackling a reconstruction. Much depends on the subject and its period, qualified by the state and amount of the recovered material and the nature of the external supporting evidence. For the purpose of this paper it will be taken that there is enough of both to merit a full scale effort; whereas there may be not much of one and none of the other, so limiting the extent of the reconstruction. There are a number of processes which can be used, some generally, others only in special situations. They can be taken roughly in the order they might be employed.

The pipe dream is probably the initial process. Forming mental pictures of what the vessel might have been like should be a continuous development. Unless one has a most orderly mind equipped with a reliable recall capability,

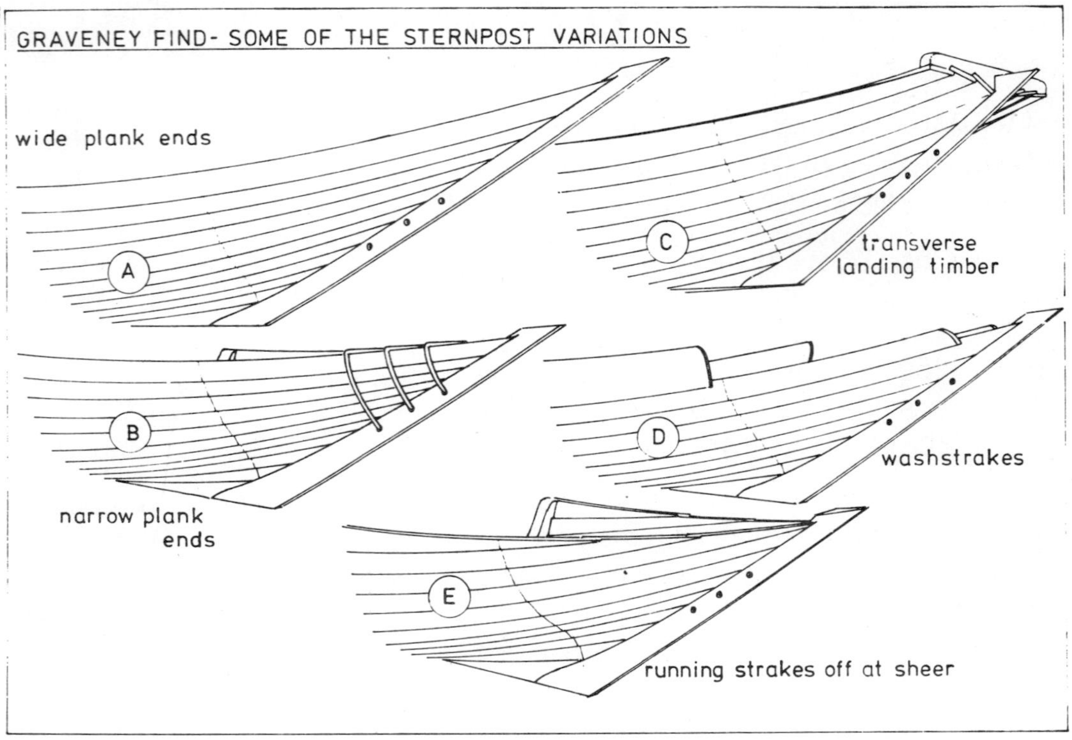

Fig. 15.1 Some forms of stern considered during the hypothetical reconstruction of the Graveney Boat.

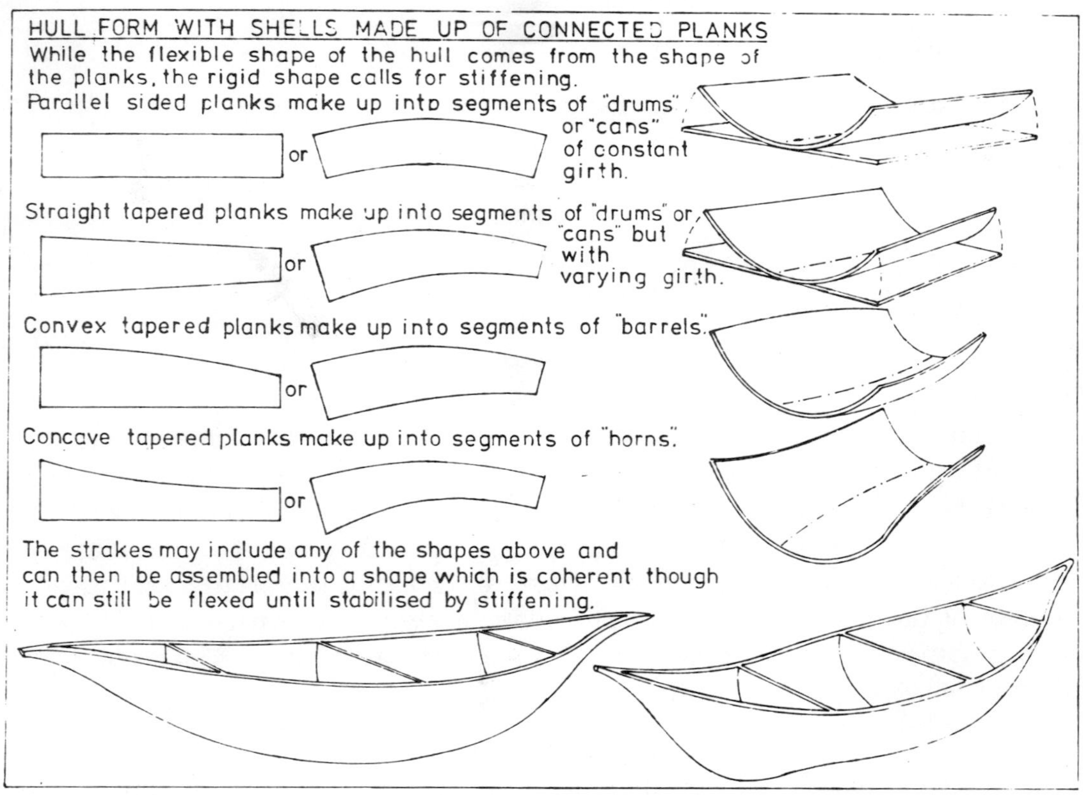

Fig. 15.2 Hull forms from different shaped plank edges.

it is suggested that the results of each brain session should be jotted down. Difficult, but less frustrating than to nurse a suspicion that one is forgetting some important earlier conclusion.

Working out all the possible variations, however far fetched, is the next step on from pipe dreams. This needs to be done systematically on paper, so as to leave no gaps in the list of starters. It might help to show what is meant, if I give an example from the Graveney Boat (The Graveney Boat edited by V.H. Fenwick, forthcoming). Actually the working sketches were only enough to define the variations. This particular problem was to work out the form of the stern, the upper part of which was absent (Fig. 15.1). The sternpost raked a lot and the remaining few plank ends were rather wide. Ten to a dozen strakes had to be accommodated, and if this was to be done with wide ends all the way up, the sternpost needed to be very long. As there were reservations about overhangs almost as long as the keel, these are some of the straight sternpost variations that were thought out and considered:-

A. Wide plank ends as described above,

B. Narrow plank ends with rope woolding,

C. Transverse landing timber for upper strakes,

D. Washstrakes to gain freeboard amidships, and

E. Running the upper strakes off the sheer line.

Some of these were dropped early on, while others were kept as still workable. As like processes were being applied to the other structures, it was found some variations had to be ruled out as they did not fit elsewhere. This showed the value of a full list of variations at the start; as without it, the elimination of non-runners takes longer.

The conservation and reassembly of what has been recovered is an invaluable step towards reconstruction. The find and the records made during and after the excavation are not only the archival starting point of one's own, but will be of other investigations. Happily most old ships are held together by imperfectly spaced fastenings, which thus offer an exclusive solution for reassembly. Distortion, before and after the excavation, lowers the value of the re-assembly until the dendrological and conservation experts detect and quantify it. This and making allowances for shrink and twist will take time and delay progress. However, an early assessment can be made from the loss of shape in bored holes, and from pitch errors between paired runs of fastening holes.

After a shell has been stitched together it is still not a single-shaped rigid structure (Fig. 15.2). Surfaces made up of pieces whose edges are straight or an exact match are either flat, cylindrical or conical in form. Forms that are built of pieces that have convergent or divergent curves become more rigid as they approach a complete sphere. Opened topped forms like the shell of a ship remain flexible until fixed by some dimension across or along its length. As drawing techniques for doing this are ponderous and probably subjectives, models are preferred.

Extrapolation by some means or other may be used to discover the shape of the missing part of the hull. Firstly a lot depends on what can be deduced about the original builder's methods, and then, whether or not he went on using them. For certain periods in the past we have examples of underwater parts of a vessel as well as pictures of the upperworks of contemporary ships. Scrutiny will suggest whether a match has been found, and if it has, how much more hull building is needed before the top hamper starts. External features call for local internal stiffening, and the converse is also true. Relating the find to pictorial evidence can both explain what you have and suggest where the topsides, capstans and spars that you haven't, were originally.

Wood bends into straightforward shapes and resists complex ones, by breaking sooner or later. Sweet wooden curves and musical harmonics follow mathematical rules. While shipwrights no more knew these rules than a boy humming a tune; they can help with extrapolation. The spline is an analogue for doing this, but it is not as objective as doing sums.

The use of mathematics often generates resistance; maybe from a suspicion it is an attempt to hoodwink. In the field of archaeology, one has to depend again and again on the findings of a specialist, whose expertise one could not match. What then is so bad about the mathematician? He has to be honest, if only to get the right answer. More use should be made of their talents. If it is thought that mathematical proofs will be unacceptable, then it may be wiser to keep them under the blotter. They will be a fine guide, but you will have to find another way of unpholding your solution.

Here are two cases where sums help. One curve is often seen in boats. Though it has a rather grand name, rectangular hyperbola, it can be drawn; first by a series of rectangles all the same area with a common corner, then running a line through the opposite corners. If you find a line that is curving like this, and you consider that the nature of it was not changing; you can work out the rest of it. The second aid is a table of differences, and the hand held calculator has taken the drudgery out of using it. Here you start by measuring something at equal intervals, say, the height of a strake every metre. Written down, you have a row of figures. Now in turn take each figure from the one next to it, and this gives the row of first differences. If they come out the same or tolerably so, then the strake is straight but sloping. If not the same, you work out the row of second differences and so on till they are. If after the third or fourth time this doesn't happen the curve isn't fair, but you can make it so by adjustments in the last row of differences and adding back. If or when the row of differences is the same you can extend the curve or fill in any gaps. You can read more about this versatile technique (Mathematics with a Desk Calculator by R.H. Cobb, G. Bell & Sons, London, 1970. Chap. V. Tabulation of Equal Differences.).

Establishing building methods is more about the size and sort of the structure than the overall shape just discussed. My own thoughts go as follows. Traditional British boatbuilders use simple rules of thumb to get their sizes. They may use relationships such as one thing being twice as long as another, or conversions such as the circumference of a spar being the same number of half inches as its length in feet, and there are rules for scantlings, spacing of fastenings etc.. Examining a find may reveal some of these rules. For instance, if the builder was spacing his lower fastenings at six times the plank thickness, he might well use the same rule for thinner planking higher up.

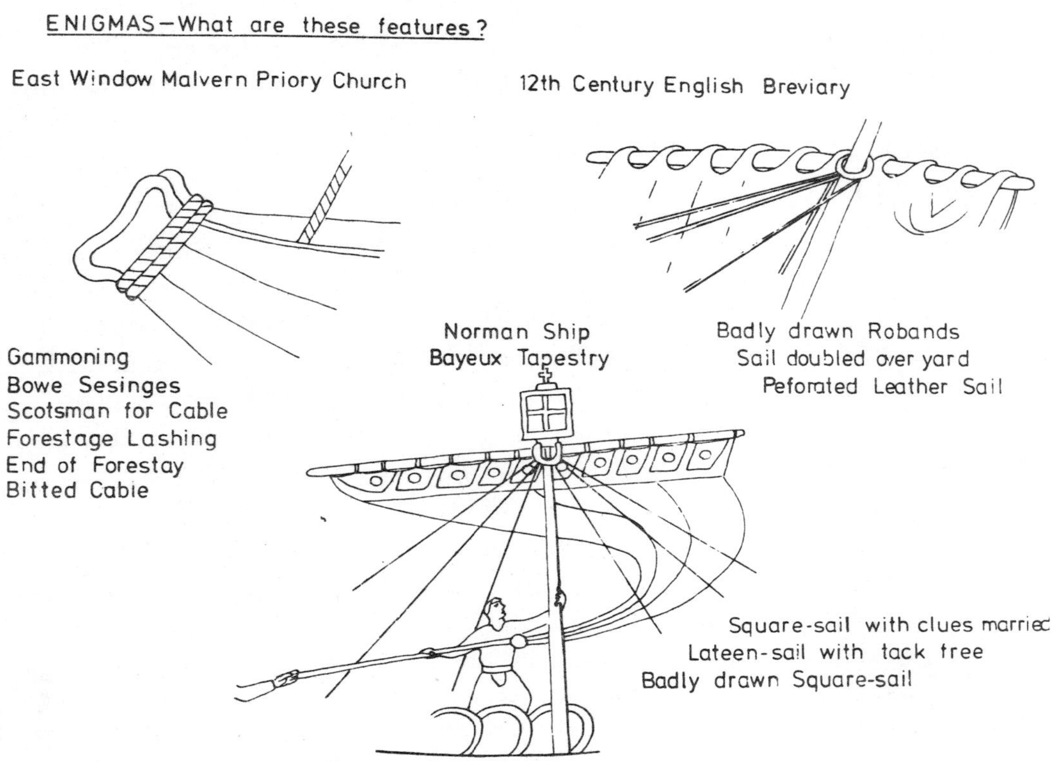

Fig. 15.3 Enigmas.

One often finds that to-day's rule fits an ancient boat, but it would be wrong to assume this was always so.

Text books and other contemporary sources become more reliable after the 16th century. By then more and more of the ship's dimensions were fractions of others, so, in theory, one can work out a complete design from very little initial evidence. One has, however, to be certain that one has the right text book and be convinced that the builder had followed it; two rather stiff conditions. W.A. Baker's article about his design for Mayflower II. (American Neptune, Jan. & Oct. 1954) is an able account of this method. Contemporary models and drafts must be checked to see if they are only projected designs, or cock shies themselves. Such models were constructed to a convention rather different from real practice.

Before the text books, there were formal reports, building accounts and ship inventories. These call for some degree of scholarship to unravel, but there are fine examples to be found in the Mariner's Mirror. Much can be gleaned by these means; what little is known about the size of the Golden Hind and Athenian Triremes is based on how they were housed. (Mariner's Mirror, 35 p. 56, and Anderson, Oared Fighting Ships, Chap. 2).

Contemporary pictures could tell a lot if one was sure what the ship was, as well as where and when she existed. It is hard to say as these illustrations were copied and used again quite out of the original context, just like to-day. They need to be looked at more for the queries they pose than the answers they provide. Here are three well known details, as yet I do not know of a satisfactory explanation for them (Fig. 15.3). The triangular sails in the Bayeux Tapestry; the ropes round stem heads; and this way of securing sails to their yards. Whether the answers to such questions are needed before one starts to reconstruct, or will come out as a result of it, is a bit like asking if the hen or the egg came first.

Interpolation of existing knowledge is helpful when the designs of vessels before and after have been well established. This still calls for a stated assumption, as one could easily arrive at a scale of spurious typological dates this way.

A vessel must stay afloat and upright long enough to make each trip. These abilities can be worked out, but we have to assume that the safety standards were probably less then than now. These calculations can be used to develop alternatives as well as testing them, and are essential in every reconstruction.

Eventually one arrives at a short list of reasoned solutions. Whereas, up till this stage much of the work will have involved various individual's selection ought to be by a group of the most unbiased persons one can assemble. In addition to making their choice, they should be asked to recommend how far the reconstruction should go on from there. As the time they will have to spare will be limited, it is necessary to take trouble over presenting the results of the investigation to them.

Models make fine cock shies, as they can be assimilated more easily than any other representation. They can be handled and compared side by side with alternative models. To me, conventional drawings are hard to interpret; and as much as artistic work pleases, it can also mislead. A model must present an alternative clearly and accurately, but go on further. Like all good forms of communication it will take longer to make than understand; but it will get the useful response one seldom gets by other methods.

By the time that a decision has been reached to embark on a full size replica the remaining alternatives should have been narrowed down enough to be tried out in a single hull, using mock-ups before finalising at times. The lessons learnt from the building and trials should not show one has frozen the design too soon. At this or some earlier stage it may be decided to publish the reconstruction, either as part of a report, plans for a model or in the course of raising funds. As in medieval times such drawings are sure to reappear out of context. So, this is a plea that any reconstruction drawings should indicate how seriously they should be taken; terms like Tentative, Second of Four Interim Possibilities or Preferred Design for Replica together with a date might meet the case.

You can see that there is a choice of methods. Appropriately used and in the right order they go towards developing a reconstruction, but only if the initial record and examination was properly done. It would not be fair to let you leave with the impression that I have done all this, but I do wish I had. It is thought that all the possible Graveney alternatives were considered, but it would not be easy to say how each was sorted out.

The opening story has one last point; even a poorly presented reconstruction may stimulate further reseach. This is not said to encourage more ill-considered reconstructions. There are enough of these, with damage still to be undone. Instead, one would like to hear of more work being directed towards a higher degree of hypothetical reconstruction. Once such a hypothesis has been worked out, presented, discussed and accepted, reconstruction should follow. Then, if funds have to be found, it should be possible to show that the ground work has been sound.

16. HYPOTHETICAL RECONSTRUCTIONS AND THE NAVAL ARCHITECT

J. F. Coates

Summary

Engineers look at artifacts and design problems in terms of purpose to be served, principle and circumstances of operation, cost, working life and repair. A naval architect tends to approach problems of boat and ship reconstruction in a similar way with particular concern about loads carried, strength, watertightness, stability, propulsion, speed, range and seaworthiness. Lack of data can prevent any one reconstruction from being proven correct, when several should be developed and sustained in parallel.

Some Bronze Age boat finds are taken to illustrate the variety of forms and questions that could occur to a naval architect in pondering the reconstruction of scant remains.

Because boats have been evolved, remains found are likely to be of successful designs, but hypothetical elements of reconstructions need assessment by numerical analyses of strength, stability, speed, power and seakeeping. Loads should be postulated and watertight strength of joints checked.

No reconstructions of an unfamiliar craft can be assumed to be satisfactory without examination of weight and stability, including stability in face of leakage. The performance of a hypothetical reconstruction is uncertain unless a replica has been actually built and tested.

Being a naval architect I hope to be forgiven if I look at the creation of hypothetical reconstructions from the viewpoint of an engineer rather than that of an archaeologist. Engineers tend to see artifacts and design problems in terms of questions like:-

What is it for?

How will it work? Where?

How to make it?

How much will it cost? Can it pay its way?

How to operate it? Who is to operate it?

How long must it last? Can it be repaired?

What will the neighbours say about it?

Those rather simple, even obvious, questions could be dressed up in higher flown terms, like system, environment, technology, and so on, but however expressed, they are unavoidable in every project today. They were no less unavoidable in the past, and certainly in projects like boats.

In thinking therefore about any reconstructions, including reconstructions of boats, one has to find good engineering answers to the same questions - albeit in a different tense.

Each raises a collection of interwoven issues about any reconstruction:-

What was it for?	Transport War	Goods People	Type of Society Geography Economy	Threat Prestige Trade
Operation?	Working life Value and cost of operation		Economics of Ownership	Society
How was it built?	Materials - forming, origin, geography			
How repaired?	Tools - skills - Society Effort - cost - time taken			
How will it work?	Loads Stability Seaworthiness Watertightness	Strength Control Crew	Working life Handling ashore Beaching Propulsion	

A naval architect should perhaps confine himself mainly to the last three questions. All the issues raised by them are his professional concern as they apply to modern ships and boats. Why not to ancient boats too? But to avoid anachronisms he naturally needs help from archaeologists particularly in understanding contemporary society and deciding what materials and tools may be assumed to be available for making the original. Equally, the archaeologist may need a little help from the naval architect in settling, or at least putting some limits of variation upon a boat's fundamental features. Questions may occur to one that benefit the other.

Load, strength, watertightness, stability, propulsion power, speed, range and seaworthiness are all closely connected factors in almost any design of boat. They are perhaps its nine vital attributes, and are determined sometimes quite sensitively, by the boat's main proportions, form, scantling and material (including fastenings). Very often a boat find consists only of the bottom of the hull, and one, if not both ends may be missing too. The reconstructor is then faced with more unknowns than equations; he therefore faces a problem which, strictly speaking, is indeterminate. One proposition immediately follows in virtually all such cases: no one reconstruction can be proved to be uniquely correct, to the exclusion of all others. Generally therefore, a reconstruction can only be more or less likely - one can sometimes do no more than aim to work out some of the more likely ones. And that leads to another proposition already advanced in the previous paper: that more than one hypothetical reconstruction of a boat should be developed if necessary to cover possibilities which may be both very different and equally valid. Investigators should try to avoid championing any one until all others have to be rejected. May I take an example?

Fig. 16.1 A model of the remains of Ferriby boat 1. National Maritime Museum.

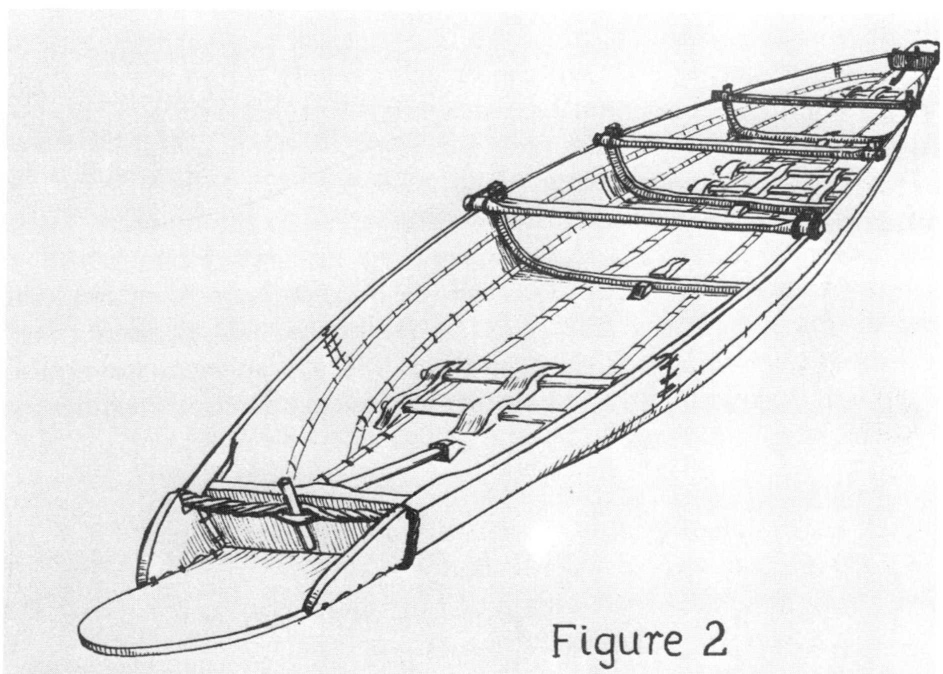

Fig. 16.2 A hypothetical reconstruction of a Ferriby boat - minimum solution with two side strakes. National Maritime Museum.

The Bronze Age boats found nearly 40 years ago by Mr. E.V. Wright at North Ferriby on the River Humber in Britain present reconstructors with many puzzles. The remains date from the 2nd Millenium B.C. and are among the oldest plank built boats of which any well preserved remains have been found. They define nearly the whole extent of the main pieces forming the bottom and a bit of one end of one lower side strake but little else. Fig. 16.1 shows a model made of the remains of the first of these boats to be found. Though built, to the extent shown by the remains, with massive pieces of hewn oak, sewn together with yew stitches, the 15 metre long boat is of slender form, of length about 8 times the likely maximum breadth. The height of the sides and how much they flared out, in other words the loaded water-line breadth, are suggested but not defined by the remains. Rough stability and strength calculations taken together would lead a naval architect to ponder a variety of forms for the complete boat. Some could no doubt be rejected on archaeological arguements, but conversely others might not in the first place have occurred to an archaeologist.

To illustrate, but not to prove, the need in many cases for plural reconstructions I offer three out of many other possibles, of the North Ferriby Boats.

The first (Fig. 16.2) is an effort to design a boat with the least material in it, that is additional to what was actually found in the remains, and necessary to form a workable vessel with a useful purpose. Such a boat would be sufficient I believe to ferry loads of about $1\frac{1}{2}$ tonnes or say 10 people and 8 paddlers across a river like the Humber, 1 or 2 kilometres wide, with adequate stability, just enough freeboard and without too great a rate of leakage through straining. There is considerable force in not straying from a rather parsimonious Principle of Sufficiency in reconstructing utilitarian objects, whatever their age. It is true that in war equipment and vessels for whose total purpose visual display may be as important as their primary function, decoration or purely stylistic features have often extended beyond relatively minor additions of finials, fashion pieces, or carving. But such vessels, though prestigious, have usually been rarities in their own times relative to the mass of workaday craft, so any reconstruction whose main form has had to be altered funamentally to include substantial features of no utilitarian purpose should attract some doubts. I suggest that it is the dull truth that ordinary, useful boats have generally been designed, or rather evolved, so that, given the technology of their time, they could be built as cheaply as their required and acceptable performance allowed.

A very substantial feature of the North Ferriby boats is their length and slender form. Extra length calls for greater longitudinal strength to avoid working and leakage at seams. Narrow beam greatly reduces stability and load carrying capacity. However length and slender form do make for speed, besides providing space for more paddlers or oarsmen.

These boats could probably be driven by 8 paddlers fairly easily at about 4 knots for a mile or so, and at 5 knots with some effort. A naval architect must ask some questions about the need for speed because one way or another, speed is expensive. Was it needed to breast the Humber tide stream of 4 knots or so? What ferry traffic in those days could have made the labour of building and maintaining a long fast boat worthwhile? Could a ferry have landed and embarked passengers except near the top of the tide anyway? The reasons for the length and slenderness of the North Ferriby boats should be sought in such questions. Reasons should be found for features so fundamental to its whole design. In ignorance of such reasons the reconstructor must accept that he has not travelled very far towards seeing the boat from the builders' and owners' point of view and must meanwhile keep an open mind about more detailed suppositions.

The second reconstruction illustration (Fig. 16.3) represents a more ambitious and higher quality craft whose materials and detailed design keep closely to the style of the actual North Ferriby remains. It could carry a larger load, about 4 tonnes, provided that the load was kept pretty low. It is true that a deep boat could carry a greater load without being swamped, but, being rather narrow for her length, she would become unstable unless the added load were kept remarkably low. At the limit it would have to become ballast on the bottom. Unless quite extraordinarily flaring sides could have been built to widen the waterplane as she was loaded, the boat's length to beam ratio of about 8 gives her, if at all deep, all the vices of the so called "Coffin Ships" of that unfortunate period in the last century when tonnage and therefore harbour dues were decided by a formula which penalised ships with healthy beams.

In reconstructing boat remains one must be wary of proportions natural in an age of luxuriously strong materials and fastenings. One may be forgetting for instance, to take leakage into account. Strains increase with load and weight, so the second illustrated reconstruction may be asking a lot of its moss caulking to withstand being rubbed as the joints "work" under load. We can easily forget, because it is a dull subject, the past servitude of sailors to baling and, later, pumping. The seams of the North Ferriby boats were most ingeniously contrived and caulked, but without practical experience of them, their watertightness should not lightly be overestimated particularly as their hulls were so long and narrow. Bilge water does more than wet the feet of passengers and crew. In flat bottomed boats it can have a profound effect on the craft's behaviour particularly as regards stability. Leakage increases with draught and could also limit the draught to which it would be sensible to load a boat. Indeed chronic presence of bilge water could be advanced as reason for the need in the North Ferriby boats for a thick floor of wood staves and twigs to reduce the effect of free surface in reducing stability: such a floor would also protect the seam stitches from damage by kicking or slipping feet or cargo. But one needs first to find out how badly the boats leaked. A replica is called for to provide a convincing assessment. I would judge that No. 2 is quite as far as that line of reconstructive thought could possibly be taken. There are certainly many possible hypotheses about the detailed construction of the ends of this kind of hull. Any need to bend wood of the thickness used must be set aside: on the other hand it would take little if any longer to hack a curved slab of wood out of half a tree trunk as a flat one, but such heavy material above water would be a liability to the design if its additional strength and solidity would be unnecessary. The boat in Fig. 3 would weigh about 3 tons: hauling it out of the water for repairs could have presented some problems.

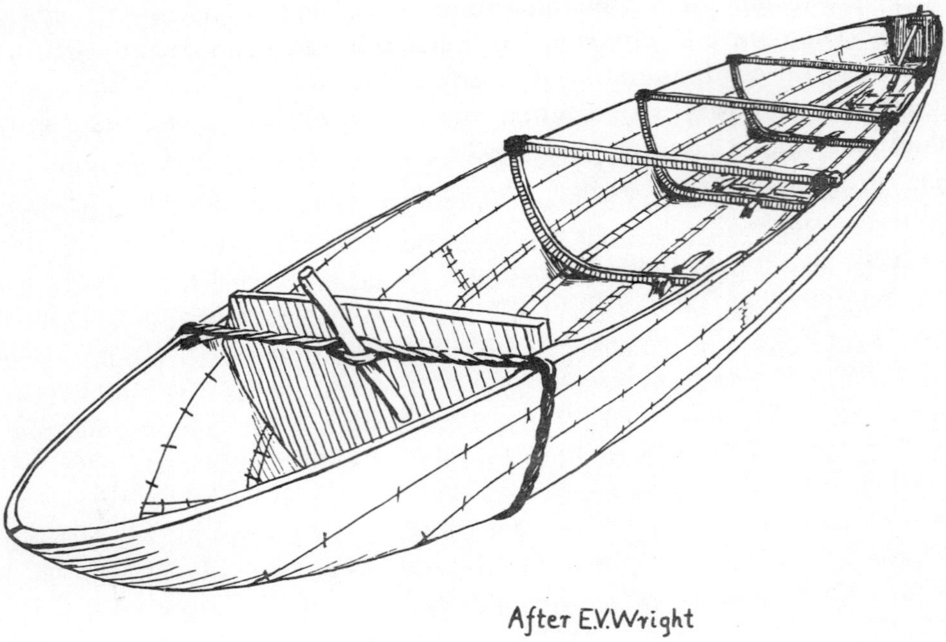

Fig. 16.3 Another possible reconstruction - with three strakes and closed ends. National Maritime Museum.

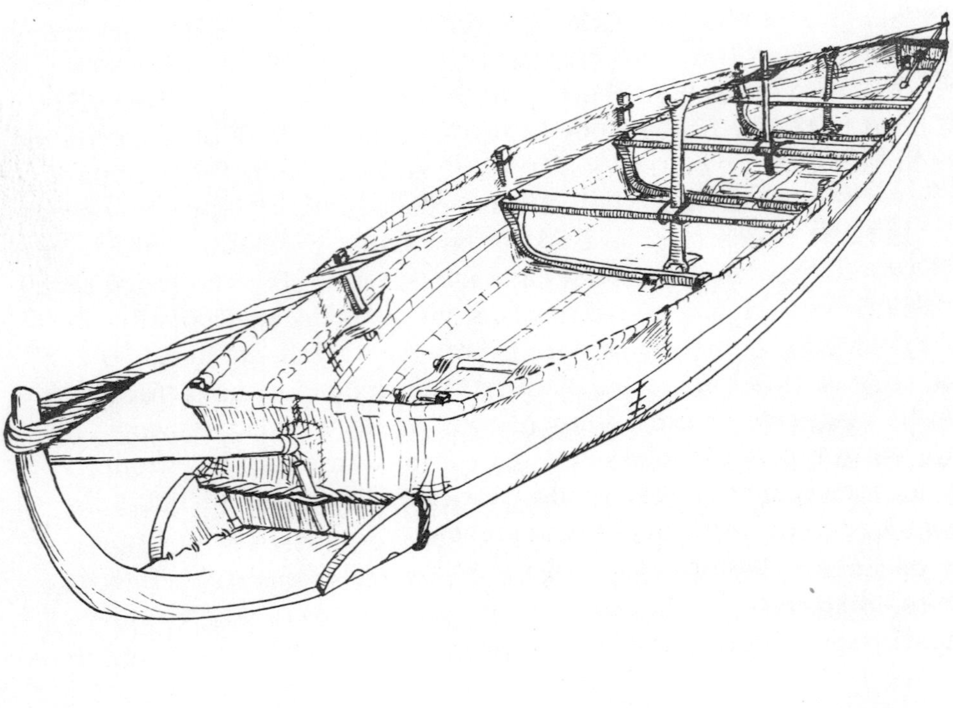

Fig. 16.4 A third reconstruction - two timber strakes with a hide sheer-strake and a hogging truss. National Maritime Museum.

The point leads to my third example (Fig. 16.4) which is one of a variety of designs that one could be led to contemplate as possible reconstructions if other sorts of construction could be admitted to have formed any or, even all, of the boat beyond what is firmly defined or implied by the remains. Here I have followed the rather obvious course of combining the very heavy oak pieces forming the lower shell with a much lighter skin construction for the upper parts and providing a hogging truss to prestress the hull permanently in sag, so that the joints will not suffer reversing forces in service. They will not therefore be so prone to leak. Skin boats had presumeably developed into an established and widespead means of navigation round the coast of North West Europe in the course of Neolithic times. Skin boats however could never have been very long lived, especially when used as a ferry, necessarily taking the ground in a tidal river at the end of every crossing. What more useful development to overcome that shortcoming, than to provide such a geographically important ferry not only with a hard bottom just as soon as it became possible to hew wood on the scale required to do so but also the strength needed to enable a slender form to be possible and hence greater speed to be achieved? It appears that the adze marks on the North Ferriby boats are of the maximum size possible with the available tools. Those boats could therefore have been, as we say today, at the forefront of technology - the Concordes of their time - in proportion to Gross Tribal Product they could have been as expensive but arguably more useful. The ingenuity and skill of the Humberside boat builders (demonstrated by the way they joined the oak strakes together) would surely have been quite adequate to solve the problem of making sufficiently watertight joints between the wooden and the flexible unstressed skin parts of a composite boat. It is not difficult to conceive designs in which no part of that joint is below the loaded waterline but, because even the clever seams between the wooden strakes would have been by no means watertight after the strains of service, immersion of some part of the skin covered region could well have been acceptable.

The three reconstructions I have illustrated stand for a large number of related practical possibilities within the confines of practical naval architectural concepts. It is natural to want to reduce that number if possible to one - the single authoritative version as it were and much effort in archaeological argument seems on occasion to be expended to that end in fields besides that of boats. In the maritime field there is the famous and wellnigh perennial case of the trireme, which only recently erupted into The Times newspaper. The ensuing correspondence would not have been very conclusive had not some facts about power, speed and resistance of ships been brought to bear upon a variety of suggestions some of which had lacked the support of physical laws or engineering expertise.

That the serious archaeological study of the boat is a relatively recent development may, I suggest, be due first to the generally subordinate place of maritime affairs in history: the Wars between Greece and Persia, the exploitation of the New World by Spain, and the period of British sea power are some of the exceptions that prove the rule. Secondly, ships and boats are technically perhaps rather trickier than the general run of archaelogical remains like pots, ornaments and buildings and the scholar studying high matters of dynasties and the ebb and flow of peoples and cultures has not been able to glean much about the thoughts and techniques of the relatively rude and unlettered mechanics, architects or engineers who contrived the wherewithal of civilised life, travel, trade and war. Neither have historians usually been engineers, nor engineers historians, so I suggest that sound study of historical ships and boats must include consideration of the naval architectural feasibility of reconstructions with as much rigour as purely archaeological considerations such as stratigraphy, context, origins and affinities.

Ships and boats have evolved by a continuity of experience so that, as in the world of living organisms, only acceptably successful designs have multiplied to form the boat population of their time. So it is statistically likely that remains of boats that are discovered now are of successful designs. Exceptions to that proposition, which may be thought to prove the rule, are cases where the very unsoundness of a craft may have caused it to be abandoned or lost in such a way as to result in its safe preservation. The Wasa is a spectacular example. Apart however from those peculiar cases, which must surely be a small minority, soundly worked out reconstructions of remains would now be found to be adequate structurally, mechanically, as regards stability and fluid dynamics, and in energy conversion if that were significantly involved in their operation. Such naval architectural considerations can eliminate as well as create hypotheses, refining and completing the set of possible reconstructions of any one boat which should all be regarded as valid. To reduce possible reconstructions to one prematurely is, I suggest, a great mistake.

I have said that, in the past, ships and boats evolved. The naval architect now has some scientific methods to help his creative endeavours. Nevertheless for ship's structure and stability only relatively recently have methods of theoretical assessment been developed. Naturally those methods have concentrated on technically, economically and militarily interesting contemporary ship designs, rather than the appraisal of say Bronze Age boats, or triremes, but such as they are, useful naval architectural methods of analysis of strength, stability, speed, power and seakeeping are available to help in making reconstructions. I may have over emphasised the novelty of numerical methods for the assessment of stability: for ships these have been available for the statical consideration of stability for well over a century now, but theortical understanding of dynamic behaviour in waves is still developing actively and for boats dynamic behaviour is more important relative to static stability than for larger ships. Structural analysis of shells, which to a structures man is what ships are, has developed in directions most applicable to metal welded ships: no systematic study of the behaviour of shells made of wood whose serviceable strength is mainly determined by relatively weak and crushable fastenings by stitches or skewers (or by the clamping action of rivets, bolts or spikes) has to my knowledge yet been carried out to help the archaeologists.

In any hypothetical reconstruction a naval architect is bound to ask questions about strength, particularly the strength of joints and what would have been their likely leakage which must have been acceptable. In every reconstruction that is a crucial question not always easy to answer.

Some likely maximum service loadings on the boat can usually be postulated, indeed that must be done if a reconstruction is to make any engineering sense. Corresponding forces in at least some members of the hull, and which would have to be transmitted by affected joints can usually be calculated, even if only approximately. If the strength of the joint or, more to the point, the repeated load that the joint can bear while remaining watertight is known and if that strength matches the calculated load, the assumptions underlying the postulated load on the hull are much supported (but by no means proved). If they do not match, then those assumptions are almost certainly disproved by that test alone - and surely the corresponding hypothetical reconstruction must then be abandoned for such a boat would be scarcely satisfactory; she would strain, leak, or even split open somewhere under heavy loading in service.

If the physical behaviour of the joints in a boat find, for which estimates of service loads can be made, is not known, then I suggest that one of the first things one should do by way of foundation work for hypothetical reconstructions is to make some test specimens of the most likely versions of such joints and to test them to discover their qualities as joints between boat hull members. The most important quality of a hull joint is not so much its strength to carry the required load without coming apart, but its strength under repeated loading while remaining watertight. The very word "watertight" is most apt, for no loose joint, however strong it may be, will keep water out. That is patently obvious in the case of wood joints whose watertightness depends upon the close mating and clamping together of their component pieces. Where neither fit nor clamping can be achieved, boatbuilders as at North Ferriby and indeed builders everywhere of sewn plank built boats as well as carvel boats have had to rely on stuffing their joints and poulticing them with materials to make what is essentially a lightly packed labyrinth of fibres to reduce the flow of water through the seam or joint. If very tightly packed, greasy fibres can by surface tension actually prevent water from penetrating the joint, but most builders and seamen have had to be content in practice with appreciable seepage. Neither a fitted nor a caulked joint remains tight for long if it "works" under service loads. It must not open or slip under the loads applied in service, which therefore can in practice be only a fraction of the actual failure load of the joint. Until the arrival in our own times of truly waterproof glues, it was nearly always the required volume of wood or required bearing areas in a joint that determined the scantlings of the members, the style of joints and therefore the style of the whole structure. The strength of wood itself has virtually never been fully exploited in structures - at the crucial parts in the joints high crushing stresses are concentrated over quite small areas while the members joined are yet carrying low stress.

In joining one piece of wooden hull to another with stitches, treenails, rivets and the rest, the problem has always been to obtain sufficient clamping force or bearing area to carry service loads by friction without slipping, and yet avoid too massive scantlings to hold the size and number of fastenings needed. It is not surprising that under such constraints the number of types of joints employed by boat builders the world over and down the years has not been so very great.

Before leaving joints the anisotropic nature of wood must be mentioned because boat builders have for a long time exploited the fact that wood swells in water across the grain more than along it. That property must be responsible for a large part of the tension in fasteners and hence clamping force in

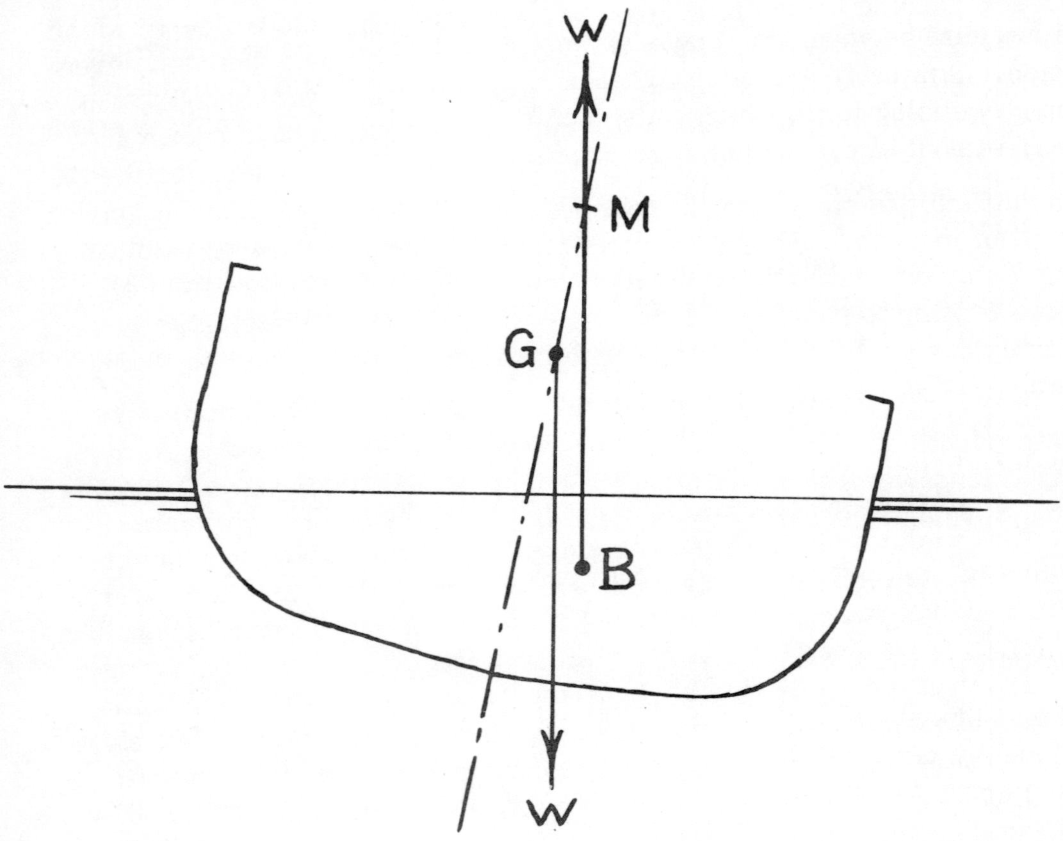

Fig. 16.5 Transverse stability diagram:
G centre of gravity of ship
B centroid of the volume of water displaced by the ship (centre of buoyancy)
W displacement of the ship
M metacentre
$\overline{GM}$ metacentric height - a measure of initial stability
National Maritime Museum.

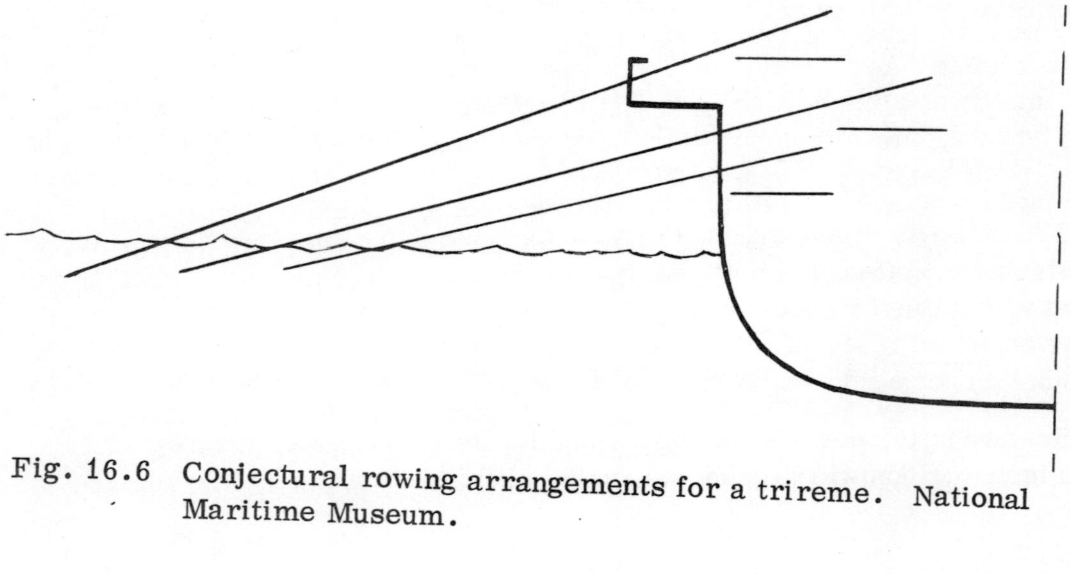

Fig. 16.6 Conjectural rowing arrangements for a trireme. National Maritime Museum.

a riveted lap joint between two pieces of wood, as for instance in clinker construction. Similarly it enables treenails, especially if split and wedged, to develop a clamping force which adds usefully to the load that can be carried by friction, without slippage, and therefore leakage.

It is as difficult to stop a naval architect from talking about stability as a gardener about manure. As important as making tolerably watertight joints is forming a ship or boat to float the right way up and sufficiently resist all those forces tending to push her over. A racing eight is a balancing act, so is a battleship due to its load of deck armour and guns and so was the Wasa, but she fell over.

Old shipwrights stuck to rules of proportion and very wise they were to do so until at least statical stability was understood mathematically. The need for those rules is sufficiently explained by 3 factors:

First that in any novel design one may assume that the centre of gravity of the ship, G, (Fig. 16.5) will be a bit higher than you think it is going to be.

Second that its weight will be greater than you think.

Third that the height of the metacentre (M) (which must be above the centre of gravity and by more than some amount if the ship is to be sufficiently stable for service) above the centroid (B) of the volume (V) of the water displaced by the ship is

a. proportional to the cube of the beam, and
b. inversely proportional to the displacement of the ship (V).

To naval architects, $BM = \frac{I}{V} \propto \frac{\text{Beam}^3}{\text{Displacement}}$

(I is the second moment of inertia of the water plane about its fore and aft axis of symmetry.)

That formula explains why beam is so powerful a factor in stability, why a deep ship (relative to beam) is tender, and why it is quite usual for a ship to be less stable when loaded than empty even if in being loaded her centre of gravity does fall: to maintain stability it has to fall enough to cope with the reduction in BM, due to increase in displacement, even though the centre of buoyancy rises relative to the keel as the ship's draught increases.

It follows that a little increase in beam, say 5%, could well convert a crank ship into a good one because BM could increase by about 15% which, depending upon how desperately close G was to M originally, could multiply GM, the measure of stability, manyfold.

The stability of any unfamiliar ship or craft can never be assumed to be satisfactory, so there can be few reconstructions that can be assumed to be sound without some check on stability. Without possession of her builders' experience and rules of proportion, no reconstructor can be sure that his hypothetical creation would be stable, except by calculation.

But no matter how well a reconstruction is worked out by theory and applied technical knowledge, no wise naval architect would be too confident about its performance and practical behaviour, even of a small boat, unless a replica of it had actually been made and tested at sea. The National Maritime Museum has pioneered this form of experimental archaeology in its practical research with a replica of the Gokstad Faering. Much has also been learned, as could have come to light in no other way, from replicas of ships such as Mayflower, Golden Hind and Non Such, even though the design of such relatively recent vessels is fairly well established. Ideas about Greek triremes would have been sorted out better if even only one of the several possible reconstructions which have been advanced had actually been reconstructed: indeed until that is done I suspect that little further real progress can be made on the subject of triremes. Preliminary practical trials of rowing arrangements said to have been made some time ago by Kopecky (Fig. 16.6) and Alexanderson need to be extended to the whole ship to put a number of shelved but essential qualities to the question. A few of those questions address the naval architect's bread and butter subjects of:

Stability	:	What weather could triremes stand and how were they worked in heavy seas?
Weight (about 60 tons)	:	Were they drawn up on beaches?
Strength	:	Did they need an encircling rope under tension to prevent hogging failure?

All three affect their naval operation and therefore the feasible limits of sea power that could be exercised by them, to an extent that should be of interest to historians of the period.

It is rarely practical to build a true replica. Many materials once easily available have become scarce either in the right form or even in any form. It is not easy to be sure of the old techniques by which materials were formed prepared or assembled - and limitations of technique have ever been a compelling discipline to architects and craftsmen: they can determine the whole style of a design in ways unimaginable until one tries to do it oneself - that is in itself a complete justification of the view, that until one has made and worked any artifact (except it be of really quite extreme simplicity or lack of novelty, and no boat is either of those), one knows very little about it. Talking and theorising may both be necessary preliminary stages to understanding, but no matter how plausible or intellectually attractive the results may be, they can do no more than form a set of ideas about a boat which may be worth testing in a workshop, on the beach and at sea. After that real experience, and getting perhaps cut, bruised and wet, one's understanding suffers a literal sea change. A new perspective of practical questions arises and one can return to those I started with armed with knowledge of an altogether deeper and more convincing kind which is unobtainable in any other way.

REFERENCES

McGrail, S. and McKee, E., 1974, Building and Trials of the replica of an ancient boat: the Gokstad faering. N.M.M. Monograph No. 11.

Wright, E. V., 1976, The North Ferriby Boats. N.M.M. Monograph No. 23.

DISCUSSION

Corlett

The Graveney Boat is a good example of using several disciplines to try and get a real idea of what a boat was built for, but, although naval architects, mathematicians and so on, can help, they must not trespass on the field of the archaeologist. In the case of the Graveney Boat the naval architect can estimate very closely the power required to drive her at various speeds, what limitations on speed applied, what type of stability she had and hence possibly the cargo she would have been likely to carry. As Mr. Coates has indicated you can also get a pretty good idea of the speed potential of the Ferriby Boats. This is helpful information to give to the archaeologist without in any way trespassing on his preserves.

At the moment I am working on triremes in some depth, and we are already involved in structural investigations. A great deal is coming to light which explains what some peculiar looking representations on vases, paintings, and coins, really were. There are several hypotheses for some of the features, but there can only be one or two real technical solutions and in some cases only one. For example, the bronze ram fitted to triremes is shown on coins to be not parallel to the waterline. There is a very good technical reason for this: it must point through the centre of gravity, otherwise when the trireme rammed another ship it would be ripped off by a force which can be calculated, something like three tons vertically at the stem.

McGrail

John Coates has shown us three variants of the Ferriby boat and they appear to have significant differences in performance. Are there any variant solutions to the Graveney boat reconstruction which would have significant differences in performance? I think Eric McKee has maximum and minimum solutions but do they differ significantly?

McKee

I don't think there would be a lot of difference in the hydrostatic shape of either solution. I didn't have much choice really as the bottom of the boat was recovered almost up to the waterline. The performance of a boat with a very high topsides would be very different from one with no topsides both in a sea when under oars or when sailing against the wind. The Admiralty experimental laboratory at Haslar ran a programme to see how she would behave in a seaway, and these figures show that she corrects herself so fast that in bad weather she might break herself up there would be so much reserve buoyancy. There is no windward capability on a boat like that, or very little, so stability in the sense of holding the boat up to the wind, isn't anything like as important as it would be in a modern fore and aft sailing vessel. This sort of data is very interesting after you have done the reconstruction: it does not tell you the reconstruction is right but it does tell you something about what you have conjured up.

McGrail May I clarify that? Is the Graveney Boat a special case rather like Sutton Hoo where there is probably only one solution to the problem of reconstruction?

McKee I can give you several solutions, but the requirement at that time was to produce a solution for the building of a replica and obviously the Museum was not going to build six full sized replicas. But I don't think that with any solution you are going to change the underwater shape very much.

Corlett I think this is true. In the case of the Ferriby Boats you could have several appreciably different hydrodynamic solutions, with measurable difference between them. In the case of the Graveney Boat the hydrodynamic solution is unique: it is above the waterline, that the possible difference appear to be.

I think another very useful help the naval architect can give is in comparisons with existing vessels. For example, you can learn a great deal about ancient Egyptian vessels from looking at the design and construction of modern Bangladesh craft. There are distinct similarities in both construction and design. Why? The answer is, that they both operate in similar environments. Examination of some of the Bangladesh craft I think could also have a distinct bearing on the North Ferriby Boats, for there again you are doing the same thing, ferrying goods and people across wide and sometimes fast flowing rivers. I have seen boats in East Pakistan that are very obvious long distance cousins of the Ferriby Boats, distant both in time and geography.

Christensen In the various stern reconstructions of the Graveney Boat shown by Eric McKee there was one with planks narrowing very much towards the stern posts and you had put some nice lashings around it to keep everything together. Why not try to use the solution we know from Scandinavia where many thin planks are assembled into one piece which runs to the stem, a stepped hood end picking up several planks?

McKee This solution was considered but was not included because I didn't reckon that the level of shipwrightry in the Graveney Boat was high enough to achieve that excellence of workmanship.

Wright Can John Coates tell us more about the sort of performance which we might expect from a two strake, minimum solution Ferriby boat with its massively heavy structure and with a very low freeboard.

Coates — If the remains as they are known were put into the water with nothing added, they would float with the upper surface of the bottom slightly above water. So one would have a workable platform in smooth water which wouldn't carry very much. I believe there is great merit in any reconstruction of an ancient boat find in seeking the minimum solution which will give an acceptable performance. Defining the ancient standard of acceptable performance is difficult, but that is why one must look most carefully at the circumstances in which it was built, the likely purpose it served. The Ferriby boats were found on the Humber, and were probably used there as ferries. We should consider whether there might have been a requirement to operate this ferry when it was a bit choppy or when the tide was running high, or whether it could wait until the wind dropped. In the latter case good freeboard would not be required. We should then consider the length of this boat: that must have cost a tremendous amount; the lowest side strakes were colossal chunks of oak. There must have been some very good reason to make it as long as I think it was - perhaps the ferry service was urgent. Perhaps also they did want the boat to cross the river when a full tide was running, in which case the need for higher sides to provide freeboard begins then to appear. I am suggesting that perhaps we should stop at two strakes each side.

Corlett — Length is the most expensive dimension in a ship, it is very expensive in any terms you like to name, money, man hours, material. I don't believe that the Ferriby Boat would have been built long and narrow in that way unless they were interested to some extent in speed. There would be pressures to obtain more regularity for the ferry service, to carry a bit more - four cows instead of three and so on. I think one has got to be careful of applying too rigorously the least effort solution, it is one solution but I don't think it is necessarily the preferred one.

Lovell — How do we know that these rather bare finds are in fact boats at all? Could they possibly be some form of vessel or tank in which you can stow things? Have there in fact been finds which were first thought to be boats and subsequently turned out to be something else?

Wright — There is no doubt in my mind that we have examples of three boats at Ferriby. I would ask Sean McGrail who has studied all the dugouts recorded in this country and many of the Irish ones as well, to tell us what percentage of them on examination look as if they aren't boats at all.

McGrail — I think the great bulk of those artifacts called dugout canoes are obviously boats of some sort. On the other hand some of the smaller objects are coffins, and there are troughs for brewing and the like. The circumstances of the find, the length and the features can often help to determine whether a find is a logboat of not. There are some finds of which I am uncertain: these are the fragmentary finds whose circumstances of recovery are not well recorded. Captain Lovell's remarks remind me that the Brigg 'raft' was once publicised as possibly a pontoon bridge, although there were obviously attempts to make it watertight. I see several possible reconstructions of the Brigg 'raft', most of them of the punt or flat bottomed barge type.

Wilson — May I comment on the statements about length? First of all long trees were very much more common in antiquity. Secondly when you are converting an oak tree with primitive instruments, the longer the plank the better because you are splitting.

Corlett — Yes, dugout canoes of course are an obvious example, they were very long in relation to their breadth and depth. But when you start putting joints into keel planks as in the Ferriby boats, I think perhaps you are getting into a slightly different field. It is a very difficult thing to do and I think you will only do it if you really want the extra length.

Windsor — Speakers have stated that the Ferriby boats were ferries. But is this not anachronistic for the Bronze Age, because a ferry is part of a land route.

Wright — North and South Ferriby are close to the chalk escarpment of the Lincolnshire Yorkshire Wolds, one of the two major prehistoric trackways along which North/South travel is fairly well substantiated. A little to the west there is another escarpment, the Jurassic escarpment, Brough-on-Humber to the north and Wintringham to the south which became more important in Romano-British times as a crossing point. In the Middle Ages there were ferries across that route and they are well substantiated in Doomsday.

Crossing major water obstacles must in ancient times have been one of the main purposes for which ships were built. At Brigg on the River Ancholme you have a comparatively narrow river where something like a flat bottomed punt would have been an ideal ferry perhaps with a rope across and you hauled it over, cattle and all the rest of it. The Ferriby crossing is a fairly substantial obstacle, frequently with bad weather where one would need something more elaborate.

Coates — Would it not be right to say that the Humber at that time would have presented really quite a long obstacle because upstream there would have been a morass from where the mouth of the Trent now is extending up to the Pennines. The Ferriby crossing would thus be a most important route.

Ellmers — In Germany, France and in Poland, ferries consisted of dugouts ties together side by side, having a platform on them making a broad craft which could carry, for instance, wagons, cattle or the like. (Die Kunde NS24 (1973):23-62.)

Naish — One does wonder who wished to cross the Humber at that point at that time, not one person but quite a lot of people in order to keep the service going. I have always thought they must have gone across the North Sea rather than just the Humber.

McGrail — We should remember that the environment, the climate and the sea level of Ferriby boat times were not necessarily as they are today. There is much research to be done here, but we must consider these factors when we speculate about possible uses of these boats. I think it can be shown with fair certainty, for example, that there was no formal River Ancholme at the time of the Brigg 'raft', but a wide shallow creek open to the Humber.

May I ask about Commander McKee's last sentence on page 1 of his paper when he appears to differentiate between recording a find as a catalogue of facts and recording it in the knowledge that he is going to have to reconstruct. Should we not have the same thoroughness, and should we not also make a strict division between observation and interpretation?

McKee — The Graveney Boat has been recorded twice, once in a totally objective way largely by photograph which were tremendously useful in the reconstruction. However, the photographer will take a picture of everything, but not necessarily from the angle the reconstructor wants. Now with a certain amount of boat-building experience behind me I concentrated my recording on certain things which I knew would help me to put it together again. For example, the bevel angle on the laps. The photography had recorded without measuring the bevel.

Coles — There is evidence for continuing if sporadic contact across the North Sea in the Bronze Age, and not just the shortest sea route but close intimate relationships between the Yorkshire and the Lincolnshire coast direct to the continent. Perhaps by treating the Ferriby finds, which show many aspects of Bronze Age social organisation, as purely inland vessels, we are constricting our views on reconstructions. We should perhaps look more widely and say could these have been ocean going vessels? If so what would they carry, and how would they stand up to North Sea conditions? Based on the relatively little work published about Bronze Age sea levels and natural conditions, the Humber would probably be as nasty then as it is today. Ferriby might not have been a terribly appropriate place for a ferry that was expected to cross once an hour or three times a day or five times a week. Could this not have been a much more major affair, that these might have been North Sea craft? This would no doubt affect the reconstructions and it would be very interesting to see a model or some projection of what such craft would have looked like if in fact it was going to be used in this way.

Coates — The Ferriby remains could be built up into a sea going version, but I suggest that it would have two quite serious drawbacks. One, it would have inadequate stability to carry a worthwhile load; and second it would be a most expensive and laborious technical solution to providing a sea going craft. If I were a boatbuilder then and I was asked to build a boat to cross the North Sea, I would certainly build a skin boat. It would be a tenth of the cost, and much more easily repairable. The advantage of the North Ferriby boat's hard bottom could be so well sacrificed for such a great reduction in cost.

Corlett — Could a boat with the proportions of the Ferriby boat be other than oar propelled? Perhaps a square sail in a following wind?

Coates — It would have to be a very pure following wind.

Farrell — Commander McKee dismissed the possibility of a pre-formed stem for the Graveney boat suggested by Christensen on the grounds that it would have been a fairly difficult sort of carpentry. Did he consider the possible use of stealers for the plank ends? I would have thought they would have been fairly simple carpentry.

McKee — I think is the quick answer to that.

17. EXPERIMENTAL ARCHAEOLOGY THEORY AND PRINCIPLES

J. M. Coles

Summary

Experiments with archaeological materials have been conducted for over 100 years, yet there has been little attempt, until recently, to consider the theory behind experiments, and the principles that must guide the work. Three levels of experiment can be distinguished, display, production and function, each of which has its own requirements. The value of repetition and series-testing can be documented, but the use of scale-models as substitutes for life-size experiments is generally hazardous. The reliability of experimental results must be critically examined, and absolute proof is as unlikely to be achieved as in other archaeological fields. The value of the approach, however, cannot be disputed, because experimental work forces the archaeologist to think much more deeply than hitherto about the materials, the technologies and the activities of past human societies.

This paper will be concerned more with the principles and theories of experiment than with boat archaeology, although particular aspects of the problem of wooden reconstruction will be noted, in the belief that they are relevant to the subject of the symposium. Although antiquarians and archaeologists have been aware of the experimental approach for over a century, there has been no concerted attempt to devise a set or sets of rules and procedures which should be followed in carrying out experiments with ancient materials. The result of such a lack of organisation has been, on the one hand, a chaotic series of subjective and objective observations, and, on the other, a wide variety of opinions about the reliability and validity of experimental results. Many observations have been made and recorded, but their context and relevance is often dubious in the absence of procedural standards. Conclusions drawn from experiments have rarely been subjected to critical scrutiny by impartial observers, and although simple results, a 'yes' or 'no' situation, are often publicised in popular media, the range of answers to experimental problems is far greater than this, and is very rarely explored.

THEORY

Experimental archaeology has come to mean the examination and testing of particular theories about the composition, the assembly, and the function of ancient material remains.

The experiments may be, and have been, single events or multiple series of tests, and they may deal with minute objects or enormous structures, stationary or portable remains, domestic consumables or weapons of war. They may be concerned with the raw materials that went to make an ancient artifact, the handling and the fitting together of the materials to make the artifact, and the use to which the artifact could have been put. The range is vast, and is barely touched by the bulk of experimental work so far done. Why this should be so is a complex question but it probably is in part an unconscious reflection of the unease with which archaeologists view their materials, and those who actually excavate are best placed to understand this unease.

SURVIVAL

Few ancient structures or objects have survived completely intact, and those that have, have suffered through burial, or exposure, so that in no case can we be sure that the extant remains accurately represent all of the composition and character of the original at its time of manufacture and use. The point is self evident, yet probably needs emphasis because it immediately opens up the question of validity. An experiment may be devised and conducted with attention paid to every observable detail, yet the model upon which the experiment is based may be inadequate. It will have altered since its abandonment, and therefore the entire experiment may be based upon false premises, unless the excavating archaeologist has been scrupulous in his recording, and the experimenter has been critical and suspicious of the records and the surviving remains. This point is examined again in the section dealing with experimental proof.

If inadequate data is among the greatest dangers facing the experimental archaeologist, then those works dealing with the most perishable substances, such as wood, are likely to suffer the most. Stones and flints are basically unaltered by many conditions of burial, fired clay often survives, as do gold, copper and bronze, but iron and bone are often poorly-preserved, and wood comes near the bottom of the list. This is no doubt the reason for the extraordinary attention paid to wooden objects, such as parts of ards, bows and boats, that have survived in peats and muds. Such attention is deserved, because of the presumed heavy reliance upon wood for all manner of objects made by most societies in the past. Yet even where wood has survived, has been recognised and recovered by the excavator, a further problem may arise unless immediate steps are taken. Such organic material is fragile, and will often have a high percentage of moisture in its body weight, sometimes as much as 95% water. The smallest amount of drying out may well alter the dimensions and distort the wood, and even if conservation is started at once, the finished product may suffer slightly in the actual process. The solution, again, is for immediate recording of the piece by its discoverer, recording of such detail and accuracy that even if the piece perishes completely, the information for reconstruction or simulation will be adequate. Wooden ards, for example, have survived and have been conserved, but copies made on the basis of the conserved ards are not exactly like the originals and therefore may not perform in the same manner. The experimental archaeologist must be aware of this potential hazard. He can do little, however, other than to estimate the original dimensions of an object if he has no accurate records of it upon its discovery, nor can he do much to account for alteration in size before discovery, through burial in peat or mud. For individual objects, these matters may be of only slight importance, but for an object such as a cart, an ard, or a boat, with several or many pieces fitted together, the problems can be of very considerable importance. The hope must be for increased vigilance on the part of excavators, and the readiness of potential experimenters to act immediately. We all know of unique objects lying beneath back-filled mud, or polythene sheets, awaiting more opportune moments for recording and examination; this is not good enough.

PRINCIPLES

The principle behind experimental reconstructions is that an object may be handled freely, may be broken if required and may otherwise be subjected to all sorts of indignities. An axe-like tool can be used as an adze, a spade, or a knife, to test its possible functions. A house can be built, lived in, pushed over, or burned down, to see how its components react. And a boat can be built, propelled in different ways, loaded to over-capacity, run aground or even sunk, simply to observe its capabilities. These examples are not frivolous; most of them have actually occurred, and the others should, but they serve to illustrate the potential for experiment. Almost every piece of work done with prethought has yielded useful and valuable information. The information may be, and often is, entirely negative, in that a presumptive implement or process or function does not succeed, but even so this is of value. A flint flake may not work as an engraving tool, even though it has been called a burin by archaeologists, and the demonstration of this non-function is of considerable use to those attempting to reconstruct Stone Age tool kits. A sheet bronze shield may not resist a slashing sword, and, in so doing, other suggestions for the use of such impressive armour may be strengthened. The suitability of most archaeological materials for experimental testing cannot be doubted.

The number of actual experiments that have been carried out over the past 100 years is very great, perhaps 250 fully recorded pieces of work, and many more ad hoc experiments barely mentioned in passing. The history of experimenters extends back in the 19th century from Pitt-Rivers, studying earthworks, to John Evans, working with stone tools, to Robert Ball, testing musical instruments with conspicuous, unique, misfortune. Within this corpus of work done, several patterns of endeavour can be seen, and it may be useful to indicate these here.

LEVELS OF EXPERIMENT. DISPLAY

Experiments can be broadly grouped into three separate sections, better described as levels of experiment. The lowest level is that of simulation, wherein a copy is made of an original artifact with attention paid only to its visual appearance for display purposes. The materials used may vary from the original, the technology employed in making the copy can be modern, and the copy itself is not tested for its function or purpose. Museum displays are a fair description of this level of experiment, and some would totally exclude the category from experimental archaeology. Yet visual effects can be useful, essentially to give a third dimension to material culture, to provide a scale, and notably to lead on to further consideration and work. Above this, such experiments cannot rise, and their greatest value at the moment must be in their public appeal which brings appreciation and support for further work. Examples of this level of experiment are numerous and include rebuilt prehistoric tombs, Celtic chariots, Roman forts, and plenty of houses.

The educational value of such simulations is important. A number of rather impressive experimental structures have been built to act as focal points for educational programmes generally involving young people who can be given the chance to appreciate the scale and character of former houses or forts, or who can actually assist in the work and thereby appreciate, in retrospect, the tasks of their ancestors. Some experiments of this kind have approached and attained the second level of research. The participation of Television authorities in work at this level is welcome, although their apparent need to provide yes or no solutions may be irritating to some experimental archaeologists. The interest of the public in such programmes is important to sustain.

LEVELS OF EXPERIMENT. PRODUCTION

A second level of experiment, which all work should in theory aim to reach, is concerned with testing for the processes and production methods used in the past. Here the experimenter is involved, not only in making a copy or replica which looks like the original model, but also in manufacturing it correctly. Appropriate materials must be used, and here the value of analyses of ancient objects, to determine their composition, cannot be exaggerated. Experimental casting of bronze axes, or the flaking of stone axes, can only be judged if the raw materials have exactly the same characteristics as the originals. The point is obvious, and again for wood it is of paramount importance that where possible the same type of wood, its age, grain and condition should be used in any attempted reconstruction. Experiments concerned with pottery production have probably suffered from a lack of care, not in building the kilns or in the firing, but in the clays and grits chosen as raw materials for the pottery vessels and for the kilns. However, it is not possible to state catagorically that the experiments have suffered in this way, because these features are rarely mentioned in published reports; there are notable exceptions. For wood, the necessity for care in choosing appropriate materials is clear.

A second element in experiments concerned with production methods is that of an appropriate technology. The greatest care may be taken in collecting the raw materials, but the working or assembly or such materials should be entirely in keeping with a level of technological ability appropriate to the society which made the original objects. We do not know how much effect a bulldozer will have upon a reconstructed earthwork, of an electric drill upon a wooden framework for a Roman gate; the first has a more visually substantial effect upon the reconstruction, but friction cracking and fraying all have a part to play in the build-up of pressures upon wooden timbers during drilling and hammering of pegs and nails. This problem of technology is perhaps the most difficult for experimenters to solve, because there is a limit to the degree of our knowledge about prehistoric or early technology and beyond this, experimenters are in dangerous territory, where subjective unsubstantive opinions are rife. The experiment must obviously be conducted with as close an approximation to ancient techniques as may reasonably be assumed and expected, and as long as the reported results clearly signify the procedures used, then appropriate assessment may be made by others as to their suitability and significance.

If it is possible to employ only those procedures and equipment likely to have been available in the contemporary society, then further experimental advances may be made. On the one hand, the implements used can themselves be observed for their own capabilities, and the wear or breakages that occur. On the other, the more difficult question of time can be posed. In experimental production of stone tools, for example, the hammer-stones or wooden mallets can be recorded for bruising or chipping, and the time taken to produce a stone axe can be reasonably estimated. In casting a bronze implement, or forging an iron tool, the equipment used, furnaces, crucibles, moulds, hammer, all can and should be carefully recorded before, during and after use. This point again is self-evident, but its reliability is only acceptable if appropriate equipment and technologies are used. The elapsed times for such procedures, whether measuring the rate of excavation of chalk, or the time taken to fell trees with a stone axe, are not so easily acceptable unless the experimenters are entirely familiar with their materials and technologies, and whether this is ever completely possible is a matter for debate. Here, however, is a point where an extremely valuable source of information appears.

ETHNOGRAPHY

It can legitimately be said that ethnography is one of the most neglected fields of experimental archaeology, because a large body of data exists both in literature and in the field which could be of considerable value. Societies which today are totally isolated from the main technological advances of the world are rare, and becoming rarer, but much has already been recorded. Detailed records of stone tool manufacture, iron smelting, forest clearance, house building, pottery firing, and many more, provide unique guides to the technologies of the past. They are unique because the people practising them know, or knew, no other ways and were entirely familiar with their materials and their procedures, and the precise reasons for selecting them all. It is not legitimate to accept all the details of such observable work as representing exactly those of the past, but as a general guide to likely actions ethnographic records are invaluable. The criticism often levelled at such records, that they represent situations far removed in both time and space from the milieu of the ancient society under examination is a legitimate criticism yet cannot totally serve to dismiss the use of such an information source. In certain cases it may be reasonable to overlook environmental differences; small processes such as iron working may be a case in point. In others, differences in environment will have been of major effect; simple forms of agriculture in tropical or sub-tropical regions should not be used as guides to prehistoric practices in northern latitudes. With care, the experimenter will find a vast source of 'experimental' data already available , and one might hope for a compilation of such material from an ethnographer rather than a social anthropologist.

LEVELS OF EXPERIMENT. FUNCTION

A third and final level of experiment is concerned with the function of the artifact, the use or uses to which the object is presumed to have been put. This level is a logical step up from that dealing with the production methods, and there are few experiments that can proceed directly to function-testing without the necessity for accuracy and relevance in manufacture. It would be worthless to test an earthen bank for erosion and weathering if it has been heaped up by earthmoving machine, but it is little better to test an accurate copy of a wooden ard dragged by chain and tractor; both aspects must be related in concept and approach. The visual appearance of a wooden boat might not be affected by the use of modern technologies in its making, but its handling and possibly its seaworthiness might be altered in some small ways that could combine to affect the conclusions drawn from the experiments. The question of the extent of alteration can probably never be resolved unless comparative tests are done, and this would be an expensive process. The general point, however, must be clear, that for function testing, the object must be made in an appropriate manner. That there is scope for some choice in this will be emphasised later when the approach to experiment is discussed.

There are two aspects in this functional work that are vital, yet sometimes ignored or neglected. The first is that of manipulation and operation; the second is environmental. The actual methods by which objects are used to carry out certain functions require more thought than untutored volunteer labour is likely to give. The reason why stone axes sometimes compare badly with steel axes, in chopping trees, is in their mode of use, because the angle of blow, and the back lift, of a stone axe is totally unlike that of a steel axe to which modern man is accustomed. The lack of success in breaking up and transporting chalk and earth, using antler picks, bone shovels and wooden baskets is in part due to our unfamiliarity with these unusual implements. The same difficulties were present in efforts to cultivate soils with ancient forms of ard, which involved entirely theoretical and subjective methods of training the draught animals, devising the harness, and manipulating the ard; the results were conspicuously successful as experiments because of the care taken in recording and publishing the observations and problems.

The ard experiments were also concerned with the environmental conditions that always affect work of this sort. Here the questions of cleared ground, old turf, soils and climate were vital if the experiments were to have any relevance in terms of past situations: other tests with other equipment are not so dependent upon physical environmental conditions which match those of the past, but even so this is an aspect well worth careful consideration. Problems of navigation in the past may have been different from those at the present, certainly in areas where major physical changes have occurred, but it is likely that experiments concerned with early cultivation require the greatest attention to matters of environmental conditions. This should not ignore or exclude the general concept of the social background within which experimenters operate, with outlooks on economies, time, motivation, ingenuity and general philosophy of life, all of which may well differ from those of earlier people and which may affect the experimental work in subtle and unconscious ways.

The three levels of experiment suggested above are not mutually exclusive, and some few experiments have been approached at an intermediate level, generally between Display and Production. The three levels may be summarized:

	Experimental Level		
	Display	Production	Function
Requirements			
Appearance	+	+	+
Materials		+	+
Technology		+	+
Environment			+
Use			+

REPETITION AND SERIES TESTING

Among the experiments concerned with the functional testing of ancient remains, several may be distinguised because they achieved far greater reliability through repetition. Repetition is often boring, whether more so in writing than talking is uncertain, but in experimental archaeology the ability to sustain repeated tests has almost always led to augmented results. Repetition in experiments concerned with the underground storage of grain have shown the relative ease by which seed-grain could be preserved by Iron Age man, but it is only by a long sequence of tests that this facility has become apparent; it was a question of gaining expertise in the work and in the small refinements in detail, only possible through a trial and error approach.

Series-tests are basically the same as repetition except that here a conscious effort is made to lead from one result, repeated if required, to a second or to a redirection of effort. From an initial interest in basic archery, Saxon T. Pope extended his experiments to include a great variety of bows, arrows, bowstrings, arrowheads and targets, each test building upon previous tests and suggesting further work. Another programme relevant here is that concerned with earthwork erosion and decay, where a series of examinations over a long period of time should augment our knowledge of the processes by which prehistoric earthworks achieved their appearance today. Repetition and series of experiments have one other element probably worth mentioning here, and this concerns the personalities involved in the work. Although it is true to say that a new approach and fresh insight in an experiment may ensue from a change in the personnel team, the maintenance of both continuity and under standing in experimental work is extremely important and can best be achieved by the constant presence of at least the instigator and/or chief technician. Personal involvement in experimental archaeology is generally maintained throughout the work, through sheer interest and dedication, and it is probably notable that those few cases where the series tests have lagged behind the planned programme have concerned rather larger teams of specialists with no one single leader or director. Personal involvment is important, and excellent demonstrators of dedication and total immersion in the projects are not hard to find at places such as Little Butzer and Lejre, nor are they less abundant in experiments not involving full-time salaried staff, in situations of extreme hardship and some danger.

SCALE MODELS

Not all who wish to experiment with archaeological material can afford the time or money to conduct major programmes of work which may involve heavy and abundant raw materials, massive expenditure of effort in the technology, and lengthy periods of function tests. An experiment concerned with a megalithic tomb, for example, would involve quarrying and dressing stone, transportation, a very large work force, and the organisation of a site to house the experiment. An ingenious way out of this impasse is to reduce the experiment in one of two ways. The first, as employed by Erasmus in Mexico, was to conduct short but full scale exercises as samples of the entire procedures; observations were made of short-term quarrying and digging, transportation, building and facing of stone and earth walls, and these time and energy records were then used to calculate the drain on a prehistoric society engaged in erecting a massive, partly surviving monument. Even so, the organisation on the part of the experimenter was considerable.

The alternative way to lower this cost-factor is to reduce the entire experiment to a small scale, 1/10 or even less. Such reductions have been attempted for earthworks and for megaliths, and with varying results. At first sight, to attempt to understand such structures by miniature models seems absurd, and there can be little doubt about the uncertainty of observations conerned with erosion and weathering of small domes of earth. Yet one scaled experiment on the megalithic monument Stonehenge has generally been accepted as demonstrating precisely how the large uprights were erected on the site; the engineering work involved in the experiment consisted of exact scaled humans, ropes, timbers and stones, with much practical and theoretical reasoning attached. It would seem, however, that many experiments are of such a character as to remove much hope of a successful scale model approach; for boats, models have often been used in tests for seaworthiness, but beyond this it is hard to visualize a dominant role for miniatures.

A final stage in scaled experiments, if costs and opportunities are unrealised, is of course to do no actual work of reconstruction or simulation, but to devise and conduct the experiment purely on theoretical grounds. It should be possible to predict exactly how a structure could be put together, or how it would react to certain conditions, given all of the properties of the materials and the strength of the agencies acting upon them. The erection of parts of Stonehenge can be visualised in this way, just as the current, wind and navigation requirements for Polynesian colonisation can be calculated. The proof of the theory, however, must still lie in the actual event, where all the other small and seemingly unimportant factors, such as temperament and variations in equipment, as well as exceptional conditions, may be inadvertently applied.

RELIABILITY

From the foregoing it should be clear that a variety of experiments have already been carried out, using different approaches and producing results of varying acceptability. There are few hard rules in experimental archaeology except those of honesty, self-criticism and determination. Yet with all this, there still remains the problem of how to assess the results of experiments, how much reliance to place upon the answers that emerge. First, the results should be repeatable, in order to demonstrate that chance alone has not determined the result. A single successful firing of a particular type of pottery vessel must be repeated so that the process can be shown to provide a consistent result. A sophisticated stone implement, such as a Palaeoindian projectile head of flint, must be produced time and time again in order to eliminate a freak percussion blow or unusual raw material that might have yielded a first successful copy. A single sea trial of a boat should not be considered enough to prove, even disprove, its capabilities.

Second, in assessing results it is important to be aware of the options thus produced. One solitary answer may not emerge from a test, and there may well be several possible solutions to an experimental problem. A stone axe-like tool may not function as an axe, but may work well as an adze; it could also serve as a hoe, or spade, or mattock, or meat cleaver and if all of the possible functions are not explored, then the range of optional answers remains wide. Only by series-testing can the list be reduced by elimination. And it will be more often a question of elimination rather than of substantiation; it is easier to prove that an object does not work at a particular task than it is to show that the object did serve one and only one function in the past. Here, microscopic examination for traces of wear on implements can help. Fortunately, for many ancient objects and structures only one or two possible functions need be entertained; a boat is presumably meant for water and should therefore float. Likewise, a house is likely to have been built to shelter something beneath its roof, if it had a roof.

QUESTIONS

One way by which experimental archaeologists can reduce the number of optional answers is through a careful approach to particular problems. By setting out to test if a house is capable of withstanding fire in the roof, or if it can be occupied in severe winter, the archaeologist can pose specific questions, set up his experiment and recording procedures, and establish his answers within relatively narrow limits. If he does not pose particular questions, he will not know which level of experiment he hopes to achieve, and will be left with a vast range of potential problems, some of which he will not have even considered to exist. The preliminary posing of questions, the questioning approach, is one of the most important aspects in the theory of experimental archaeology.

As an example, house building demonstrates the potential of the questioning approach, and the relevenace it has to the procedures adopted. If an experiment is to test the materials, time and motion studies, in the erection of the building, then great care in recording such details will have to be expended, and only appropriate methods can be used. If the building is to be destroyed by fire, its refinements perhaps need not be so extensive, but its exact assembly and the position of each of its members must be noted, and its physical condition must be recorded before the conflagration. If the building is to serve only as a display area, or to house museum exhibits, then its precision in erection need not be so carefully controlled. The same general conditions will apply to other experimental subjects, including boats, stone tools, and earthworks, although the details will naturally differ. The three levels of experiment noted above are appropriate here, although they are perhaps too rigidly exclusive at the production and function levels.

PROOF

Having carefully devised a problem-orientated approach to an experiment and having religiously observed all of the principles, the experimental archaeologist will be left with a set of results or, rather, records of the observations made. The question of proof then has to be faced. How much reliance can be placed on a set of results? Archaeological proof is always hard to demonstrate, and the progressive reduction in information is well known:

ancient evidence (object or activity) >

total buried evidence >

surviving evidence in the ground >

archaeological recovery of evidence >

archaeological understanding of evidence.

The experimental archaeologist is probably no better and no worse than this. All of the available evidence must be used, and the experimental results subjected to a balanced opinion which will state the uncertainties as well as the achievements. Where a test has eliminated a possible answer, where a presumed function does not work, then the 'negative' answer is almost certainly a positive one, and that ancient function can be said not to have been performed by that tool. On the other hand, if a certain implement does function successfully in performing a certain act, it need not necessarily mean that the tool was actually used in this way in the past. It may have performed another function, not considered by the archaeologist. The answers, the reliability of results, clearly will not be simple yes or no situations, but presumptive and possible answers. It is not often that proof absolute is claimed, yet there have been some; the work of Townley Johnson on Bushman rock art and their paints and media quite clearly indicated only one positive answer to the media problem, and therefore that answer was the correct one so far as the evidence suggested. Work on Bronze Age shields was less conclusive, but it demonstrated the unsuitability of metal shields for actual physical battle, and the satisfactory qualities of leather shields for battle; the negative result (metal) seems conclusive, but the positive result (leather) could be only one of a number of uses for these shields. Archaeologists should not be dismayed by these uncertainties; they are inherent in prehistory and ill-recorded early histories. The work of Heyerdahl is another case in point because although he has been able to show that voyages in reed boats across the Atlantic are possible, he has not claimed that this proves that such voyages did take place in the past. Not all experimenters have been so modest about their results.

VALUES

The value of experiments is that they force the archaeologist to think about all of the characteristics and qualities of his material, and not just those that are superficially evident. The essential features of artifacts, whether small portable tools or large permanent structures, are not always those most readily apparent, and experiments are the only way by which the archaeologist can come to understand the practical reasons for the selection and manipulation of materials in the past. Experiments, then, create opportunities for a new approach to archaeology, one concerned with the economic requirements of ancient man, and therefore basic to our understanding of past human societies.

ACKNOWLEDGEMENTS

The history of experimental archaeology is long, and it is only by observation of the range and character of this work that trends can be seen, principles determined, and levels of experiment deduced. What has been said in this paper represents only one viewpoint of the total work already done, and the author is therefore entirely dependent upon the practical experiments of others. Individual workers have not generally been singled out here, nor references listed to particular experiments; many of both are noted in the book Archaeology by Experiment (Hutchinson, London, 1973).

18. ASPECTS OF EXPERIMENTAL BOAT ARCHAEOLOGY

Sean McGrail

Summary

The authenticity of a boat replica should be investigated before deductions are made about the capabilities of the original vessel. This point is illustrated by reference to the 1893 replica of the Gokstad ship, and to the recent Greenwich faering replica.

Boat replicas represent a heavy investment and the decision to build should be preceded by cost/benefit analysis which should include consideration of the replica's long term use.

In general, there are several solutions to the problem of hypothetically reconstructing an incomplete boat find, but full size boat replicas have only limited adaptability once their main design parameters are established. An intermediate stage of theoretical model building is therefore recommended, in which inter-disciplinary techniques are used to investigate the variant forms.

I very much regret that it proved impossible for Alan Hinks to present a paper at this Symposium. He is chairman of J..Hinks and Son, the North Devon shipbuilders (Farr, 1976: 17, 18), and unexpectedly has to carry out sea trials on a recently completed ship. We would have learned much from his experience of commercially building replicas of ancient boats. I hope that one day we may hear him at another forum - possibly on site in his boatyard, as recommended in another context by Hasslöf (1972). I felt it impossible to invite others to prepare a paper on this topic at short notice, and therefore I reluctantly undertook this task myself.

BOAT REPLICAS

Later in this Symposium Timothy Severin will describe the building of his skin boat Brendan and how he sailed her this year from Ireland to Iceland. Brendan is the latest in a long line of boat replicas, though not all could be classified as archaeological experiments. Paul Johnstone of the BBC "Chronicle" programme, who but for his untimely death would have presented a paper at this Symposium, had traced details of a replica trireme built for Napoleon III in 1861 (Eveillard, n.d. 233); this is possibly the earliest replica of which we have records. Johnstone himself was involved in the building and filming of several experimental craft: firstly Atkinson's Stonehenge experiment in 1954, when a replica bluestone was transported by a multiple

Fig. 18.1 A replica of the Nydam boat built in 1935 for use in a German film. H. Åkerlund.

Fig. 18.2 Viking, Magnus Andersen's 1893 replica of the Gokstad ship. Norsk Sjøfartsmuseum, Oslo.

Fig. 18.3 Viking in the North Sea, with a foresail in addition to her square mainsail. Magnus Andersen.

Fig. 18.4 Hugin. The 1949 Danish replica of the Gokstad ship. Danish Tourist Board.

Fig. 18.5 The Danish National Museum's Norland Ottring.

"dugout" (Atkinson, 1960: 113-4); secondly the building of a tarred canvas curragh and its trial crossing of the Irish Sea in the 1960s; and more recently he worked with Professor Marstrander on his Bronze Age skin boat experiment (Johnstone, 1972; Marstrander, 1976).

A not so well publicised replica is that of the Nydam boat shown in Fig. 18.1 (Åkerlund, 1963:45-7). Paul Johnstone noted in a summary of his projected paper that this was the first boat replica trial to be recorded in film and he wondered whether this film could now be traced.

A better known replica is Viking, Captain Magnus Andersen's 1893 replica of the Gokstad ship, (Fig. 18.2) which he sailed from Bergen to Newfoundland in 28 days (Andersen, 1895). Viking was built as close to the original form as possible, with the ribs lashed to cleats on the planking, and therefore Andersen's account of her voyage is of great importance for the understanding of the sea worthiness and the sea keeping qualities of this form of hull. But there are doubts about what can be learned from her performance under sail. Her square mainsail was similar in shape and material to those used in late 19th century Norway: that is it was made of hemp or flax, and of high aspect ratio, tall and narrow. Christensen (1968: 39) considers however, that Viking Age sails were probably twice as wide as they were tall, that is of low aspect ratio, and the consensus of opinion is that they were made of wadmal, a coarse woollen cloth, possibly strengthened by strips of linen or leather (Haasum, 1974:113). Furthermore, Viking sometimes carried a topsail, and sometimes a foresail (Fig. 18.3). Thus great care must be taken in drawing any conclusions about the sailing capabilities of Viking Age ships from this replica's performance.

Another replica of the Gokstad ship, Hugin, (Fig. 18.4) was built in 1949 and sailed from Denmark to Kent (Røjel, 1949). The English language account of this replica does not give sufficient information for us to judge her value as an archaeological experiment, although one interesting fact is recorded: 20th century Danes, 2 metres tall, could not easily row in the space between adjacent rowing stations. One may conclude from this that Viking Age oarsmen were not as tall as their 20th century descendents, or that rowing styles have changed. The remains of this replica can still be seen on the beach at Pegwell Bay, Kent.

There has recently been increasing Danish interest in building boat replicas: Egon Hansen (1974) has recorded six built since 1963. Apart from three notes by Crumlin-Pedersen (1966: 256-8; 1969; 1970:9, 18) describing aspects of the Imme Gramme replica of the Ladby ship, there is no readily accessible publication giving details of these experiments. Crumlin-Pedersen (1975: 130) has stated that these replicas have not "been rigged and sailed properly, nor have they been subjected to serious trials". Nevertheless it would be of great value if he could give an authoritative summary of past Scandinavian experience with boat replicas so that future research workers can learn from it. Possibly the forthcoming report on his current work evaluating the problems of sailing a Norland Ottring (Fig. 18.5) with a single square sail, could become the occasion for a comprehensive paper.

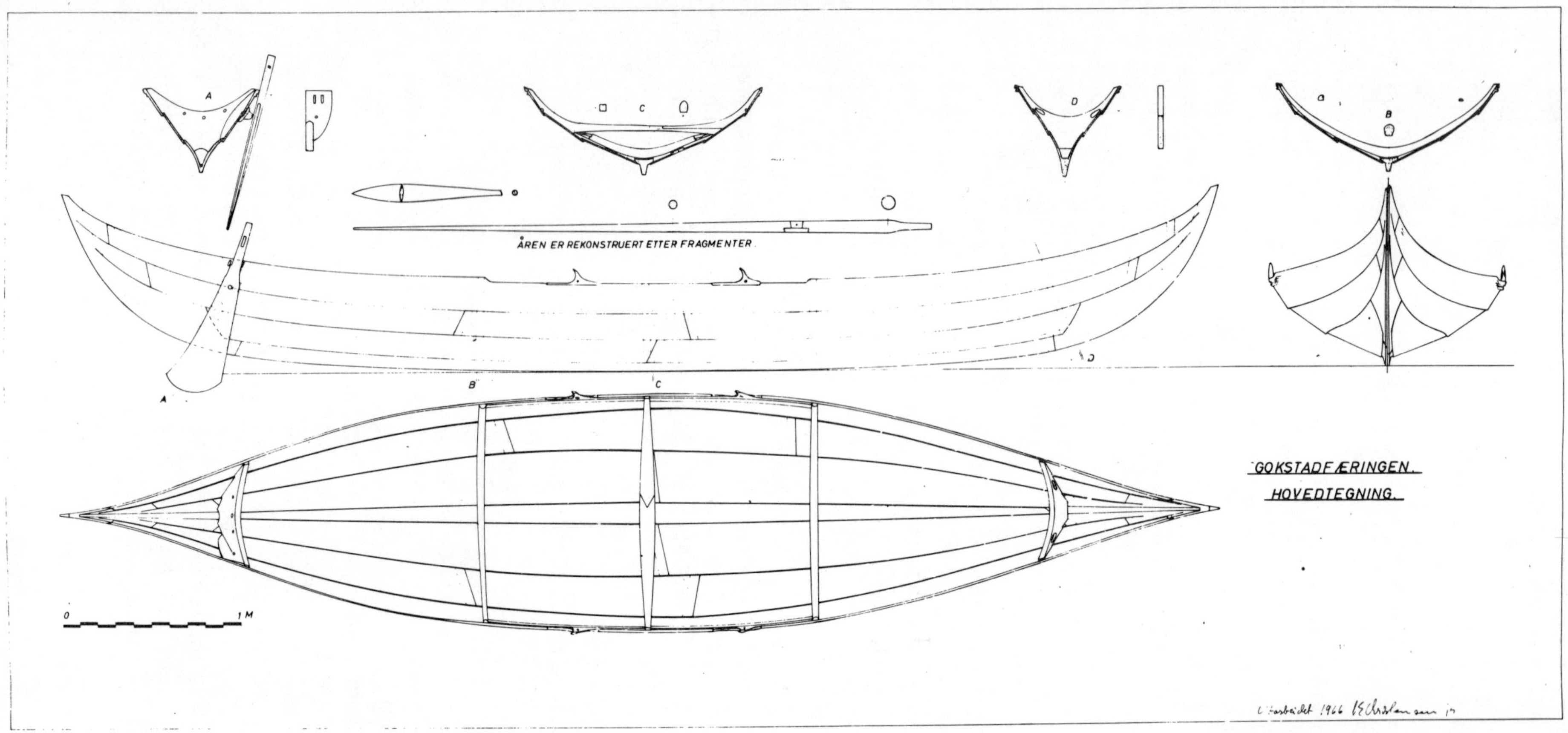

Fig. 18.6 The Gokstad <u>faering</u> construction plan by Arne Emil Christensen. Universitetets Oldsaksamling, Oslo.

THE GREENWICH FAERING REPLICA

The National Maritime Museum's direct experience of experimental boat archaeology is limited to the building and trials of a replica of the faering from Gokstad (Figs. 18.6,10,11). This was not an ideal experiment (McGrail, 1975A: 5), but we learned much from it, and gained valuable experience in the theory and practice of replica building. As with the 1893 Gokstad replica, there are problems in attempting to use our experimental results to draw conclusions about Viking Age seamanship. Some of these difficulties have been considered elsewhere (McGrail, 1974: 34), but I would like to discuss further aspects here, to illustrate some of the problems inherent in boat replica work.

Fig. 18.7 shows the forward inclined frame fitted to the Greenwich faering, with two holes piercing it. The dimensions of these holes had been scaled up from Christensen's 1:10 drawing (Fig. 18.6) and their longest dimension was c. 6 cm that is, about three fingers breadth. The function of these holes was not clear to us until we operated the boat from the beach, when it became natural to attempt to use them as handholes when hauling the boat into and out of the water. Inaccurate scaling meant that they were in fact too small, although those on the original faering in Oslo could be so used. When building his faering replica, Egon Hansen of the Prehistoric Museum, Moesgård, Aarhus, Denmark obviously appreciated this use (Fig. 18.8), although his handholes are somewhat bigger than those on the original faering.

Christensen's construction plan for the faering (Fig. 18.6) shows an oar, reconstructed from a fragment, which is c. 3.29 m (10ft.9in.) long. It is now next to impossible to say which of the fragmented oar remains belonged to which boat, and it may be that Christensen based his drawing on fragments from the larger, 6-oared boats, also found in the Gokstad ship (Christensen, personal communication). In static trials at Greenwich an oar of this length seemed to be too long for the faering, and Harold Kimber, the master boatbuilder in charge of building the Greenwich replica, made two oars of length c 2.52 m (8ft. 3in.), calculated as twice the beam of the boat at the tholes. This pair was used in the rowing trials together with 20th century pairs of oars of 2.46 m and 3.05 m length (McKee, 1974: 18). In retrospect, it might have been wiser and more in keeping with experimental methods for us to have done more research in the Bygdøy archives before building the replica, or to have used a pair of 3.29 m long oars, and attempted to find a rowing technique and an operating environment suitable to them. The fact that we also experienced difficulties during our trials when using 20th century rowing methods in the seemingly inadequate distance between the thwarts (.95 m = 3ft. $1\frac{1}{2}$in.), emphasises the latter point.

McKee (1974: 13-17) has described some of the problems encountered in evaluating the steering arrangements of the faering. It may not have been sufficiently emphasised however that the effectiveness of the side rudder probably depends to a marked degree on the shape of the rudder boss (Fig. 18.9), and the fit of the rudder to that boss. There are obviously great difficulties in getting the precise three dimensional shape from Christensen's drawings, as can be seen by comparing the Greenwich boss (Fig. 18.9) with the drawing (Fig.18.6). Furthermore, at the time of the Greenwich experiment it was not

Fig. 18.7 The fore stem and the forward inclined frame of the Greenwich faering replica. National Maritime Museum.

Fig. 18.8 A similar view of the faering replica built by Egon Hansen at Moesgård. National Maritime Museum.

Fig. 18.9 The rudder boss on the Greenwich faering replica. National Maritime Museum.

Fig. 18.10 The Greenwich faering replica under oars. National Maritime Museum.

Fig. 18.11 The Greenwich faering replica under sail. Basil Greenhill.

appreciated that the inboard surface of the original rudder may have been dished to match the shape of the boss. The surviving evidence is not unambiguous and the depression on the inside of the faering rudder may be due to wear in use or pressure damage in the ground (Christensen, personal communication), but if the depression is intentional it would effect the rudder efficiency. In addition, we did not persist in our attempts to find the best method of binding the rudder by withy thongs through the boss to the aft inclined frame (McKee, 1974: 14), but used hemp rope for the recorded trials. With these doubts and discrepancies it is not possible to draw conclusions from the Greenwich experiment about the efficiency of the original faering's steering gear. Nor need we yet consider McKee's tentative hypothesis (1974: 26), that since the steering gear of the Greenwich faering is only effective for small angles of helm, this may have been 'a significant obstacle to the development' in Viking times 'of a sailing boat that could change tacks'. More research and experimental work are required.

LONG TERM USES OF BOAT REPLICAS

During 1975 and 1976 Eric McKee sailed the faering replica with a single square sail (Fig.18.11) to gain some understanding of the problems of handling a small boat with this type of rig. Further trials in the Spring of 1977 will be the final ones in the first series. Would a further series of trials be productive? We are planning longer term trials to assess wear in use and other effects. But is there more we can get from replicas such as this one, which represent a heavy investment? Time and money may be justified by the knowledge gained during the building and the trials. We may also gain an additional return by using the replica as a test vehicle to learn the art of sailing with a single square sail; or to evaluate a prototype trials programme; or to act as a control by providing feedback to theoretical models used in the naval architectural evaluations. But can a boat replica be archaeologically useful right to the end of her life? I have elsewhere suggested (McGrail, 1974: 37) - possibly lightheartedly - that the faering replica might be interred in a burial mound to be excavated in future years on the lines of the experimental earth works at Overton Down (Jewell and Dimbleby, 1966) and Wareham (Evans and Limbrey, 1974). There would be problems of finance and logistics, of continuity and repeatability - but is this suggestion worth considering seriously? Would boat archaeologists and others gain sufficient to justify the effort? If not, are there better ways of using a boat replica in the long term, to the greater gain of Archaeology? I believe that questions such as these should be asked as part of the cost/benefit analysis that must precede any decision to build a boat replica.

THE PRINCIPLES AND METHODS OF EXPERIMENTAL BOAT ARCHAEOLOGY

The theory of archaeological experiments expounded by John Coles in the prededing paper generally applies to boat replicas. Elsewhere (1975A) I have discussed procedures which may ensure maximum archaeological value from a boat replica experiment. These include the choice of a worthwhile subject for the experiment; the selection of the project team; the choice of the environmental conditions for building and testing the replica; the choice of tools and of constructional methods and the way to "translate" the original into a replica; the design of a trials programme; and finally the necessity to record the experiment in detail, including the reasoning behind decisions to adopt or reject particular courses of action, so that others can assess the validity of the experiment.

Since 1974 the National Maritime Museum has been planning the building of a replica of the Graveney Boat, a mid 10th century A.D. cargo boat whose fragmentary remains were excavated by the National Maritime Museum and the British Museum in 1970 (Fenwick, forthcoming). This preliminary planning has led me to believe that greater emphasis should be placed on the early stages of an experiment. In general there will be several possible solutions to the problem of hypothetically reconstructing the full original form of an incomplete boat find: some of these solutions may vary only because of minor differences in fittings and "trimmings", but others may have significantly different hydrostatic and hydrodynamic characteristics, and operational capabilities. I believe that these variants should be thoroughly investigated in a variety of ways <u>before</u> a decision is taken which version the replica builder is to use in his practical experiment. Thus, by small scale model building, by naval architectural calculations and the plotting of hydrostatic curves, by computer simulation, and by tank tests, these theoretical alternatives can be evaluated for practicability of building, for stability and load carrying effectiveness, and for performance with different methods of propulsion under a range of operational and environmental conditions. These are experiments which can be repeated using many different values of the parameters, and they can be repeated by others. These theoretical models can be 'dismantled' and 'rebuilt' many times before we need crystallise our ideas on the version to be built as a full size replica, which will have only limited adaptability once its main design parameters are established. Much should be learned during these theoretical evaluations: negative findings and the rejection of hypotheses can give valuable insight, often as valuable as positive information; inconsistancies in reasoning may be highlighted and directives reformulated accordingly; and the criteria by which to choose a preferred solution should be clarified. In certain cases this form of analysis may show that it is not yet the time to proceed to the full size replica phase - the decision possibly being deferred until more evidence becomes available.

There must however be controls for these theoretical investigations, and indeed as John Coates points out (This volume p. 226) naval architects prefer to have practical confirmation of calculated assessments, especially those of speed and of sailing performance. There should be feedback to theoretical models from both the building and the trials phases of a boat replica experiment, enabling the theoretical models to be amended, for example, by the addition of new variables whose existence was not suspected until the practical work revealed them. The first boat replica investigated by these methods will undoubtedly take longer to build than anticipated, and there will be much discussion whilst the theoretical and practical aspects are synchronised and made consistant, but future experiments should benefit from the refined analytical techniques which emerge.

Haasum (1974), has used some of the interdisciplinary techniques I have described above in a theoretical investigation of Viking ships under sail. Her work has been criticised by Crumlin-Pedersen (1975: 130) because of her uncritical use of sources and methods. Whilst agreeing that this criticism may be well founded, I would nevertheless applaud Haasum for her pioneering spirit in attempting to use hydrodynamic and aerodynamic analysis to gain insight into the possible capabilities of ships of the Viking Age. There is a place for this type of theoretical assessment in any study of an ancient boat, providing that the data is sound. Additionally, there is a requirement to validate the theoretical model by practical trials, but this may have been beyond the resources available to Haasum.

The National Maritime Museum will probably use this intermediate stage of theoretical model building to investigate not only the Graveney boat but also the Brigg 'Raft' (McGrail, 1975b) and the Kentmere boat (Wilson, 1966). E.V. Wright and John Coates are now evaluating several alternative hypothetical re-reconstructions of Ferriby boat 1, (Wright, 1976) and it is conceivable that they too may use these theoretical procedures to explore their reconstructions. I believe that time spent on this stage of any research programme will be of tremendous benefit not only because of the knowledge immediately gained, but also in the ultimate value of the full size replica, should it be decided to build one.

Acknowledgement I am grateful to Arne Emil Christensen for discussions on aspects of this paper.

REFERENCES

Andersen, M., 1895. Vikingefaerden, Kristiania.

Atkinson, R.J.C., 1960. Stonehenge, London.

Åkerlund, H., 1963. Nydamskeppen, Goteborg.

Christensen, A.E., 1968. Boats of the North, Oslo.

Crumlin-Pedersen, O., 1966. 'Two Danish Side Rudders' Mariner's Mirror, 52: 251-261.

Crumlin-Pedersen, O., 1969. 'Kopi af et Vikingeskib' Skalk 1969 (2): 26-27.

Crumlin-Pedersen, O., 1970. 'Viking Ships of Roskilde' in National Maritime Museum Monograph No. 1 Aspects of the History of Wooden Shipbuilding: 7-23, Greenwich.

Crumlin-Pedersen, O., 1975. 'Viking Seamanship Questioned' Mariner's Mirror 61: 127-131.

Evans, J.G., and Limbrey, S., 1974. 'Experimental Earthwork on Morden Bog', Wareham, Dorset, England: 1963-1972' Proceedings of the Prehistoric Society, 40: 170-202.

Eveillard, E., n.d. 'Navigation de plaisance' fascicle No. 70 in Lacroix, E. (ed) Nouvelle technologie des arts et metiers Volume 4, tome 7. Paris.

Farr, G., 1976. Shipbuilding in North Devon, National Maritime Museum Monograph No. 22, Greenwich.

Fenwick, V.H., (ed), forthcoming. Graveney Boat.

Haasum, S., 1974. Vikingatidens Segling och Navigation, University of Stockholm, Theses and Papers in North European Archaeology 4. Stockholm.

Hansen, E.H., 1974. Unpublished paper on archaeological replica building. Moesgaård.

Hasslöf, O., 1972. 'Concept of Living Tradition' in Hasslöf, O., et. al (ed) Ships and Shipyards, Sailors and Fishermen: 20-26, Copenhagen.

Jewell, P.A., and Dimbleby, G.W. (ed), 1966. 'Experimental Earthwork on Overton Down, Wiltshire, England: the first four years.' Proceedings of the Prehistoric Society, 32: 313-42.

Johnstone, P., 1972. 'Bronze Age sea trial' Antiquity, 46: 269-274.

McGrail, S., 1974. Building and trials of a replica of an ancient boat: the Gokstad faering. Part 1. National Maritime Museum Monograph No. 11. Greenwich.

McGrail, S., 1975A. 'Models, Replicas and Experiments in Nautical Archaeology' Mariner's Mirror, 61: 3-8.

McGrail, S., 1975B. 'Brigg Raft re-excavated' Lincolnshire History and Archaeology, 10: 5-13.

McKee, E., 1974. Building and trials of a replica of an ancient boat: the Gokstad faering. Part 2, National Maritime Museum Monograph No. 11, Greenwich.

Marstrander, S., 1976. 'Building a hide boat. An archaeological experiment' International Journal of Nautical Archaeology, 5: 13-22.

Røjel, J., 1949. 1949, Cruise of the Viking Ship Hugin. Copenhagen.

Wilson, D.M., 1966. 'Medieval boat from Kentmere, Westmorland' Medieval Archaeology, 10: 81-88.

Wright, E.V., 1976. North Ferriby Boats, National Maritime Museum Monograph No. 23, Greenwich.

19. PRELIMINARY REPORT ON THE *BRENDAN* PROJECT

Timothy Severin

Summary

A 36ft. skin boat replica has been built of the West Irish type attributed to the late 5th century A.D. The design was based on surviving Irish curraghs and literary references. Research was conducted into the properties of oak-bark tanned leather for the hull skin, on an oak and ash frame lashed with leather thongs. Leather dressing was grease from sheep wool. Sea trials indicate nil windward performance, 2-3 knots cruising speed, but considerable stability and seaworthiness. On passage from West Ireland to Iceland there was no appreciable degradation of the leather skin. The project continues.

The main purpose of the Brendan Voyage, which is still unfinished, is to learn by practical trial whether a large skin boat of the type which may have been in use in the west of Ireland between the 5th and 9th centuries A.D., can survive a major ocean passage. In particular it is hoped to learn whether such a boat could follow the 'stepping stone' route by way of the Hebrides, Faeroes and Iceland to Greenland and North America which has been postulated by a number of scholars. If so, I believe that first-hand knowledge of this route is contained within the *Navigatio* of St. Brendan,[1] a Latin text now dated to about 800 A.D. which seems to describe the geography of this part of the North Atlantic.

The method of the project has been to find out as much as possible about the type of skin vessel which was in use at the time, build as close a replica as the facts would allow, and then attempt to navigate the craft along the same route.

The primary historical source for the construction of the replica skin boat was the text of the *Navigatio* itself. Though terse, the *Navigatio* is clear about the manner of boatbuilding, the type of leather which is stated to be an oak bark tanned leather, and the use of some sort of butter or fat to dress the leather. It was possible to cross-check some of these details with reference to other early Christian texts which mention the skin boats used by Irish monk-navigators, and also to glean some clues as to size, rig, and ship handling.

In practical terms the ethnographic evidence derived from the skin boat descendants in the west of Ireland was invaluable in laying down the guidelines for design and construction of the replica. The skin boat descendant of the west of Ireland is, of course, the curragh; and it is perhaps significant that the most elegant and seaworthy of the various Irish curragh designs, as Hornell[2] pointed out, is the curragh of the Dingle peninsula in County Kerry, a small area which is also the home of St. Brendan and the place from which he is said to have embarked on his westward voyage.

Clearly the initial step towards deciding to build a replica was to check the potential seaworthiness of a surviving curragh. These boats are still in use in the Dingle, though their numbers are decreasing rapidly. Perhaps 40 of them are still in regular use, most with an outboard motor 'box' to replace the mast and sail which is no longer used on the working boats when lobster fishing. They do, however, conform in size, design and construction to the Dingle curraghs which Hornell described.

There is only one full-time curraghmaker surviving, John Godwin of Maharees; and to learn something of his techniques the author ordered the smallest curragh - known locally as the 'two hand punt' - and assisted with its covering with the hot tarred canvas coat that now replaces the leather skin. The completed two hand curragh then served as a useful test bed for practical experiments, rigging and sailing these boats.

Additional information about contemporary materials and technical skill was obtained by examining other early Christian artifacts. The available types of wood, for example, could be determined from earlier work on charcoal analysis of hearthfires, and most important of all it was possible to examine contemporary Irish leatherwork in the form of the book satchels or budgets,[3] some of the very few surviving examples of early medieval leatherwork and done - conveniently enough - in association with monastic communities.

The most fruitful line of materials research was undoubtedly into the properties of oak bark leather as mentioned in the Navigatio. Parallel to such straightforward tests as hanging squares of the leather in the sea and periodically seeing what had happened to them, the laboratories of the British Leather Manufacturers' Research Association very kindly carried out a research programme, using the specialised equipment available to them, into the properties of oak bark leather to try to determine its suitability for the hull membrane of a boat. Using wet and dry samples, samples treated with different dressings, and samples under dynamic and static conditions, the laboratories checked tensile strength, weight, extension, and impermeability as well as the suitability of various fatty dressings such as cod oil, tallow, butter, and the wool grease from sheep. These tests were conducted so that the oak bark leather was 'in competition' with selected special-purpose modern leathers as well as with rawhide and alum tannages. To the great interest of the research team - and my own gratification - the medieval oak bark leather proved far superior to its rivals as a hull cover. Tensile strength after prolonged soaking, for example, was far stronger than any other type of leather. Similar tests were conducted on sail leathers and the leather used for thongs to lash the wooden frame of the replica.

On this accumulated evidence, then, Colin Mudie the naval architect produced a design study and a set of lines plans for the construction of the replica vessel (Fig. 19.1). The size of the vessel, named Brendan, was arbitrarily selected at 36ft. as being convenient for reconstruction and also sufficient to contain the numbers of passengers who are supposed to have accompanied St. Brendan and similar voyagers. The beam of the vessel was designed at

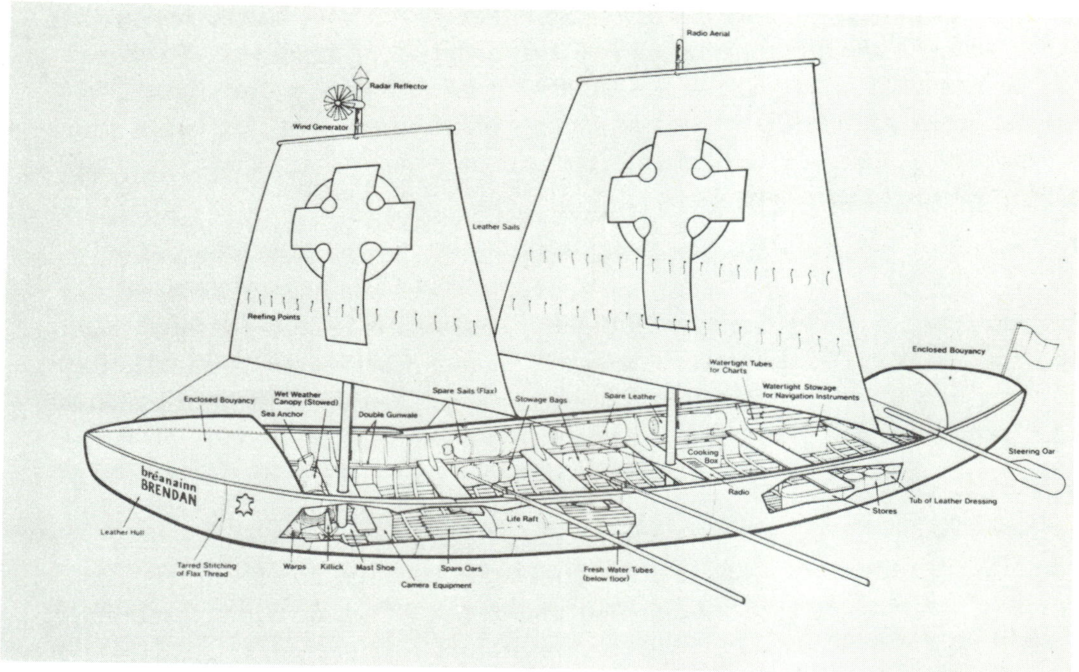

Fig. 19.1 Schematic drawing of the original planned lay-out of Brendan. Subsequently, fore and midship cabin shelters were added. The steering oar was supported on an H frame, and the unstayed masts - shown here passing through the thwarts - were held by mast partners running fore and aft at thwart level.

Fig. 19.2 The skeleton of a 1:12 scale model of Brendan, built by Arthur Saluz of Poole. This was subsequently covered with East Indian goatskin.

8ft.; her draft at approximately 1ft. and the designed weight for the bare hull was approximately 2000 lbs. though this last figures was considerably exceeded in the finished boat.

The next stage was to complete a scale model (Fig. 19.2) from Colin Mudie's plans, from which it was hoped to learn in advance some of the construction difficulties that might arise in the full scale vessel. Even in the model it was found that the logical way of building the boat was upside down, i.e. laying the gunwales first, as is still done by John Godwin.

For the initial shaping of the wooden frame the Crosshaven Boatyard in Co. Cork was selected. The shipwrights were asked to fair up a wooden frame to Colin Mudie's drawings. The only timber to be used was timber available in the early Christian era in Ireland - oak for the gunwales and thwarts, ash for the frames and stringers. To fair up the hull shape it was decided to pin it together lightly with ordinary wire nails. At this stage the professional shipwrights' work was finished, and volunteers began to lash together the frame, removing the wire nails as the work proceeded. The leather selected for the thongs was alum tanned - a known technique in Roman days. This leather in fact was unsuitable for the hull membrane, though more stable under stretch once the initial elasticity had been taken out, Copying the technique observed in the surviving leather curraghs, the leather thongs which lash together the boat were joined end to end to provide a continuous strand across each frame from one gunwale to the other. It was found that this lashing technique produced a sufficiently strong hull so that up to a dozen people could stand on the up-turned frame while they were working. To preserve the ash frame from sea-water attack, the entire structure was coated with wool grease, which also protects the leather.

42 ox hides were eventually needed to cover the hull of Brendan, each one being fitted individually in place. The hides were of young oxen, averaging 4mm in thickness, or approximately the thickness of surviving examples of early medieval leather from Ireland[3]. The hides were stitched together by hand, using either a back stitch, or a saddler's two-hand stitch. The latter involves using two needles, so the labour force had to divide into pairs, each person passing the needle through the leather to his partner. Hand-rolled flax threads of 16-cord thickness were used to stitich the hides together. By far the most difficult areas were found to be the bow and stern portions of the hide covering. The only craftsmen available to work this type of leather and in such thinkness was a carthorse collarmaker, whose leatherworking skills were found to be essential. As on the surviving fragment of an early Scottish curragh,[4] the leather hull membrane was finally attached by lapping over the upper gunwale and lashing down to the significantly convenient lower gunwale.

The oars for the Brendan are of the traditional type described in Hornell and still in use. They pivot upon a wooden bull and single thole pin. Using the harness maker's technique it was found that the simplest fixing for the side steering oar was a leather strop.

Initial tests were carried out at the Crosshaven Boatyard, with the shipwrights serving as crew. It was found that the boat was difficult to manoeuvre and particularly susceptible to windage (Figs. 19.3 and 4). Early sailing tests

Fig. 19.3 Sea trials of Brendan off County Cork, Spring 1976.

Fig. 19.4 The stern steering sweep fitted to Brendan, proved ineffective during sea trials and was later discarded; the side rudder was enlarged.

were done on the lakes of the Upper Shannon River. The boat was found to be considerably more stable than anticipated, and the unstayed masts were adequate, a previous source of some worry. It was also found that a larger steering oar was necessary, if the boat was to be steered properly (this may relate to the steering oar carved on the Bantry Boat[5]). Flexing of the hull, which was expected, is in fact noticeable. Leakage through the leather is negligible, even though the seams were neither tarred nor pitched as had previously been thought necessary.

On May 17th of this year the Brendan set out from Brandon Creek with a crew of five (and, happily, escorted by two Dingle curraghs for the first few hundred yards) and sailed around the west coast of Ireland, through the Outer Hebrides and to the Faeroe Islands, and in mid-July reached Reykjavik in Iceland. In general the weather conditions she encountered were rather poor for the time of year, and there was plenty of seaway to test the flexibility and durability of the structure, expecially during three gales, one of which prodcued quite heavy seas off the Irish coast. In my opinion the Brendan was a seaworthy craft under all these conditions, though she was slow - a cruising speed of 2-3 knots was all that one could expect in any wind strength less than a Force 4/5. Windward performance was nil, for while it was possible to point the boat 60-70 degrees off the wind, the considerable leeway produced an effective course of 90 degrees off the wind. On the other hand this course could be maintained in surprisingly rough seas without fear of swamping or capsize. Leeboards - probably anachronistic - were not as effective as leeside rowing, but the crew of five was too small for effective use of the oars beyond manouvering in and out of harbours and anchorages. On the other hand, given the size of crew attributed to St. Brendan, I believe that a genuinely useful performance of the boat under oars could be expected. This I hope to test later.

The leather itself was virtually impermeable. Much more water entered the boat from rain (of which we had a depressing amount) or from an occasional wave over the side, than ever penetrated the leather skin. In places the leather skin developed wrinkles and corrugations over the stringers after prolonged immersion. Also some of the sheer was lost, apparently due to hogging after beaching and - I believe - a fault in construction of the wooden frame. However the leather itself, when the boat was hauled up for inspection in Iceland, was found to be in excellent condition, though scuffed and scratched by collision with flotsam. Even the woolgrease was mostly in position though supporting a light growth of barnacles. In sum, this skin boat after almost six months afloat was still sound and watertight.

Brendan is now hauled up for the winter in Iceland - itself a useful practical exercise - and I hope to re-launch her at the beginning of the next sailing season, probably in late May, in order to attempt the last leg of the project, the voyage direct to North America. For this we will need favourable winds.

Thus the project continues, but it is fair to state that already I have a greatly increased respect for the accuracy of the medieval texts we have been dealing with; and I would submit that the leather-covered boat should now be regarded as having a genuine ocean-going potential. So far, at least, Brendan has not let my crew or me down.

NOTES

1. Selmer, Carl (ed) Navigatio Sancti Brendani Abbatis, Univ of Notre Dame Press, Publications in Medieval Studies Vol. XVI, 1959 is the critical edition. Translated by Webb J.F. in Lives of the Saints, Penguin Books, 1965; or O'Meara, John J., The Voyage of Saint Brendan, Dolmen Press, Dublin 1976.

2. Hornell, James 'Curraghs of Ireland' Mariner's Mirror, Jan. 1938, p. 29.

3. Breac Moedoic Budget in National Museum of Ireland, Dublin; and the Book of Armagh Budget, Trinity College, Dublin; also Waterer, John J., "Irish Book-Satchels or Budgets" Medieval Archaeology 12, (1968).

4. Fenton, Alexander "The Curragh in Scotland" Scottish Studies, 16, (1973) pp 61-81.

5. Johnstone, Paul "The Bantry Boat" Antiquity 38 (1964) pp 277-284.

DISCUSSION

Göthche: I am one of a group of Danish volunteers undertaking sailing trials of an 80 year old fishing boat from the northern part of Norway. This type of boat had a square sail until the beginning of this century and we are learning to use this type of sail. On tacking we reached a speed of about 5.5 knots and sometimes we have been to nearly 10 knots with the wind from astern. She goes to an angle of about 50 degrees to the wind which is not as good as the modern fishing boat we have in Denmark. We have also used the boat with a topsail as used to be done in Norway.

Coles: It would be of considerable value if such a boat could be used to give basic training for people who wish to learn about these boats.

McGrail: Are the Danes still learning about this boat?

Crumlin-Pedersen: The work done with this vessel is preparing our minds for the interpretation of the archaeological remains as we see them in the Viking ships. By travelling in Norway and talking to the old boatbuilders and fishermen who built and sailed these ships we have got to know practically everything about how this was done, to the point where we have detailed knowledge of handling the boat. I am not saying that this situation is the same as that of the Viking sailor. I am saying that we now know the situation as it was around the turn of the century in northern Norway, and thus we are much better suited for the job of understanding and interpreting the Viking sails and the behaviour of the Viking ships, than someone who has done only theoretical work.

Walle: I fear experiments with replicas may lead to wrong conclusions because you may bring in too much modern knowledge and there is a danger in using this to prove old methods. Experiments in a water tank would lead to better results because you have better means of measurement than when you are sitting in a yacht. In a yacht a sailor is inclined to believe that he is much closer to the wind than he is.

Crumlin-Pedersen: We have established a trials area in the Roskilde Fjord with three buoys placed at a known distance apart and we have carried out a number of sailing trials between then. In due time the results of these will be published together with a comparison between this vessel and more recent boats mostly with gaff rig. This Nordland boat was as good at tacking as the gaff riggers. Regarding the wind tunnel experiment which has been quoted by Sibella Haasum (Vikingatideus Segling och

Navigation, 1974), there is a tendency to regard wind tunnel or other test results from scientific experiments to be the truth itself, but if you carry out the wind tunnel tests with a sail you have to make some assumptions about the shape of the sail, and if this shape is not the optimum one you will get a result which is different from the one you get by sailing in the sea with the capability of adjusting your sail to give it an optimum shape: that is what many generations of sailors have been developing. Therefore I think it is important for us not just to leave it to model tests and not to be uncritical about practical experiments but to go to the material with a critical mind and find out something which can withstand criticism.

McKee: What I am trying to do with the Greenwich <u>faering</u> replica is learn how to sail her and at the same time to evolve a way of comparing different boats. I have learnt quite a lot about sailing a replica small boat. It will be sometime before I can publish anything because one is literally in the stage of designing the experiment but I have found out one thing which may be of interest. Last summer I was sailing with a rectangular sail made of a very soft material, and I managed to get this much smaller boat than the Nordlander sailing rather better than a broad reach. This summer I have had a sail which will stand very much closer to the wind, but the boat itself doesn't go so well to windward with this modern sail. I think a lot of this is due to the fact that there isn't a grip on the water from a hull like this. I am very interested in the material of the sail too and I hope to try a sail with diamond or lozenge shape pieces, because in the way I can get the warp and the weft not parallel to the head of the sail, and I have an idea that there may be something quite important in this.

Baker: Do not believe all that you read about tank testing of yachts; you cannot always believe your results. While an underwater cut off transom proved possible in the tank for the 12m yacht <u>Mariner</u> for the last American Cup trials, it just did not work out full scale in water.

Evans: Has the Greenwich <u>faering</u> replica been used by local fishermen? I think if you get people without preconceptions who are absloutely masters in their own small craft to see how the replica handles this would be extremely useful. Secondly, did you stand up and row, pushing as opposed to pulling?

McGrail: We did try pushing, pulling, and even paddling. Thank you for the suggestion about local fishermen.

McKee: Because the mast was up and I wasn't sitting on the thwart that it is convenient to row from, I recently found myself push rowing her from the after thwart which is not normally used for rowing, and I got on splendidly. Yes you can push row those boats very nicely. We have had watermen rowing her, these are men who used to work an equivalent to a Thames wherry on the Tamar, and one man can move her, using his own oars mind you, very much better than anybody else we have had rowing her so far, just one man rowing.

Christensen: Early curragh building is basketry or derived from basketry, but Severin built a large Dingle curragh of lath construction on the double gunwale.

Severin: Among the Irish curraghs I did not see this distinction in that the lath built ones transpose as you move along the coast into the basket built ones. Both techniques are being used and I can't find the dividing line between the two.

Roelfzema: You would have had such a perfect opportunity to make a double hulled boat, I don't know if this was a Celtic characteristic but a double skin looks much safer to me.

Sever I don't think there is any reason to do so, she is perfectly good as she is. One of the problems is that today the cows aren't what they used to be, and we tried to use a leather about 4mm thick, and that is quite thin by modern standards, and that seems to be sufficient. It would be very hard to do a double skin because of the stitching techniques.

Editorial Note: Mr. W.A. Baker of the Hart Nautical Museum, Massachusetts Institute of Technology, U.S.A. described the building of several replicas or reproductions in the U.S.A. These were mostly built for Museum display, although some are working boats. The boats are replicas of:

i A 30ft. boat typical of those carried by Mayflower in 1620.

ii A 16ft. logboat, used as tenders in 17th century U.S.A.

iii A 30ft. Potomac River plantation boat of *c*. 1725.

iv A 24ft. whale boat used by privateers in Long Island Sound during the Revolution.

v A 25ft. English longboat of *c*. 1775.

vi A late 19th century small fishing boat.

20. ANCIENT BOATBUILDING - A PROVISIONAL CLASSIFICATION

Arne Emil Christensen

Summary

After defining the limits of the present paper against the topic, the author discusses recent trends in the study of the subject, with special emphasis on Northern Europe. The importance of the concepts "shell building" and "skeleton building" as tools for the understanding of boatbuilding techniques is stressed. A brief survey of the main source groups is given, and a tentative classification of ancient boatbuilding methods is suggested. It is stressed that the material is still insufficient for definite results and that more research is needed on many sectors of the subject. Of particular importance is the study of Celtic shipbuilding in Western Europe, the possible transition from shell to skeleton technique indicated by late Classical and early Byzantine wrecks, and the study of living boatbuilding traditions in Western Europe and the Mediterranean.

The huge topic "Ancient boatbuilding methods" can not by any means be thoroughly covered by a symposium paper, so I have selected what I see as some of the main points.

I will define boats as craft which float because they displace water. This definition will omit, however interesting they are, the various floating devices used in mild climates to support swimmers; rafts and reed boats. I will also arbitrarily, omit the bark canoes of North America and other parts of the world. This leaves us with two main types of vessels; boats of wood, and boats of wickerwork or a wooden skeleton covered with skins, textiles or bitumen. I will try to group the material and see if this grouping can give any information of significance on the various methods used by different boatbuilding "schools". Geographically, I will mainly restrict myself to Europe, but I will draw in comparative material as I find it necessary. Finally, I will try to sum up what we know, what we think, and where we need to pursue further research.

The main aim of the paper is not to solve problems, but rather to formulate them, and provoke further research into this, in my opinion, very important branch of the history of technology. In North-Western Europe, the topic has been rather intensely studied, especially within the last 15 to 20 years. As I started work on my M.A. thesis in the late 1950s, the few persons working actively on boat archaeology that I could locate and discuss problems with, could be counted on one hand. We all worked in the clinker-building tradition, and few people seemd to take any interest in other old methods of boatbuilding.

Luckily, matters have changed greatly, Many scholars now devote time to the study of shipbuilding history, underwater archaeology has expanded greatly, to give much new material, and new museums devoted to ships have been established. A series of finds has given new information, and the re-publishing of old and insufficiently published ships has drawn attention to other and less spectacular types than the Norse Viking Ships, which have had a tendency to draw most of the interest from scholars and the general public alike. Two journals have been founded for this and related topics, Cahiers d'Archaeologie Subaquatique and The International Journal of Nautical Archaeology.

Thus the field has expanded in several directions. The material has increased, making it necessary to reformulate theories that have long been generally accepted, and to ask new questions.

Olof Hasslöf's paper "Carvel construction Technique, nature and origin" published in 1958 has supplied the theoretical background for much of the recent research.[1] Hasslof formulated the concepts "Shell building" and "Skeleton building" and demonstrated that they were better tools fo research than the terminology previously used, "Carvel" and "Clinker" building. Using carvel-clinker terminology may lead to wrong conclusions, as true carvel ships and smoothskinned shell-built vessels look alike when finished. Especially when iconographic material is used, this pitfall is evident. Shell technique as used in clinker building is still practised, so modern comparative material can be much more readily used in the "clinker-building" area than in other parts of the world. In other areas, the links between ancient and modern boat-building are not so obvious, and they have not been studied and documented with the same intensity as the clinker-building tradition has.

If we sum up the work of what may be called the "Hasslöf school" of research, (including the present author), much of it centres on the basic hypothesis that wooden shipbuilding in the Mediterranean, Western and Northern Europe was done in shell technique until fairly recently. In the North, boats were clinker-work, in the Mediterranean, smoothskinned tenon and mortise technique was used. When Hasslöf wrote his paper, the Mediterranean material was scanty. Apart from the Dashur and Nemi ships, very little was accessible in properly published form. In addition to Hornell's work on the connection between the modern frameless boats on the Nile and the Dashur boats,[2] Hasslöf mainly used iconographic material showing surface dovetailing of planks in the "dashur manner". Hasslöf contradicted Hornell's statement that carvel building "must be an act of invention", and that "no gradual evolution of the one into the other is concievable."

For the clinker-building area, the gradual evolution can be demonstrated by practise which is still within living memory, and Hasslöf suggested that similar evolution may have taken place elsewhere. Hasslöf assembled some material pointing to Western Europe in the late 15th century as a possible place for the transition into carvel work. Important in this context is the English practise of giving old clinker hulls a new skin in carvel technique.

The study oi recent boatbuilding within the clinker-building area has played an important role in much of the work published by the "Hasslöf school", and it has been proved that this is a source of prime importance for the proper understanding of archaeological boat finds. One example is Ole Crumlin-

Pedersen's paper on the hull shape and construction of the "Kog" type.[3] Concerning the other method of shell building, ships excavated under water in the Mediterranean in recent years have given numerous examples of tenon-and mortise work that must have been built shell first.

It is now generally accepted that this method was used by Egyptian, Greek Punic, Roman, and Byzantine shipwrights, but it is still not known whether this was the only method in use in the Mediterranean.[4] Other finds, however, have given indications of a much greater age for the transition from shell to skeleton work than Hasslöf suggested. Finds like the 4th and 7th century Yassi Ada ships and the Pantano Longarino ship indicate a mixed shell-skeleton technique strangely similar to the 19th century Scandinavian method of building small freighters with a clinker bottom and carvel topsides.[5] In these cases, in Scandinavia, the shipbuilder would shape the bottom of his hull in shell technique, to ensure that the shape developed properly to give good underwater lines. Floors, futtocks and top timbers are then fitted, and the remainder of the hull is carvel planked for strength and sturdiness.

In Northern Europe the work done by Marsden and Ellmers on early Celtic ships has widened our knowledge considerably and shown that the picture is more complicated than previously assumed.[6] Celtic shipwrighty had its own methods, seemingly not related to either tenon and mortise or clinker work. In a recent paper Marsden has arranged the North European material from the Roman period in six groups. One of these is Mediterranean (mortise and tenon work), one is Nordic (clinker). The remaining four are seen as different Celtic shipbuilding traditions. From important finds, Marsden labels them the Blackfriars, Zwammerdam, New Guy's House and Utrecht types.[7]

New material is constantly being reported, and the time may not yet be ready for general conclusions, but it will be worthwhile to summarize the sources available and see how they can best be utilised. Our most important sources naturally, are the actual remains of ancient vessels. The catalogues recently complied by Ellmers and Muller-Wille shows how extensive this material is in the North and Northwest of Europe,[8] and the number of finds will surely multiply in the future, due to underwater work. For the Mediterranean no complete catalogue exists, the most extensive is the one drawn up by Basch,[9] but nearly every season new finds appear with a remarkable amount of the wood intact.

Many of the Mediterranean vessels, among them highlights like the Kyrenia ship, the Punic wrecks from Sicily and the Cheops ship, are still awaiting final publication, so their full significance can only be guessed. The series of papers started by the IJNA on "Mediterranean hull types compared" will surely help to clarify matters.[10]

Ancient verbal and iconographical sources have been frequently and often brilliantly used in the study of what ancient ships looked like, and to some extent also, how they were made. However, they pose great problems of source criticism. As the finds are still lacking, the iconographic sources are still our prime source for understanding rigging, but as a source for hull shape and methods of building, they will often not be fully understood until actual remains of the ships are found.

Fig. 20.1 With 1100 years between them, roughly 45 generations of boat-builders, the two boats clearly demonstrate long, continuous lines of tradition. Above, the smallest boat found in the Gokstad burial of c. 850 A.D. Below, traditional boat built near Bergen, Western Norway around 1967 A.D. Photo Terje Olsen.

Ancient woodworking tools are a source of importance, when it can be demonstrated that they have been used in shipbuilding. Analysis of toolmarks on the ships will supplement the tools themselves as a source.

The results of modern experiments cannot be called a source in the usual definition of that word, but they are very helpful in the research process, for instance in proving that a theory is impossible or improbable in practise.[11]

Finally, I will include modern comparative material among the sources. For the clinker-building area, its value has been proved in full (Fig. 20.1). A striking example of how such material can be used across cultural borders is the paper by L. Casson on the Roman tombstone of the shipwright P. Longidienus.[12] Using comparative material from the clinker-building area and the results of underwater archaeology, Casson gives the proper explanation of the scene: the shipwright is fitting a rib to the finished shell of the boat he is building.

The variation in the sources as listed above, gives rich possibilities of attacking problems from many angles, but at the same time there are great problems involved in the many different aspects of analysis and source criticism necessary. Documenting an excavation, or judging the quality of similar work done 50 years ago makes it neccessary to use methods of source criticism far different from those needed to interprete an ancient text or sketchy ship pictures.

It is hardly to be expected that the same scholar should posess the training and experience necessary to analyse toolmarks on ancient wood, study texts in Greek, Old Norse and Celtic, wield the axe or oar in a comparative experiment, and interview a retired and probably deaf boatbuilder, equally well. Teamwork is the obvious solution to the problem, of course. During the fieldwork done by the Scandinavian Maritime History Working Group in the years 1965-67, we all felt that our different background training, in ethnology, history, naval architecture and archaeology, gave great advantages over working in a group where all had the same training, or working alone.[13] However, those of us who are archaeologists, should strive for as broad a knowledge as possible of all types of sources. It is our job to interprete mute artifacts, to analyse fragments of a lost totality. The "model" we establish to explain our finds, how the artifacts were made and used by early man, is of course limited by our own knowledge, experience and imagination. The wider the knowledge and experience is, the richer the model gets, in details or alternatives. On the other hand the search for "parallels" must be kept within the limits of common sense. There have been many sins comitted, where the "parallel" is visually striking, but has no possible cultural connection with the ancient material. Here experience will help in tempering a too rich imagination.

For this occasion, I will leave most of the sources mentioned above and concentrate on the ships themselves. On the basis of the available sources, I have tried a classification of the "ancient boatbuilding methods" known to us, within the limits stated in the introduction.

1. Wooden boatbuilding

2. Skin boatbuilding

We can divide further according to building technique.

1.1.1. Log boats, made by hollowing a log with tools, fire or both. Shape may retain the round cross section of the log, or the boat may have hard chines both inside and outsides, giving a more or less boxlike structure. Boats of this category have been in use from at least Mesolithic times to within living memory in Europe, and they can still be seen in use in many parts of the world.

1.1.2. Log boats widened after hollowing with the help of fire and water. Archaeological finds date back to the Roman period, at least; the method is still used in parts of Eastern Europe. The method is also reported from the Far East and the northwest coast of Canada.

1.2. Clinker building. Planks overlap, are joined by sewing, iron nails or wooden nails. Finds push the date of the method back to c. 350 B.C. but it may be older. Evolution of this method from either log boat or skin boat has been much discussed. The method can be studied today in the boatshops of Scandinavia, Britain and the Baltic.

1.3. Smooth skinned plank boats, the strake edges joined by mortise and tenon, lashings, iron staples or iron nails drived obliquely. In Egypt, finds go back to the Old Kingdom; the iron nailed variation has been in use well into this century. Similar edge joining methods have also been used and are still used in the Far East and the Pacific.

I have choosen to include in this group the North Ferriby and Brigg vessels, seeing them as a northern variation of Mediterranean Bronze Age boatbuilding. It is probable that the group as a whole is derived from log boats. In Egypt, reed boats definitely influenced the shape of wooden vessels, and to some extent probably also their construction.

1.4 "Celtic"[14] plank-built boats with hard chines. Finds and literary sources give a terminus ante quem of the 1st. century B.C. Modern derivatives may exist, in the "Nachen" type of rivercraft used on the Main and Rhine.

The ancestry of the type, or rather group of finds, seem to be log boats of the hard chine variety.

1.5 "Celtic" plank boats of the "Holck" type. So far, finds and linguistic evidence gives a terminus ante quem in the 2nd century A.D. Modern clinckerbuilt vessels of the "pram" type may be derived from the Holck. Ancestry seems to be log boats of the 1.1.2. type.

1.6. True carvel work, defined as ships where planks are nailed to pre-erected frames, the planks not being fastened to one another. Age unknown, probably late Classical period in early form. The method may have developed from group 1.3. Still in current use.

It will be seen that my grouping of the Celtic boats differ from that suggested by Marsden. I have choosen to group the Blackfriars and Zwammerdam types together, and my group 1.5. covers Mardens Utrecht and New Guy's House groups. This grouping places more weight on hull shape than on technical details, and may be an over-simplification.

The skin boats should be subdivided in 2 groups.

2.1 Boats where the skeleton consists mainly of longitudinal stringers, connected by ribs or bottom and side struts. All arctic skin boats fall into this group, whether they are of the "Umiak" or "Kayak" type. Kayaks generally have ribs of thin bent laths inside the stringers, resulting in multichine hull shape, while the umiaks generally have straight bottom and side battens separating the stringers, and the number of stringers is smaller than in a kayak.

2.2 Boats where the skeleton is in basketwork, or derived from basketwork. This group has a wide geographical distribution and it is uncertain whether there has been cultural contact between the various areas, India, the Middle East and Celtic Wales and Ireland. Many of the curraghs as built today have a lath skeleton, but even here the basket work ancestry seems certain when the curragh - coracle group is seen together.

It is with some hesitation that I offer this classification, and I am fully aware that any grouping of the material will force some of it into too rigid "boxes". After all, only a generation ago, wooden shipbuilding was either "carvel" or "clinker". We still need to know much more about the variations in the "repertoire" of the old shipbuilders, before our groups can have any hope of getting close to realitiy. However, classification and reclassification is a necessary tool in studying an archaeological material, and I hope my classification will make some technical differences and similarities more visible.

The separation of the skin boats in to two categories is necessary, as both have been used together in the discussion on whether the skin boat or the log boat is the ancestor of clinker-built craft. With the danger of starting that discussion again, it should be pointed out that all other groups of plank-built boats have a wooden ancestry. This is still valid if we include Polynesia and the junks of China and Japan. As I see it, the skin (and reed) boats are blind alleys. They develop within their own limits, sometimes to great perfection. They may influence the shape of wooden boats, but very seldom do they influence their construction. Egypt is the exception to the rule, the system of longitudinal beams at deck level seen in the Cheops boat and indicated on many models may have been taken over from papyrus rafts with wooden reinforcements. Generally we can say that where wood was available, primitive man make his boats of wood, and wooden technology has dominated boatbuilding ever since, until steel and fibreglass was introduced. (Exceptions do exist, notably among mobile groups of hunters where portage is often necessary, as among the Indians of North America.)

In the Mediterranean tradition of shipbuilding, an early find like the Cheops ship shows us that Egyptian shipwrights mastered the building of very large wooden vessels early in the Old Kingdom, with the wonderful reliefs from the graves of Ti and others as supplementary evidence to show us how the shipwrights worked.[15] There are huge timbers involved, the largest plank in the Cheops ship is 23m long. This, together with the Ferriby boats and the large war canoes of Oceania, shows us that iron tools are not necessary for the building of great vessels, they are within the possibilities of stone and bronze age technology.

Even with these huge timbers, all fastenings are wood and rope. In fact the use of metal nails and bolts belong to a well developed iron age technology. The bronze age never got around to the general use of nails, the material was too expensive, or the idea had not yet occurred.

When thick wood is used, for ships' timbers or other structures, dowels and treenails are natural solutions to the problem of fastenings. When thin wood is used, there will be a danger of the nails pulling through, and not enough wood thickness to cut mortises. The only possible pre-iron fastening method will be sewing or lashing with the available fibres. However, thick wood may be lashed as well as treenailed or tenon-fastened, obvious examples are the Cheops boat, and a number of the canoe constructions of Polynesia. We do not know when tenons took over completely in the Mediterranean shipbuilding tradition, and sewing was discontinued, but tenons are used in the Cheops boat, and they can be seen in some of the Old Kingdom reliefs. Until late Classical times, shell building seems to have reigned supreme in the Mediterranean, as it did in the North, within the clinker-building territory. The aids and methods used by the Mediterranean shipwrights to obtain the right hull shape are not known, but Van Doorninck has found evidence of the midship rib being erected early during the building of the 4th century Yassi Ada ship, clearly to act as a midship mould to guide shape.[16]

The Celtic building methods show great variations, even within relatively small numbers of finds known. Characteristic are the massive timbers used, the early use of heavy iron fastenings, and the high frequency of hard-chine vessels. Some of the Celtic vessels are clearly shell-built. The Utrecht ship, and probably the Zwammerdam boat 3 may be expanded dugouts with added washstrakes, while the Blackfriars type may be examples of early carvel work, I am still not quite sure.

Summing up, it must be stressed that this particular field of research is still new and there are plenty of opportunities for pioneer work. Much still remains to be done in the study of clinker built vessels, but for the time being, this is the best-studied branch of the topic. In the Mediterranean a pattern can be seen, but much of the best material is still only temporarily published and the pattern may have to be modified. We need to know much more about the Celtic tradition in wooden shipbuilding, in its various forms. The archaeological material will surely come, in the meantime, much remains to be done on recent boatbuilding within the area, and here fieldwork is urgently needed. Traditional crafts are on the way out all over Europe, and with the last of the craftsmen, a line of tradition disappears that could give both ethnologist and archaeologist information that cannot be obtained either from the soil or archival material. The recent book by Francois Beaudouin[17] on French vessels shows us the great variation in hull and rigging that can be found within one country.

As for Spain and Portugal, we know what vessels looked like, the information still existing in the yards and among retired boatbuilders has not yet been fully explored.[18] This is also the case in the Mediterranean. Mortise and tenon work is no longer is use, as far as we know, but the dovetailed surface tenons used in the Dashur boats were recorded by Admiral Paris on French river craft in the last century.[19] Who knows what one may find? Another line of research which should provide fruitful is the study of carvel practice in small yards. The use of drawings and mould loft has not yet penetrated to all districts, and the "predrawing" methods of carvel building should be recorded in as many places as possible. The differences in approach between shell builders and skeleton builders, the various aids used in shaping hulls, is information that can still be assembled in fieldwork where the spade is left at home. As has been the case in the North, living tradition may explain obscure points in verbal and iconographic sources. It will enrich our "model" for use on archaeological material. The systematic study of the Mediterranean wreck material will surely lead to a greater understanding of how hull shapes and building methods developed from the Cheops ship to late Classical and Byzantine shipbuilding, and on to the true carvel hulls of Venetian renaissance shipbuilders. But I think that the key to a deeper understanding of shipbuilding history in Western and Southern Europe lies not in the past, but in the present, in recording living boatbuilding traditions.

Fig. 20.2 Shell building for different waters. Widely separated in time and space both pictures show ships built in shell technique. The Egyptian shipwrights fasten their ship with tenons: William the Conqueror had his vessels built by the clinker method of his Scandinavian ancestors - G. Steindorff, F. Stenton.

NOTES

1 O. Hasslöf. 'Carvel construction technique, nature and origin.' Folk-Liv XXI-XXII. Stockholm 1958. An expanded and undated version of this paper is published in Ships and Shipyards, Sailors and Fishermen. Ed. Hasslöf, Henningsen, Christensen. Copenhagen,

2 J. Hornell. Water Transport. Cambridge 1946 p. 213 ff.

3 O.. Crumlin-Pedersen. 'Cog-Kogge-Kaag, træk af en frisisk skibstypes historie.' Handels og Søfartsmuseet pa Kronborg, Arbog 1965. Helsingør.

4 The most recent general survey of the material is L. Casson. Ships and Seamanship in the Ancient World. New York 1971.

5 Hasslöf 1958 p. 56.

6 D. Ellmers. 'Keltischer Schiffbau.' Jahrbuch des Romisch-Germanischen Zentralmuseums Mainz. 1969. D. Ellmers. Fruhmittelalterliche Handelsschiffahrt in Mittel- und Nordeuropa. P. Marsden. A Roman Ship from Blackfriars, London. London 1966. P. Marsden. 'A Boat of the Roman period found at Bruges, Belgium in 1899, and related types.' IJNA vol. 5-1 London. 1976.

7 Marsden 1976 p. 41 ff.

8 D. Ellmers. Fruhmittelalterliche Handelschiffahrt in Mittel- und Nordeuropa. Neumunster 1972 M. Muller-Wille: 'Bestattung im Boot. Studien zu Einer Nordeuropaischen Grabsitte.' Offa vol. 25/26. Neumunster 1970.

9 L. Basch. 'Ancient wrecks and the archaeology of ships.' IJNA vol. 1, p. 12 ff.

10 'Mediterranean hull types compared." A series of papers commencing in IJNA vol. 5 nr. 2.

11 S. McGrail. Models, replicas and experiments in nautical archaeology. The Mariner's Mirror 61. London 1975.

12 L. Casson. 'Ancient Shipbuilders, new light on an old source.' Transactions and Proceedings of the American Philological Association. XCIV. New York 1963.

13 Some of the theorectical views reached in our discussions, as well as some practical results of research, will be found in the book Ships and Shipyards, Sailors and Fishermen, ed. Hasslöf, Henningsen, Christensen, Copenhagen 1972.

14 During the Greenwich conference, the phrase "Celtic shipbuilding" met some opposition. As none of the other names suggested for this group of ship finds was entirely convincing, I prefer to follow the term used by Marsden and Ellmers.

15. Abdel Moneim Abubakr and Ahmed Youssef Mustafa. 'The funerary boat of Khufu.' Beitrage zur Agyptischen bauforschung und Altertumskunde. Heft 12. Wiesbaden 1971.

16. F.H. van Doorninck. 'The 4th century wreck at Yassi Ada.' IJNA vol. 5 nr. 2. London 1976.

17. F. Beaudoin. Bateaux des Cotes de France. Grenoble 1975.

18. The work of O.L. Filgueiras shows how much information one can get from living tradition. e.g. Raboes da Esquadra Negra. Porto 1956.

19. Admiral Paris. Souvenirs de Marine part 2 plate 64. (Bateau dragueur de l'Adour).

21. CELTIC SHIPS OF EUROPE

Peter Marsden

Summary

The recent identification of a family of shipbuilding traditions in central and north-western Europe during the Roman period and later has opened up new areas of study in the history of ship technology, particularly in making more precise interpretations possible. The vessels discussed in this paper are classified into four basic types - firstly Blackfriars, a seagoing sailing ship; secondly, New Guy's House a Thames river barge; thirdly, Zwammerdam a large river barge common on the Rhine and in Switzerland; and fourthly, Utrecht another type of river barge found on the Rhine.

Methods of cutting the timber to build the ships are considered, and the large size of the vessels, which is inconsistant with their primitive construction does suggest that prior to the Roman invasion the vessels were mostly of much smaller size, and that their large size during the Roman period was due to the sudden demands of a vastly increased volume of trade.

Between Scandinavia, and the Mediterranean basin there lies an area which includes the British Isles, northern France, Belgium, the Netherlands, West Germany and Switzerland in which very little was until recently known about early native shipbuilding traditions. The clinker tradition of Scandinavia, as illustrated by the Nydam boat of the 4th century A.D., and later by the Viking ships such as Oseberg, Gokstad, Tune and Roskilde, is famous, while far to the south the classical shipbuilding tradition of the Mediterranean as shown by the Greek ship of the 4th century B.C. from Kyrenia, and by the later Roman ships from Lake Nemi and Fiumicino in Italy, Yassi Ada off Turkey, and Grand Congloué off France are now equally well known.

But of the region between the two little was known until 1962 when part of a large seagoing ship of the second century A.D. was excavated in the bed of the Thames, at Blackfriars, in the City of London. This ship was completely different from the other discoveries and was the first recognised example of what has since been rather loosely termed the 'celtic' tradition of shipbuilding. Since then other discoveries have been made, or the significance of earlier finds recognised, and it seems that there was during the Roman period a family of 'celtic' shipbuilding traditions each having much in common with the others. I have summarised most of these discoveries in a recent publication (Marsden, 1976), but in addition to my research much valuable work in this field has been undertaken by others, particularly by Dr. Detlev Ellmers, Director of the German Maritime Museum at Bremerhaven (Ellmers,

1969), by Maarten de Weerd of Amsterdam University (de Weerd and Haalebos, 1973), and by Beat Arnold and Michel Egloff of Neuchatel Museum in Switzerland (Arnold , 1974; Egloff, 1974).

At least eighteen 'celtic' ships and boats (excluding dugouts), thirteen of Roman date, have now been identified in central and north-western Europe. and I understand that several other unpublished discoveries have occurred. Very little detail of most of these finds has been published, so at this stage it is not possible to make a detailed analysis of their significance. Not only do these vessels tell us much that is new about ancient shipbuilding technology and the use of ships, but also they must alter many long held theories about the origin and development of early ships in north-west Europe.

For example the overlapping of planks was sometimes used (Marsden, 1976, 47), so it is unlikely that the clinker method of shipbuilding that was introduced into the celtic area during the Migration Period was such a stranger, and the extensive use of clinker planking in medieval shipping south of Scandinavia may have been partly due to the fact that the native celtic shipwrights already used a form of clinker planking in their own traditions.

The range of 'celtic' ships and boats of Roman and medieval date has been found on various sites from Lake Neuchatel in Switzerland to the mouth of the Rhine in the Netherlands. In addition they have been identified at Antwerp, Bruges and Pommeroeul in Belgium, as well as in London (Marsden, 1976) and the recognition of this apparently interlinked family of early north-western European traditions opens up new vistas in our understanding of later developments in European ship technology. Indeed, as Ellmers has shown, the same basic tradition still probably lives amongst certain small boats in a tributary of the Rhine (Ellmers and Pirling 1972), and I suspect that on our own river, the Thames, the traditional swimhead Thames Barge is likely to have been a descendant of the celtic tradition.

The vessels are mostly large, shallow draught, flat-bottomed craft used on inland waterways, particularly on the Rhine. But the latest research shows that they occur probably as far west as the Somme in France where in 1808 a vessel probably with this same construction was found at Abbeville (Traulle, 1808).

Although the first identified find occurred in 1962 in London, it was already known that a celtic shipbuilding tradition had existed because it was briefly described by Julius Caesar in the first century B.C. Caesar was describing the ships of the Veneti tribe of north-western Gaul, but the lack of information in his description meant that very little could be judged regarding the significance of the tradition. It was clear, however, that the celtic ships could be large, for Caesar found them very difficult to defeat from his smaller Roman war galleys. It was only after immobilising the Veneti ships by cutting their rigging that the Romans managed to capture the enemy in what is the earliest recorded major sea battle in northern Europe. Their ships, Caesar wrote, were different from the Roman ships of the Mediterranean as they were high sided, flat bottomed craft propelled only by leather sails, while their hulls were made from thick planks and massive beams held together with iron nails as thick as a man's thumb.

It is clear that the ships of the Veneti were different from the ships of the Scandinavians where yet another tradition existed for in that area, as the Nydam boat shows, the vessels had thin overlapping planks and slender ribs, and a much more elegant form. Although a subjective comment with which not everyone will agree, I do think it fair to say that compared with the Scandinavian tradition the 'celtic' tradition, however well suited to the environment in which they worked and the job they had to do, was rather crude - their vessels being generally more like floating boxes and tubs. I shall return to this matter later as I believe it has a particular significance.

But first the term 'celtic' must be explained before the characteristics of the tradition are discussed. The term is farily loosely used, but it is based on the fact that the areas in which these vessels have been found were primarily occupied by Celtic peoples, though in the Low Countries and in south-eastern England the Belgae are believed to have been of mixed Celtic and Germanic stock. Most of the so-called 'celtic' ships have been discovered in the Belgic region near the mouth of the Rhine and in London, though the finding of two vessels in the purely Celtic region of Lake Neuchatel, Switzerland, does strengthen the belief that these were vessels built by Celtic shipwrights. It is realised that the term is perhaps not as suitable or specific as some might wish, though it does provide a useful umbrella term to show that these craft are distinctively different from those that we tend to classify under the terms 'Scandinavian' and 'Mediterranean.' The belief that 'celtic' is a suitable term is strengthened by the description of the ships of the Veneti, a purely Celtic tribe, which compares so well with the characteristics of the Blackfriars ship; but in view of all the uncertainties the term 'celtic' should be used with a small 'c'.

Prior to 1958 the only generally known discovery of a ship of the Roman period in northern Europe was the mid-section of the hull of a small trading vessel found in 1910 in London on the site of the County Hall, near Westminster. Now preserved in Lancaster House where it has long been well-known to ship historians, the significance of this find remained unknown until the 1930s. At that time discoveries in Lake Nemi in Italy of enormous ships showed that the County Hall ship was built according to the classical Mediterranean tradtion But in fact this was not the first plank built craft of the Roman period to have been found and identified in northern Europe for in 1899 a similarly constructed vessel of the first century A.D. had been found at Vechten, near Rotterdam, in the Netherlands. Unfortunately this was published in an obscure Dutch archaeological report (Muller, 1895) and the discovery remained unknown until recently.

When in the 1930s Lake Nemi, a flooded volcanic crater near Rome, was drained to expose the two great sunken Roman ships that for centuries had been known to lie there, archaeologists were at last able to have their first view of classical Roman shipbuilding technology and though the forms of the ships were not those of seagoing craft their build was similar to that of the County Hall ship. This discovery then tended to strengthen the general assumption that the usual ships to be seen in the northern province of the Roman empire were the Mediterranean types, illustrations of which dating from the Roman period have been so frequently found in the Mediterranean basin. Similarly the ships of the north have tended to be thought of as exclusively vessels of the Scandinavian tradition - simply because very little was known about the Celtic tradition of central and north-western Europe. This clear cut division was more apparent than real while the Celtic tradition was unknown, but judging from the prepond-erance of native craft amongst the discovered Roman vessels, the major ports of northern Europe in Roman times were probably mainly filled with local ships and boats, with the occasional though not rare visit of a Mediterranean ship.

Excluding dugout canoes, which, during the Roman period have very limited importance in anything other than local economy and no value in the development of shipbuilding technology because complex plank-built ships had already been developed by this time, there are at least thirteen Roman ship and boat finds of Celtic type. And for these I have proposed that there are four main types Blackfriars type, New Guy's House type, Utrecht type, and Zwammerdam type, the names being taken from the place where the best example of each type has been found (references to various types and examples discussed below are in Marsden, 1976).

The Blackfriars type was the only seagoing vessel of this series, the examples being Blackfriars ship 1 of the second century A.D., from London, and the Bruges boat which was found in 1899 and has been recently dated by Carbon 14 to A.D. 180 ± 80. These vessels probably had no keel, but had a flat plank built bottom, and carvel built sides, the planks being nailed to massive ribs or floor-timbers and not so massive side frames. A distinctive feature is the curiously shaped floor-timber into which the mast was stepped. It seems that these vessels were deep tub shaped craft, if one can judge from the Blackfriars ship. The origin of the type is uncertain but E.V. Wright and L. Basch (Basch, 1972, p. 41-2) have drawn attention to the great similarity between the Blackfriars ship and the Ferriby and Brigg boats of Late Bronze Age - Early Iron Age date. The wholly plank construction of the bottoms of these vessels perhaps suggests that they may not have been developed from the dugout canoe, but rather from the log raft.

The New Guy's House type is represented by only one find, a river barge from Guy's Hospital, near London Bridge. This was discovered in 1958, and much of it still remains to be excavated - a task which will hopefully be carried out now that a professional archaeological organisation exists in the area. This vessel was also a tub-like craft, the beam to length ratio being very great and it had carvel built sides and a flat bottom, and it seems clear that unlike the Rhine barges described below, its bottom curved around to its sides. Unlike the ships of Blackfriars type whose place of construction cannot be accurately established due to these being seagoing ships, the New Guy's House boat was a shallow draught river craft that must have been built on the Thames.

The Zwammerdam type also was a river craft whose flat plank built bottom, and low sides were connected at the chine by a plank cut to an L - shape in section. Vessels of this type could be truly enormous, the largest known being 34 m long. Examples of this type dating from the Roman period are Zwammerdam ships 2, 4 and 6, (Netherlands) the Kapel Avazaath ship (Netherlands), the Bevaix and Yverdon ships (Switzerland), probably the Druten boat (Netherlands), the Pommeroeul boat 1 (Belgium), and probably the Abbeville boat (France).

The origin of the Zwammerdam type is presumably the dugout canoe, for the L - planks are remnants of the early dugout origin, the dugout having been spilt lengthways to allow planks to be inserted to form a wider flat bottom.

The Utrecht type also no doubt developed from the dugout canoe for its bottom was formed of a large dugout the sides of which were built upwards with planking. Only one Roman examples of this type as been found, this being Zwammerdam ship 3 found in the Netherlands.

Some of these native types certainly continued in use into the Middle Ages, the box-like Zwammerdam type remaining but having the single L - shaped plank replaced by two planks joined to form nearly a right angle. Two examples of this type have been found dating from the Middle Ages, a boat from Krefeld in Germany (Ellmers and Pirling, 1972) and another vessel found in Utrecht in 1974 (Hockstra, 1975). It is perhaps from a variation of this type that the traditional swim-head Thames Barge was developed, and these continental discoveries underline how very important it is to record modern traditional boat types and their construction. Examples of the swim-head Thames Barge have not existed for a long time, but every opportunity should be siezed to excavate one should it be found as indeed three dating from the eighteenth century were found at Battersea on the Thames several years ago.

The Utrecht type also continued in use into at least the early Medieval period - to the 11th or 12th centuries - three examples of this type having been found. The first discovery occurred at Utrecht in 1930 (where it is still preserved in the Central Museum), the next at Antwerp in 1905 (Philipsen, 1965) and the last at Utrecht in 1974 (Hoekstra, 1975).

The fairly crude and simple form and construction of these Celtic vessels is, I believe, a key factor in understanding their significance in Roman times, as is also suggested by a detailed study of the choice and use of timber from which the Blackfriars Roman ship was built. This is an aspect which was not considered in my report on the Blackfriars ship and indeed the fact that there is a need for this discussion at all is an indication of the greater knowledge and understanding which is nowadays required in the study of early ships. In this paper I am briefly considering the choice of timber and the methods used in shaping them.

But first let me say a few words about the completed Blackfriars ship. She was quite a large vessel, built entirely of oak, estimated originally to have been about 55 ft. (16.8 m) long and about 21 ft. (6.4 m) broad with a depth amidships of more than 7ft. (2.1 m). The carvel built sides met the fairly flat carvel built bottom to form an angle or chine of 30-35°. There was no keel, but instead two broad longitudinal planks each 3 in (0.076 m) thick. The bottom planks and the first strake above the chines were nailed to massive floor-timbers, while the side planks were nailed to smaller side frames, the bottom ends of which were also fastened to the first strake above each chine.

The ends of the ship were not too well preserved though there is clear evidence that they were pointed. Propulsion was by means of probably a square sail on a mast stepped into a specially shaped floor-timber, in which a coin of the Emperor Domitian had been placed for luck, about one-third of the length of the ship from the bow. No evidence of the steering arrangement was found, but this was no doubt achieved by using one or two steering oars. The whole construction was heavy, with thick broad planks and massive floor-timbers which had clearly been cut from substantial oak trees.

But although a fairly accurate sketched record was made in 1962 of the tree-ring patterns shown in the sections across the timbers, I did not discuss their significance in my report on the ship published in 1965. It is thanks to the more recent work on other ships, particularly by Ole Crumlin-Pedersen of the Viking ship museum at Roskilde and of Sean McGrail of the National Maritime Museum at Greenwich that these matters have been highlighted and shown to be important. In fact, a study of the original cutting of logs to make ribs and planks will soon be made when detailed studies begin of the several Saxon and Medieval boat remains recently found in London by the Department of Urban Archaeology, of the Museum of London.

The marvellous photograph in the National Maritime Museum's monograph on the Gokstad faering, of boat-builder Harold Kimber choosing an oak log from a great pile for building a replica of the Viking boat, echoes the first step in building a ship that happened countless times down the ages (McGrail 1974: 15).

The bottom planks in the Blackfriars ship were as much as 2 ft. 7in. (0.79 m) wide and clearly of considerable length (certainly more than 18ft, 5.5 m), and a study of the tree-ring alignments across the planks, which were sectioned in several places, enables the approximate centre of the trees from which they were cut to be judged. It is clear from these that the log or logs had been cut tangentially to form the planking, and because the 'Celtic' ship-building methods were really quite primitive I susupect that the logs were most likely sawn in the simple 'through-and-through' technique (McGrail 1974, p. 42-3). Fortunately, a careful study of the planks in some of the recent ship discoveries, such as those at Zwammerdam in the Netherlands, could establish the plank cutting technique with more certainty. Nevertheless, in order to make the widest planks in the bottom of the Blackfriars ship an oak tree roughly with a minimum diameter of 7ft. 6in. (2.3 m) had to be used. Naturally this is a very rough estimate, but it does give some indication of the size of tree involved; and although it is very difficult to judge the age of such a tree, I am advised that it could have been approximately 200-300 years old when it was cut down to build the ship, probably in the early 2nd century A.D.

The ribs and stempost were cut from smaller logs, the floor-timber having mostly been cut from whole logs, while several of the small side frames had evidently been cut from the same log. Although the floor-timbers could not be sectioned to accurately record the tree ring patterns, the sizes of the original logs were estimated both from what could be seen of the rings and from the extent of the lighter, softer sapwood preserved in the ribs, as well as from parts of the external curvature of the log (the bark having been removed) which was also preserved. Thus it is clear that the tree trunks from which the floor-timbers were cut were very approximately between 12 in. (0.3 m) and 40 in. (1.01 m) in diameter. And also it is clear that most of the slightly U - shaped floor-timbers had been cut from fairly straight tree-trunks, so that when the log was lying on the ground the central part of the rib, the lowest point of the U-shape, had been cut from the bottom of a log to allow the higher ends of the rib to be cut from the top of the log.

There is little direct evidence to show how the planks were cut and the ribs shaped as no distinctive marks left by the cutting tools were observed. It seems most likely that a very large saw (at least 11 ft., 3.35 m long) was required to cut down the trees and then to cut the planks from the logs; while the curved ribs were presumably adzed or axed into shape. The ribs and planks were held to each other by long clenched iron nails; and so that these could be driven through the floor-timbers holes $\frac{3}{4}$ in. (0.02 m) in diameter were drilled out, presumably by using a spoon bit, and an oak peg was inserted before the iron nail was driven in. The timbers of the Blackfriars ship indicate the great age and extent of the oak forests in northern Europe during the first and second centuries A.D.

These 'Celtic' ships illustrate a native shipbuilding tradition of north-western Europe which, no doubt, existed before the Roman invasions. Prior to those invasions when the economy of the region was of very limited extent it is likely that the native ships, reflecting their primitive construction were generally of small size and limited number. With the coming of the Romans there suddenly arrived a sophisticated economy requiring a vastly increased waterborne traffic. Although the Mediterranean tradition was introduced into the north by the Romans as shown by the County Hall, Vechten and Zwammerdam 2a ships, the native shipwrights could have had little time to learn these new techniques, and instead, judging from the numbers found, it is likely that vessels of their native tradition were much more commonly used. Indeed it is probable that they had little time to refine their own conservative construction techniques, and instead, very big ships with simple constructions and form, more suited to smaller vessels, were built according to the primitive old methods, and it is this that we see both in the Blackfriars ship and even more clearly particularly in most of the Zwammerdam ships of Roman date recently found near Rotterdam.

Finally I would like to add that the Science Museum and the Museum of London through the initiative of Basil Bathe of the former museum most carefully had very exact scale models made of the remains of the forward part of the Blackfriars ship; while more recently the Roman gallery of the Museum of London, to be opened at the end of 1976, is featuring this ship not only with a partly reconstructed model of the vessel at a large scale, but also with a full size partly reconstructed cross-section of the ship incorporating both the original timber elements of the vessel and accurate replicas of those parts of the hull which were badly damaged on drying out.

REFERENCES

Arnold, B., 1974. 'La barque gallo-romaine de la baie de Bevaix' Cahiers d'archeologie subaqatique (Fréjus), iii: 133-50.

Basch, L., 1972. 'Ancient wrecks and the archaeology of ships' International Journal of Nautical Archaeology (London) 1: 1-58.

Egloff, M., 1974. 'La barque de Bevaix, épave gallo-romaine du lac de Neuchâtel'. Helvetia Archaeologica (Basel), 19/20: 82-91.

Ellmers, D., 1969. 'Keltischer schiffbau'. Jahrbuch des Römisch-Germanischen Zentralmuseums, (Mainz), 73-122.

Ellmers, D., and Pirling, R., 1972. 'Ein mittelalterliches schiff aus dem Rhein'. Die Heimat, Zeitschrift für niedrrheins kultur - und heimatpflege, (Krefeld), 43: 45-48.

Hockstra, T., 1975. Note on ancient ships found at Utrecht in International Journal of Nautical Archaeology (London), 4: 390-391.

Marsden, P., 1966. A Roman ship from Blackfriars, London. Guildhall Museum, London.

Marsden, P., 1976. 'A boat of the Roman period found at Bruges, Belgium, in 1899, and related types', International Journal of Nautical Archaeology (London), 5: 23-55.

McGrail, S., 1974. The Building and Trials of the Replica of an Ancient Boat: The Gokstad Faering, Part 1, Building the replica. Maritime Monographs and Reports No. 11. National Maritime Museum, Greenwich.

Muller, S., 1895. Verslag van het verhandelde in de algemeene vergadering het Provinciaal Utrechtsch Genootschap van Kunsten en Wetenschappen. Utrecht, 1895.

Philipsen, J., 1965. 'The Utrecht ship', The Mariner's Mirror 51: 35-46.

Traullé, M., 1809. Lettre adressée a M. Mongez, Membre de l'Institut, par M. Traulle....sur les débris d'un bateau deterré dens le levées de la Somme. Paris, 1809.

Weerd, M. de and Haalebos, J.K., 1973. 'Schepen voor het opscheppen'. Spiegel Historiael, (Bussum), July-August: 386-97.

22 BOAT-ARCHAEOLOGY OF THE THAMES

David Sturdy

Summary

The Thames, like every other major river, has great possibilities for research about and discoveries of early boats; a flexible and deliberate approach will produce a wealth of information. Parallel approaches are suggested: backwards from the documentary material and sideways from the evidence of riverside sites. Four traditional types of working craft from the River Thames above London Bridge are described.

This paper outlines the potential for a programme of purposeful research into the archaeology of the boat along the Thames. Important work has been done in the tidal waters and especially in London, but more work is greatly needed on the upper waters of the river. No satisfactory boat-find has yet been recorded along 150 kilometres of navigable river between Staines and Lechlade. Important finds are constantly being made in dredging the river, and boats must sooner or later be recovered in dated contexts with associated objects.

Creeks and abandoned side-channels are the most likely to contain boat-remains in substantial form: by considering archaeological sites on land we can identify both areas of general interest for specific periods and key-areas of outstanding importance for possible early boat-finds (Fig. 22.1 and 2).

The ancestry and relations of the boats known in historical times will be clarified only by the recovery and study of actual pieces by excavation. Only with a series of dated examples can we discuss possible continuity of tradition from the provincial-Roman Blackfriars ship to the river-barge of Elizabethan pictures; whether the wherry has a Scandinavian origin; or what the punt came from. It is, however, valuable to describe the traditional boat-types of the Thames and define the changes which they underwent in reasonably well-documented historical times. The very considerable changes of type, shape and function which can be discerned remind us that earlier periods probably saw equally complete change and great development.

This paper deals with the everyday working craft of the river above London Bridge. It does not discuss either the state-barges or the sailing-craft of the estuary, such as Barking-smacks, Thames sailing-barges or the earlier hoys.

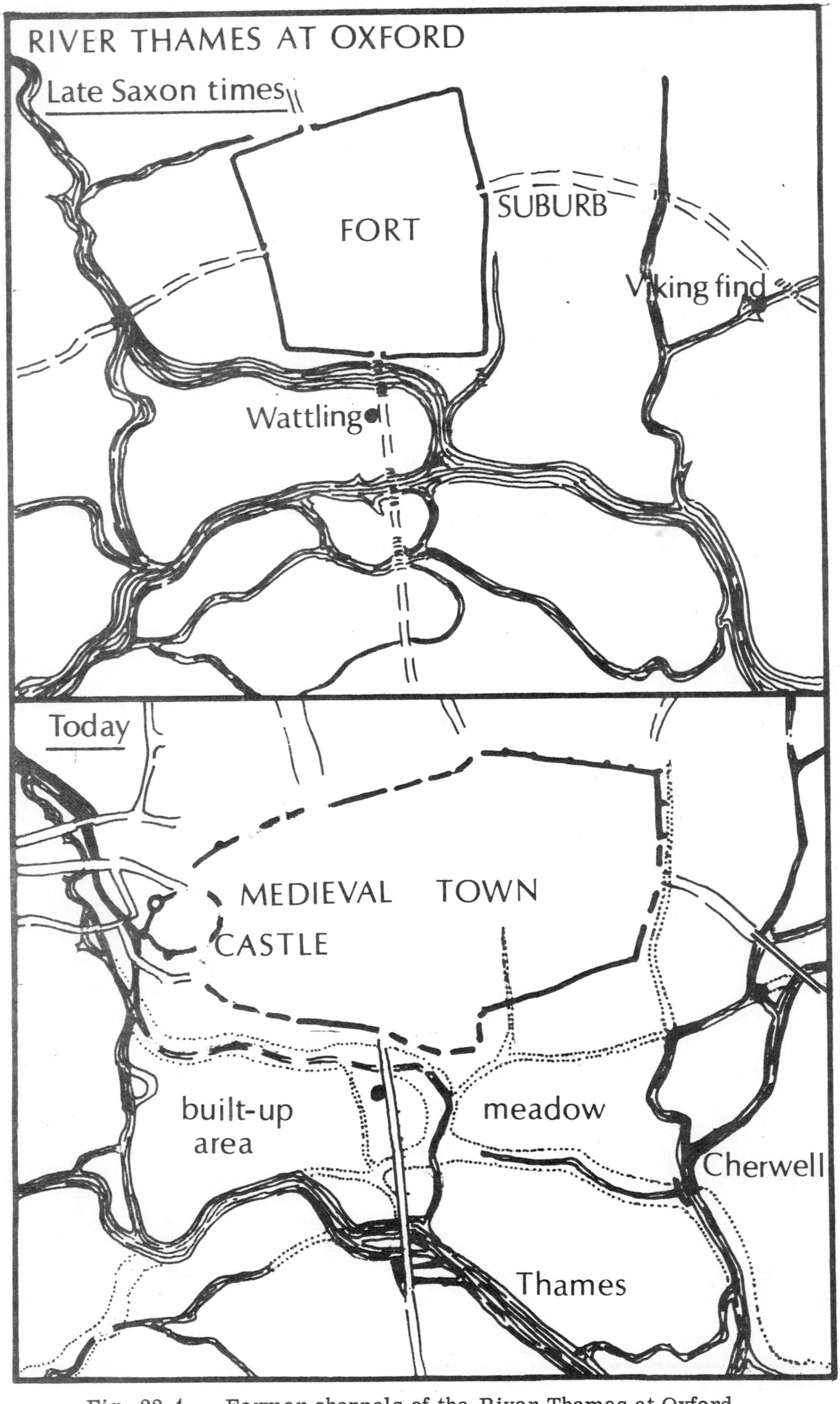

Fig. 22.1 Former channels of the River Thames at Oxford.

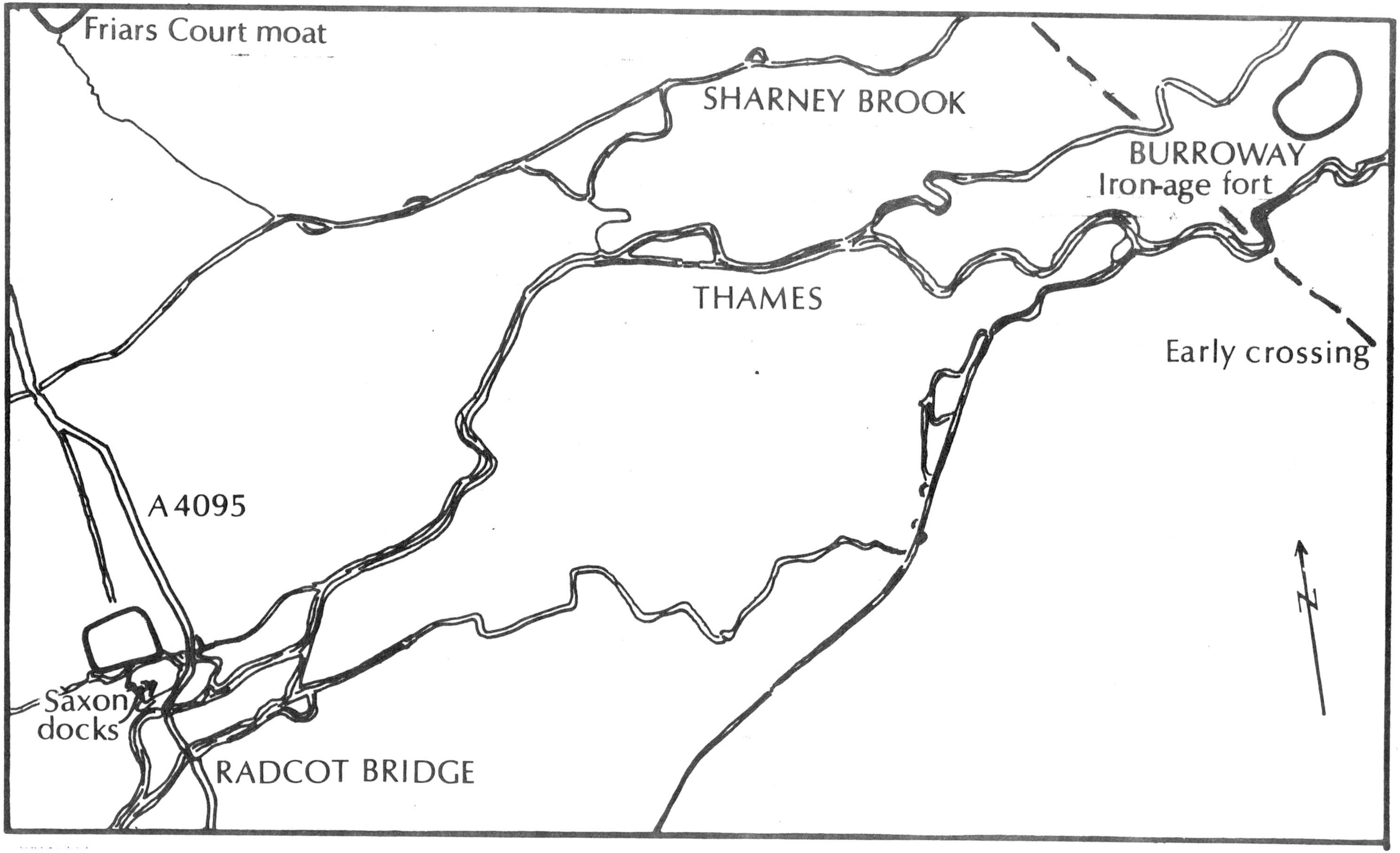

Fig. 22.2 Two possibly important areas for boat-finds on the River Thames above Oxford. Scale 1:10,000.

THE WHERRY, SKIFF OR ROWING-BOAT

The most common small craft on the Thames at London was the wherry, as familiar and as freely used as the taxi today. The word, here applied to a light rowing boat, is of obscure origin and is first recorded in A.D. 1443. The later use of 'wherry' to mean a large barge-like vessel has survived on the Norfolk broads.

A clear early representation of about A.D. 1500 is given in the well known view of the Tower drawn by a Flemish artist in England for the Poems of Charles Duke of Orleans. This shows two boats with high bow and stern, each with a group of passengers in the stern and two oarsmen in the bows pulling across the river to the Custom House.

It is clear that these craft changed considerably in about A.D. 1535. An Act of 1555 refers to 'boats and wherries' of 'the old substantial sort, and sure making of boats which was used before the space of twenty years then past'. The new kind of boat was 'so little and small in proportion and so straight and narrow in the bottom' as to be dangerous. The Act prescribed that wherries should be $22\frac{1}{2}$ ft. in length and $4\frac{1}{2}$ ft. in breadth.

Views of London in the mid 16th century show wherries both with high bow and stern, presumably the solid older sort, and with a low profile, presumably the new handy pattern.

This development is well confirmed by both documents and illustrations. The wherry's later changes in its development into the modern rowing-boat need both more reseach and actual examples to study. The earliest surviving examples, the deep bluff-prowed Gravesend Wherry of 1888, the pleasure-skiff of 1905 (both at Greenwich) and the Oxford racing eight of 1829 (in the Science Museum) differ greatly.

The Wherries shown in their thousands on 17th and 18th century illustrations were double-ended and lacked rudders; two oarsmen and up to six passengers are generally shown, and occasionally a small sprit-sail. The example of a 'Thames wherry' whose lines were published in 1768 was 26 ft. long, rather larger than the 1555 standard. They seem to have been fitted with rudders on a vestigial transom only after about A.D. 1860. By this time the expansion of London and the building of many bridges across the river meant that they were no longer the universal way of getting about the town. For this the Hansom Cab had supplanted the wherry. They had become family pleasure-boats and the old skills of steering them with the oars well up towards the bows was no longer practised.

The Peter-boat used for fishing in the estuary and upriver to Brentford in the early 19th century presents problems. This distinctive broad thwartless craft is very uncommon on 17th and 18th century illustrations, but a few can be found. They bear a marked similarity to the Dutch 'Kub-boot' from the north of the Zuider-See and to the 'Aaljolle' of the Cranz am Elbe near Hamburg, an area reclaimed and colonised by Dutch immigrants in the 15th and 16th centuries. How far the Peter-boat is derived from the wherry, and how far it is an actual Dutch type brought over to work on reclamation of the Thames is obscure.

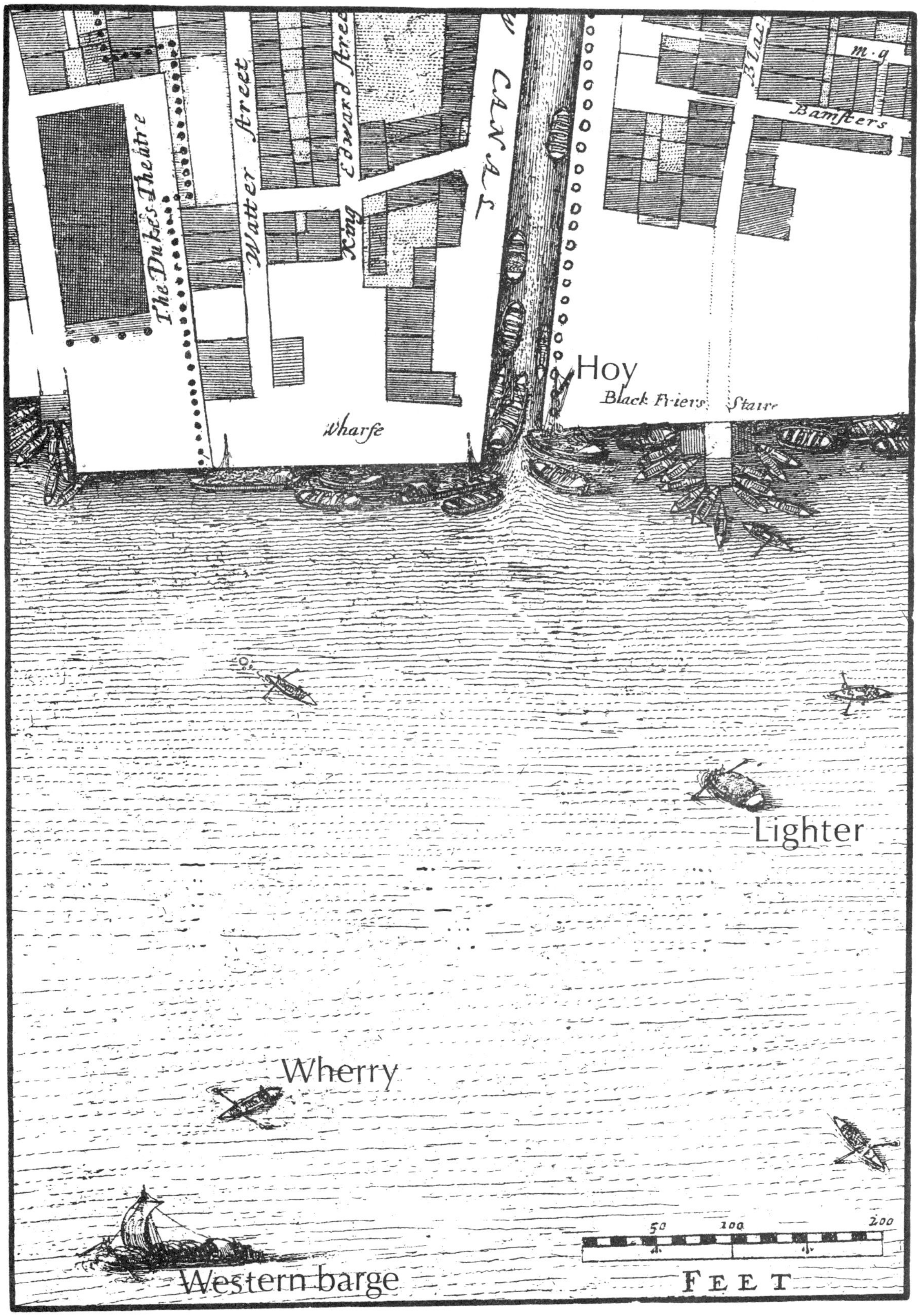

Fig. 22.3 Boats at the mouth of the Fleet River, London in 1676. A detail from Ogilby and Morgan's map. Hoys carried passengers and cargo in the outer waters of the Thames Estuary. Lighters carried cargo, and Wherries passengers, on the Thames at London. Western barges carried cargo up-river. The Hoys measured 55 to 60 ft., Lighters 40 to 50 ft., Wherries (much exaggerated in the foreground) not more than 30 ft, and Western barges 75 to 100 ft.

THE LIGHTER

As characteristic as the wherry, which was to be found all along the river, the lighter has always been restricted to London waters. The origin of the lighter is obscure. The word first occurs in A.D. 1372 and means, obviously, a vessel into which a ship is unloaded or lightened. Most vessels of this period were small enough to lie on the mud against a riverside wharf, and it seems unlikely that lighters developed their form and way of use until the great expansion in the size of ships in the late 15th century. It also seems unlikely that lighters were numerous until the early or mid 17th century. Wyngaerde's view of c. 1550 does not show any recogniseable lighters and Braun and Hoefnagle in c. 1560 show only a few possible examples lying singly at buoys. Their wherries are much oversized and may perhaps be confused with lighters. Possibly too, lighters of the late 16th century were no more than 30ft long, scarcely larger although rather deeper than wherries.

In the late 17th century, however, the lighter can be clearly distinguished from the wherry. The early lighters of this period were quite unlike the swim-headed slab-sided lighter of today. Many representatives show them as double-ended boat-shaped craft between 40 and 50 feet long, with a well marked sheer rising to fairly blunt stems. They were without mast or rudder, worked from either end with sweeps and must have handled very differently from modern lighters. With their rounded cross-section and virtual absence of keel, to judge from the solitary example found by Paul's Stairs (Marsden, 1971), they had actually to be rowed across the stream (Fig. 22.3). The Victorian lighterman could make a crude 'aerofoil' with the side of his lighter and his sweep and 'lift' his much heavier vessel across the tide. For progress along the river both sorts of lighter depended on waiting for a tide, until tug-drawn trains of lighters became commonplace in recent times.

The change from small boat-shaped lighters to larger punt-form ones seems to have occurred in the 1820s, about a generation after a similar change from cargo-punts had happened in river-barges. This makes it difficult to accept the conventional explanation that the Thames sailing-barge in its early swim-headed form developed from the swim-headed lighter. Both appear in engravings and pictures at about the same time, and the origin of the Thames barge needs more investigation.

The Humber keel, apart from being rigged, bears a considerable resemblance to the early Thames lighter.

THE RIVER BARGE

The Thames river-barge of the 16th, 17th and 18th centuries is a distinctive type, a large shallow vessel 75 to 100 feet long with little rise to bow and stern, a long shallow rudder, a single central mast for towing or a square sail, and a cabin of hooped canvas in the stern. They are shown under every possible form of propulsion. In the 1560s Braun and Hogenberg's view (1572) has two running downstream past Westminster under sail, one with four oars out. Later views show them being poled or towed by strings of men or horses. Ten or twelve horses were needed to pull upstream and seven downstream in an account of 1775. They had a crew of three to six men and a larger cargo capacity than many seagoing vessels of the period. In 1775 vessels of 140 tons were said to be able to go up to Wallingford and 70-tonners to Lechlade.

Writing in about A.D. 1700-1720 Defoe says of Reading: 'Their chief trade is by water-navigation to and from London...., particularly coals, salt, grocery wares, tobacco, oyls, and all heavy good. They send from hence to London by these barges, very great quantities of malt, and meal, and these are the two principal articles of their loadings, of which, so large are those barges that some of them, as I was told, bring a thousand or twelve hundred quarters of malt at a time, which according to the ordinary computation of tonnage in the freight of other vessels, is from a hundred, to a hundred and twenty ton, dead weight'.

Of London he says 'the vessels which bring this malt and meal to Queen Hith, are worth the observation of any stranger that understands such things. They are remarkable for the length of the vessel, and the burden they carry, and yet the little water they draw; in a word, some of those barges carry above a thousand quarter of malt at a time, and yet do not draw two foot of water. Some of these large barges come as far as from Abingdon.'

The city water-baliff estimated in 1746 'that 300 Barges pass and repass at least 6times in a year'. Many of their cargoes must have been very small, since only 56,000 tons of cargo was recorded as passing Marlow in 1767. This could have been carried on only a third the number of voyages estimated. However all these Georgian statistics seem thoroughly vague and muddled, and Defoe's capacity-measures seem to give double his tonnage. The river-barge of this type, whose origins may go back to medieval or even Anglo-Saxon times, disappeared in the late 18th century. For a time in about A.D. 1770-1800 pictures show the river-barges with much greater sheer and then from about A.D. 1800 big swim-headed cargo-punts replace boat-shaped barges completely on the upper Thames but the older form remained on they Wey. This rapid development of a long-established type may have been caused by the Act of 1770 which established the Thames Commissioners, who soon started building pound-locks. Before this there were no locks for 50 miles upstream to Maidenhead, above which there were many flash-locks, or removeable weirs. Barges going upstream had to be hauled or winched over them, and downstream they had to 'shoot' the weir. Only near Oxford was there a group of pound-locks built in the early 17th century.

These river-barges were very like the early Severn trows and generally similar to early Norfolk keels.

THE PUNT

Above London the small punt 6-7 m long has probably always been the characteristic working-craft of watermen, millers and fishermen. A number are shown on Loggan's view of Oxford in 1675, but the type is likely to be much earlier; a small punt of the 11th century A.D. has been found in Denmark. Before the late 19th century they were workaday vessels. The Victorians took them up for racing and for family boating and hundreds are still in use today.

Larger punts of 10 to 18 m long are also shown on the 1675 view. They are still quite commonly used for riverside maintenance-work and, in all kinds of widths, for ferries. One of the very few views of a punt in London is the Horseferry at Lambeth shown, also in 1675, by Hollar in his Prospect of London and Westminster. This has a coach and pair abroad and seems to be worked by a pair of sweeps at the forward end.

From this period punts were carried about on waggons in military 'pontoon-trains' and frequently used for improvised river-crossings. A fine view of a practise-crossing in July and August 1780 at Gravesend shows strings of them, with drawbridges for disembarking being hauled across the Thames on chains, rather belated practice for the American war.

Fairly soon after this large-scale demonstration of the value of the punt, and a repetition of the exercise in 1799, the river-barges described above were replaced by very large cargo-punts up to 30 m in length. From about A.D. 1800 until railways killed the river-traffic in the mid 19th century these cargo-punts were in general use on the Thames, and are shown in many illustrations. Within a generation of their introduction on the upper river similar punt-like forms were adopted both for lighters in London waters and for sailing-barges in the Thames estuary. It would be attractive to suggest that these developments were stimulated by the military pontoons of 1780 or by the large cargo-punts on the river, but the lines of the 'Chalk barge' published in 1788 shows a tendency towards the form.

The most likely explanation is that the ever-growing demands for heavy transport in London led to the simplest design and construction for river-barges, lighters and sailing barges alike.

The largest size of Thames cargo-punt has lately been matched by the group of Roman stone-carrying punts found in Holland, and the origin and early history of the punt is of the greatest interest and merits further research.

NOTE

Illustrations of boats in the Thames near Oxford may be found in the Card-Index of local views in Bodleian Library; Malchair sketch-books of c. 1770-80 in Western Art Drawings collection, Ashmolean Museum; Taunt photographs and other collections in Oxford City Library.

Illustrations of boats in the Thames near London may be found in the GLC Map and Print Collection, County Hall, and the Guildhall Library Print Collection. Most boroughs along the river have useful views in their reference libraries.

REFERENCES

Braun, G. and Mogenberg, F. Civitates Orbis Terrarum (1572) 3, facsimile by GLC is widely available.

Griffiths, R. An Essay to prove that the Jurisdiction and Conservacy of the River Thames &c..... (1746) 159.

Loggan, D. Oxonia Depicta (1675).

Marsden, P. "A Seventeenth-century Boat found in London", Post-Medieval Archaeology 5 (1971) 88-98.

Ogilby J., and Morgan W. A Large and Accurate Map of the City of London scale 1: 1200 (1676), facsimile edited by C. Welch for London and Middlesex Archaeological Society (1894).

DISCUSSION

de Weerd: An illustrated preliminary report on the three dugouts, the three barges and the rudder from Zwammerdam will be published soon. It will include a discussion on Roman period ship construction along the Rhine. The reference is: M.D. de Weerd (forthcoming) 'Romerzeitliche Transportschiffe und Einbaume aus Nigrum Pullum/Zwammerdam, Provinz Zuidholland, Nierderlande' Akta X. Limeskongresses (1974) Xanten/Nijmegen.

Arnold: In the different classifications proposed, we have too much tendency to examine only the planks and occasionally the frames, but these represent only part of the boat. We should also consider the caulking, the mast step and other fittings for sailing, and other details. If the size of the boats can vary in proportion to the intensity of trade in an area, the detailed techniques in the shipbuilding may be similarly influenced. It is by the study of these small details (but are they really details?) that we established an original hypothesis for the construction of the Bevaix boat (Arnold, 1974; 139 Fig. 5). We may compare the complex caulking of the North Ferriby boats (lashing-moss-wooden lath-lashing fastening the planks together) with those of Bevaix or Yverdon (rope-moss-wooden lath-small nails).

Ellmers: I originated the term 'Celtic boat' and I did so to show that the Celts had their own boats which were not in any way related to Mediterranean or to Scandinavian boats. (Jahrbuch des Romisch - Germanischen Zentralmuseums Mainz (1966), 16: 73-122). But perhaps we should name this different style of boatbuilding "Continental"?

Wilson: The word Celtic describes languages, there is no other possible way it can be used, it should not be used to describe these boats. Lets try to use a name like 'Rhenish' or 'Western European Rhenish'.

McGrail: Why not call them Ellmers type 1 or Arnold's type 2 and so on?

Marsden: I suggested in my paper that it ought to be 'celtic' with a small 'c', but 'continental' might be appropriate.

de Weerd: The Zwammerdam boats, and perhaps those in Switzerland too, are Roman provincial barges, and this may be the term to use.

Upham: These boats operated in a tide, whereas the Romans had more or less tideless seas, a very small range of tides. These so called celtic boats are nearly all flat bottomed and they had to beach. Is the word 'tide' relevant in your type name?

Wright: We might think of 'Gallo-Roman' or 'native Gallic' for these boatbuilding traditions.

Wilson: We do not know that these boats have their origin in a Gallo-Roman or Celto-Gallo-Roman or Anglo-Roman or Brito-Roman area. We do know that they are generally found in rivers, and that they may have something to do with tides. I think though the best rule in these circumstances is to use a place name for a type, the Zwammerdam type or the Blackfriars type.

Crumlin-Pedersen: There are now sufficient boat finds throughout north west Europe to permit specialists to investigate in some detail the shipwrightery and the basic boatbuilding traditions lying behind the individual finds. This will bring us to ask whether these are developments within a particular area. We should record details of the boatbuilder's trade, the way he tackles constructional problems, and the supply of timber available to him; and these features, together with more archaeological facts, will enable us to place the find in a boatbuilding tradition. The overall picture seems to me to be one of a number of different lines of boatbuilding existing at the same time. At a given time certain of these basic lines are developed into the larger ship of the period. But this doesn't mean that all the other lines disappear; they remain as small boats, and one hundred years later another of these is brought forward because economical, political, and technical possibilities encourage building in that construction.

Ellmers: Christensen's classification considers some important contructional details and takes these as guide lines to separate boats into seven or eight groups.

Christensen: Small details are also extremely important and especially for people who go on apprenticeship and learn by copying what the master does. You may change the shape and size of the ships but the builder will stick to his details of construction. The finger print of the boatwright is in the small details of boatbuilding and these must be studied. We now know some of the repetoire used by ancient boatbuilders in Europe. We don't know all, but I am perfectly sure that new finds may be very broadly lumped into the old categories. If we look carefully at each new find we will be able to refine our groups and make more groups.

Marsden: There are two main aspects in any ship or boat, one is the form and the other is the detailed construction. And although form may change according to the environment a boat works in and the uses to which it is put, the details, the nail fastenings, and the caulking methods, tend to live on into the more recent types. The caulking is generally moss on the Rhine barges; at Guy's Hospital and Blackfriars the caulking is hazel twigs. The size of the Guy's Hospital craft suggests it was a river craft, and therefore it must have been built on the Thames or the estuary round here, it can't be from the continent. Blackfriars is a sea going vessel, with teredo borings in the planking. But it may be that although we have two craft of very different shape the fine details of hazel twig caulking suggest that Blackfriars may have also been built on the Thames or in south east England. The absence of hazel twigs on the Rhine land craft could mean that these were not built in south east England.

Robb: You have a problem rather similar to that when alchemy moved to chemistry and I suggest that you take the characteristic features you have used to describe the ships, whether it is the caulking, or the way that the planks were joined together or whatever, and describe these boats using a vocabulary with some kind of shorthand such as the initial letter as was used in chemistry, C for carbon O for oxygen and so on, so that you get a coded description. Once you put things into classes such as Celtic boats you begin to suggest then that there is a neccessary connection between them, but there is not neccessarily a technical connection between them, simply the third connection.

Coates: Sea going boats move about and they all need mending so it is quite likely that a boat which was built at A needs to be mended at B quite a long way away, by a man who served his apprenticeship in B but perhaps had also been to C. Ships also have to be re-caulked. There must be a great mixing of classifications within one boat, so a Danish built ship might well end up its life with Rhineland caulking and somebody else's waterline planking, so when the boat finally comes to rest and is dug up by the archaeologist what is its classification?

McGrail: I recommend quantitative methods to Peter Marsden and Arne Emil Christensen. If they record everything on a find, and providing they have a sufficiently large and reliable data base, they will then be able to search for the optimum classification scheme, either by hand or by the dreaded computer.

Christensen mentioned the logboat from Valaar Moor which Crumlin-Pedersen has said was expanded. (in Hasslöf, O. (ed.) 1972, Ships, and Shipyards, Sailors, and Fishermen: 208-234). I am not absolutely convinced by his demonstration. Are there any ethnographic parallels for oak logboats being expanded?

Christensen: This has not been researched at all. We know that eastern Europe uses aspen, poplar and occasionally spruce for expanded dugouts, and the western coast of the United States uses red wood. The Valaar Moor boat seems to be an expanded dugout, the bottom plank of the Hjortspring boat may also be an expanded dugout. These should be investigated further.

Crumlin-Pedersen: I recommend close study of the relations between the Graveney boat and the medieval shipwrecks found at Blackfriars and Customs House. It will be worthwhile going into shipwrightery here to see whether you have features indicating a local Thames tradition during this period. A number of specific features we see in the Graveney Boat; some sections in the frames, the way that the hood ends come up to the stem and stern, and the keel section, are also found in the medieval vessel found at Blackfriars (Marsden, P., 1972, 'Blackfriars Wreck 3'. International Journal of Nautical Archaeology, 1: 130-132.)

I think the interaction between navigation, settlement and fortification is very important, and these factors offer a clue to understanding many of the more general problems in archaeology.

Howard: There are two very promising sites that are in possible danger and I think should be watched. One is the old port at Hedon about six miles from Hull, where one of the havens has already been built over. That was silted up somewhere about 1600 and there must be quite a lot of interesting stuff in the bottom there. The other one is the old harbour of Castle Rising in Norfolk. The village with its market square is now largely grass; it is on the slope of a hill overlooking a flat place which in the 1500s was still washed by the sea, according to a statement in the church that the waves broke the church windows. I think we should watch these sites before the builders get at them.

23. SUMMARY AND CONCLUDING REMARKS

Basil Greenhill

I see that I am down on the programme to give a brief summary of the proceedings of this Symposium to be followed by further discussion. Now this assignment, it seems to me, is almost impossible and I attempt it only under the protection of my carefully guarded amateur status. You may wonder why the summary should come now when there are further sessions tomorrow. I think that those responsible for drafting this programme were quite right to put the summary in here because it is clear when you look at the programme that it was designed to bring out certain themes and I shall enumerate them in a moment as I see them. In the first three days of discussion these themes were brought out and on the fourth day, having built the vessel as it were, as a very special reward we get the carved and gilded work, the presentation in one form or another for the first time in this country of one of the most important ancient boat finds ever, the Cheops boat.

Now I think personally that on the whole the programme has worked out pretty satisfactorily, and that several important points have emerged in the discussions. One thing that is clear is the implicit confirmation of a definition of boat archaeology which we have tended to adopt at this Museum - and let's not get involved at this stage in arguments about the definition of boat or ship in this context, but use the term boat archaeology without commitment. Now in setting up the Archaeological Research Centre (ARC) we had to define our objectives, and we did this in broad terms, as the study of boats before the technical revolution of the 15th century. This apparently widespread revolution appears to have taken place at a fantastic speed and to have had a profound effect upon the history of man. We have for our purposes decided to treat the study of this revolution as a separate subject, and to treat the subsequent development of the ship as an historical matter, available information of which is increasingly reinforced by evidence of an archaeological nature largely as a result of the work of divers. This topic then merges into maritime industrial archaeology with our work on the ketch, Shamrock at Cotehele in the west country, which has close parallels with the work being done at the Danish National Museum on their galleas now rebuilt and at sea this summer. The discussions in this symposium have, I think, implicitly tended to confirm this broad brush kind of definition of the field of interest we cover.

The pattern of discussion which has developed has been one of increasing seriousness and depth and I would like to thank the speakers who have taken such immense trouble to present their fields of interest to us so clearly and specifically, speaking in my language whatever their language of origin. Speaking again purely personally I found the addresses of Dr. Levy and Dr. de Jong gave a new dimension to a subject with which I had been previously unfamiliar, and that Professor Filgueiras's fascinating presentation opened a whole new field of detailed study of a group of European boat types. I have no doubt at all that each of you has received similar fresh stimulus through one or more of these formal presentations and from the interventions in the often intensive discussion which has taken place.

Now I think at this stage I might take the occasion of my reference to Professor Filgueiras to draw out what seems to me to be the first of the main strands which I have detected as coming out of the complex intellectual spinning process that we have all been going through in the last three days. This has been reinforced particularly by interventions from Arne Emil Christensen and Eric McKee. It is that we should do all we can to encourage and further, while there is still time, the collection of the most detailed records that circumstances permit - and this can be anything from a few photographs to a thesis study - of surviving traditional boat types. I should say "so called" traditional boat types because one aspect of this study must be the degree to which they are traditional. I think it has emerged clearly also that we must treat this material not as evidence of what was done in the past but as evidence of what was possible or might have been possible within the framework of limited technologies at different periods in different places, and to provide us also with possible solutions which could from time to time have been adopted in the past. At Greenwich we have taken this particular message very much to heart and at a meeting as recently as this lunch time, we have begun to lay down the framework of a policy for the acquisition of ethnographic boat material in this Museum, there having been a recent increase in accomodation which makes this more possible than it has been in the past.

The second theme I think is that there seems to be a general feeling that there is a real necessity for some form of glossary in the English language which might be the beginnings of a generally agreed terminology to use in this new and developing study of the archaeology of boats. Now I sense the feeling, and no doubt the discussion after this summary will correct me if I am wrong, that the time is ripe for some sort of initiative. If this is so it is possible that you may care to consider requesting the Trustees of the National Maritime Museum to consider ways in which this Museum can move towards the preparation and publication of a first illustrated glossary of terms for use by boat archaeologists. I have chosen my words carefully so that the commitment if it is asked for, and if undertaken, can be as broad as possible. I have reason to believe that should this Symposium feel such a course of a action desirable, the Trustees may well feel that this is an appropriate further addition to the work we are already undertaking in the archaeological field.

The third strand emerges less clearly, but it is nevertheless there and it comes out particularly in the presentations yesterday by Mr. Graham and by Sean McGrail, the Noah of our ARC, if I can return the happy compliment he paid me yesterday. This strand centres round the controversial matter of quantitative methods and the establishment of internationally agreed attributes for recording boat finds and their analysis, possibly with electronic aids. Now here we are getting into pretty deep water. I said at the beginning of this Symposium, and also at the beginning of this address, that I have been very careful to maintain my amateur status. Perhpas now I can take the liberty as a non-professional to say that, looking in from the outside, archaeology seems, perhaps even more than any other forms of human activity, to depend for the considerable resources of nervous energy required of those who participate in it successfully, upon intense personal involvement and this involvement means that even more than most people, archaeologist, professional and fringe amateurs, do not wish the annihilation of being merged into Peer Gynt's button moulder's crucible of boiling brass. And that perhaps is how some people, - and I deliberately chose the Scandinavian simile, - very understandably see a computer. Nevertheless for the long term development of the study it seems to me that there must emerge a quantitative branch with agreed international standards and agreed selected data. But I think this will be a very slow growth. From Sean McGrail's work there may emerge the possible basis of a British list of attributes for recording logboats. It is likely that there will be merit perhaps in the extension of this in two directions slowly. One, for it to be the basis of an international system of logboat recording; and second for a gradual extension of the system into plank built boats. This must grow of its own accord spontaneously. These systems may well be capable of future computerisation as a natural development, happening reasonably painlessly but in parallel with the continuation of the complex of more traditional methods which have been summarised as "intuitive" in these discussions. As the international store of data grows so it may well transform itself into something which Jonathan Cutbill may be prepared to accept as amenable for development as a bank, but the growth must be slow and gentle and natural, and it cannot overload people whose existing commitments are fundamental to the development of the study. Now here I must make a further qualifying point: however much we struggle, the button moulder's crucible waits for us all and inevitably the study is going to become more scientific. The process really began with the introduction of naval architects into the field, and Ole Crumlin-Pedersen himself was a great pioneer. Then came Eric McKee with the engineer's point of view. And you will have realised from the papers presented by him and John Coates under the chairmanship of Ewan Corlett, how much such specialists can contribute to our understanding of ancient boats and their capabilities. The presence in this room of such a formidable combination as Corlett, Coates and McKee indicates our intentions at Greenwich. But let us make one point clear, as far as we are concerned, the archaeologist must remain in charge, naval architects, engineers, and similar specialists are advisers, and the archaeological staff will move more and more into the general professional field in the broader sense. The key word is archaeology, and professionalism in archaeology is paramount.

Now the last of these main strands as I see them, emerged particularly clearly in the discussion after the two papers on recording finds and presenting information. Does the symposium feel that there is a case for the publication of a simple guide, an elementary first-aid handbook beginning with some such phrase as "if humanly possible do not disturb the casualty", for those who become involved with an ancient boat find? I am told that the British Museum of Natural History has a publication entitled What to do with a stranded Whale. Now if this is so we might look at this with advantage, perhaps the parallel might be closer than we think. But to be serious do we feel that there is an occasion for such a publication and that it would serve a useful purpose ? What I have in mind would be strictly for those who do not know the sharp end from the blunt, boat archaeology speaking.

I think it has also very clearly emerged from the discussion that we can all of us only approximate, only do the best possible in the circumstances with each individual find. It is the function of a professional institution such as those which have emerged or are emerging in Scandinavia, Germany and elsewhere, and in this country, to set standards, to develop disciplines and to develop methods. Limited and inadequate as we consider our resources still to be, and I am sure our colleagues have the same feeling about theirs, most people faced with boat archaeological finds will do so with even more limited resources. Compromise will continue to be the order of the day, but it should be a compromise of developing international professionalism and steadily developing standards to which people will strive to approximate.

Now a suggestion - it isn't a theme, it is a suggestion - which has been made more than once in our deliberations is that the Scandinavian working group or something like it might be revived or a parallel started elsewhere, and I would like to add my tribute to those who have praised the work of this group and pointed out that their publication in English "Ships and Shipyards, Sailors and Fishermen" is required reading for anybody who gets involved in this subject. We in this Museum have something similar going already, which was actually inspired by the Scandinavian example. While archaeological work is going forward in the ARC, Eric McKee does his ethnographic recording, if we can use the term; Campbell MacMurray is beavering away on the sociological themes from oral sources; Ian Merry and others work with us on regional maritime history; and other people are working for us on ship registration material.

In conclusion, I think that this Symposium has been very important for a number of reasons, and the contibutors have illuminated the various branches of the study. The symposium has clearly revealed the cross connections and the ways in which the various interlocked disciplines are developing towards a new branch of archaeology. Perhaps this Symposium marks our coming of age in this country. It is, after all, 40 years since the great landmark of Ted Wright's stroll along the temporarily revealed hard on the north shore of the Humber, and about the same time since the first opening of the grave at Sutton Hoo, and twelve years or more since the great pioneer struggles of Peter Marsden in the mud at Blackfriars with the contractors breathing down his neck. I think the study of the archaeology of the boat has now arrived internationally and the level of discussion here has been such that the published proceedings of this symposium is likely to stand as a useful basic text book for quite some time.

I hope I have fairly summarised some of the main themes of our deliberations during the past three intensive days. There are no doubt other themes and strands, some of them very important to which people will wish to draw attention in the discussion in the next few minutes and these will be duly recorded by the invisible bugs with which this Lecture Theatre is infested. The discussion will be conducted by the Chairman of this session and since I claim no merit for this summary, except that it may help to focus this final session, I now draw my carefully preserved cloak of amateur status around me and return to the anonymity of the main body of the hall chanting as I do so, publish, publish, publish, for God's sake publish, because that's what it is all about.

DISCUSSION

Wilson: I believe that the first theme mentioned by Mr. Greenhill that is to encourage the recording and collection of traditional boats and so on, is obviously a national concern for anybody who is represented here. I think we can all agree with that and press on with the necessary action. Now can I have comments on the proposal for a glossary in English ?

Wright: There is a clear need for this glossary with illustrations. I believe that a development of Eric McKee's pioneer work the Glossary in his Clenched lap or Clinker with some illustrations would be an extremely handy tool for anybody working on this subject in the English language.

Ellmers: We are excavating so many things for which there is no real expression in the available dictionaries that we need a glossary. I think that it would be a good thing to have a central address where we could send reports on features which occur and which are not in common dictionaries.

Baker: Having worked with some maritime historians and been greatly disturbed by their misuse of the English language, I propose to the future editor that such a glossary be made for their use and that it be kept simple.

Corlett: The glossary should use standard terms, whenever possible, I have heard the word 'ribs' used frequently, but no naval architect or ship-builder normally uses the word 'ribs'; he calls them frames. We don't want two words, and if you are going to make a scientific naval architectural analysis it is much better to have proper naval architectural terms. Where there are items for which there is no modern equivalent then obviously a new term has to be devised.

Christensen: In Scandinavia, if you go to a carvel builder's yard and ask him he will use the word "spant" for rib, but if you go to his neighbour the clinker builder and ask him he will use the word "band" for the same fitting, and they are both very positive that the word is right for their particular boats. Our problem is that we run into so many things where no proper naval architectural term exists. Most of our work is with non-carvel ships, and the naval architectural terminology is for the standard carvel ship.

Wilson: I think that if this is to be done it is to be done for wooden boats, say before 1500, and a terminology developed on the basis of what has been found. This is what is most needed by archaeologists.

The third question is the problem of quantitative methods, agreed international standards, and agreed methods of computing. Information retrieval is the big problem facing all archaeological and museum work and it is a thing which has obviously got to be thought out in great detail. Can we usefully here as a body go any further than we have, other than perhaps to give it our general blessing ?

de Weerd: Information retrieval is not so useful without a standard glossary.

Wilson: What do you feel about the fourth idea of a guide for those finding boats ?

Coles: Such a guide should not be devised for those people here, it should be for the other 9,500 archaeologists in this country who whilst digging may find a boat, and may think it is a wooden box or any old thing. This guide would have to be in layman's language and would have to be widely publicised, not in the nautical journals but in the journals that all archaeologists read. Such a thing would be of very great value to all archaeological units in this country.

Wilson: Could I suggest in that case that possibly the best compilers of this are the National Maritime Museum and that they should then get together with the Council of British Archaeology to distribute it.

McGrail: We have told regional units about the research facilities available at Greenwich and of our own research interest, and we have offered them some general advice emphasising that a boat find is a problem, there are difficulties in recording and conserving it, and perhaps they should seek advice. There may well be a case though for more detailed advice being published.

Wilson: In the field the general archaeologist may not have time to call in the specialist and I think it is for this particular situation that this booklet is needed. On the other hand contractors and farmers and so on are not the people we want to give this handbook. It is the job of the Council for British Archaeology and its various local groups to tell such people to report anything they find to the local museum.

van de Merwe: The Wreck Act protects scheduled wreck sites under water, but boats found above High Water cannot be protected under this act; nor can they be protected under the Ancient Monuments legislation because they are not considered monuments, they are considered to be chattels, movable possessions.

Croome: We want the Ancient Monuments legislation to include boats and, as it were, wrecks on land.

Evans: After the discovery of the Graveney Boat we were given a time limit by the Kent Water Authority before they flooded the dyke. Dr. Rupert Bruce Mitford took some rather weak archaeological evidence to the Ancient Monuments Board in an attempt to schedule the land underneath the boat as a harbour, to be used in a last ditch stand to protect the boat, if necessary.

Marstrander: During the discussions here several members have rightly emphasised the necessity for international co-operation in important fields of research. We seem to be generally agreed that the recording and thorough investigation of dugouts in the north and west European area ought to follow common lines to reach satisfactory results. In this connection it was also proposed to establish an international data bank. There were also other important proposals including a glossary. There seems to be a case for establishing an international maritime committee which could co-ordinate and organise this sort of research, and which could maintain continuity between meetings, and have the responsibility for planning and arranging future symposia.

Editorial Note. After a thorough discussion of Professor Marstander's proposal it was agreed to form an *ad hoc* working party of 12 members to plan the formation of a body to be known as the International Symposia on Boat and Ship Archaeology. Details of this organisation may be obtained from the Archaeological Research Centre, National Maritime Museum, Greenwich, London SE10 9NF.

GLOSSARY

This Glossary is reprinted from Archaeology of the Boat by Basil Greenhill. A. & C. Black, 1976.

Boatbuilding terms are not standardised in Britain: Essex will interpret a word differently from Wessex. In addition, the same parts of a boat or the same boatbuilding technique may be known by different names in different counties. This glossary is based on the glossaries in Eric McKee's Clenched Lap or Clinker, and Sean McGrail's Building and Trials of the Replica of an Ancient Boat.

baulk a tree trunk which has been roughly squared.

bevel a surface which has been angled to make a fit with another.

blind fastening one in which the point of the nail does not protrude through the timber.

bottom boards lengths of timber fastened together and laid over the bottom of a boat as flooring.

brail rope used to bundle a fore-and-aft sail rapidly. It may be rigged singly or with others, generally on boomless sails that have their peaks extended by a sprit or a standing gaff, which are not readily lowered.

breasthook, stemlock, fore hook. In wooden ships a stout knee fitted internally across the bows, holding the sides together In boats this knee is at gunwale level.

carvel build there are several definitions in use; in particular, this term is sometimes taken to be synonymous with skeleton build with flush-laid strakes. The term is confused and ill defined.

caulk to insert material between two members after they have been assembled, and thus make the junction watertight (see luting).

chine beam longitudinal strength member fitted at the turn of the bilge and into which the planking is rabbeted and fastened above and below. If an angle is formed, this is known as a hard chine, and is usual in 'V'-bottom designs.

clench, rivet To deform the end of a fastening so that it will not draw out - usually done over a rove.

clinker built a form of boatbuilding in which the strakes are placed so that they partly overlap one another - usually upper strake outboard of lower strake, but occasionally the reverse arrangement is found.

construction plan a scale drawing of a boat with a longitudinal section, horizontal plan, and several transverse sections. The position and nature of the scarfs, and other important constructional details and scanlings, may also be given.

cove, scotia. A hollow shaped moulding. May be used to hold the luting between two strakes.

cramp, clamp, gripe. A device for holding elements of a boat together temporarily.

crook a curved piece of wood which has grown into a shape useful for boatbuilding.

dagger plate a metal retractable device on the centreline for combating leeway.

dolly a metal billet held against the head of a boat nail whilst it is being clenched.

double-ended a boat which is (nearly) symmetrical about the transverse axis - pointed at both ends. It would not be normal usage to describe a pram as double-ended.

drop-keel, centre board or centre plate. A wood or metal retractable device on the centre-line for combating leeway.

fair (n) a line is fair when it passes through its guide marks without any abrupt changes in direction. The lines of a boat are fair when the level lines, half breadths, buttock lines and diagonals, being themselves fair, all correspond.

fair (v) to render a set of lines eyesweet and mutually true on a ship's draught.

feather-edge tapering to nothing.

floor a transverse member - often a crook - extending from turn of bilge to turn of bilge, and set against the planking.

frame a transverse member made up of more than one piece of timber, usually extending from sheer to sheer and set against the planking (see rib and timber).

futtock any of the timbers used to make up a frame in a wooden ship. More exactly, not the floor or top timbers.

garboard the strake next to the keel or keel plank.

grommet strand(s) of rope layed up in the form of a ring.

gunwale a longitudinal member fitted round the inside top edge of the sheer strake of an open boat. In open boats, mainly positional - the timber or arrangement of timbers that make up the sheer line.

hog a longitudinal strength member of a boat fastened to the top of the keel to provide a landing for the inner edge of the garboard between the forward and aft deadwoods. Also a form of distortion which causes the ends of a boat to drop.

horn to check the squareness of a mould, relative to the boat's centre-line.

joggle to cut out a notch in a piece of timber so that it will fit close against another member.

keel the main longitudinal strength member, scarfed to the stempost forward and the post aft.

knee a naturally grown crook used as a bracket between two members set at about right angles to each other or more.

land that part of a strake which is overlapped by the strake immediately above it.

lay-off to draw out the lines of a boat full size.

lines the interrelation of sections in different planes which show the shape of a boat's hull. They usually consist of (a) sheer plan with longitudinal sections, (b) half breadth plan with waterlines or horizontal sections, (c) body plan with transverse sections. Diagonal lines, longitudinal section lines on the half-breadth plan, and waterlines on the sheer plan, enable the three plans to be related to each other and checked for fairness. Lines converted to numbers are known as a Table of Offsets.

loom that part or an oar inboard of the point of pivot; it includes the grip. The section of an oar between the loom and the blade is called the shaft.

luting traditionally, luting is a plastic substance such as paint used between two adjacent members. In this text the term is used to describe any material inserted between two members before they are assembled. (see caulk).

mast step fitting used to locate the heel of a mast.

moisture content the weight of water in a specimen of wood expressed as a percentage of the weight of oven dry-wood. Thus the figure can be greater than 100%.

moulds transverse wooden patterns taking their shape from the body plan. (see lines).

mould loft loft with a levelled floor on which the lines of a vessel are drawn out full size and faired.

moulding a pattern or linear decoration cut into a length of timber.

pay cover caulking in seams with a layer of hot pitch. Loosely but commonly used - to coat a ship's bottom with tar.

plank a component of a strake that is not all in one piece.

rabbet, rabet, rebate. A groove or channel worked in a member to accept another, without a lip being formed.

rays layers of parenchyma cells in horizontal strands running out from the centre of a tree towards the circumference.

rib a simple form of frame. This term may be more appropriate than frame, when applied to small open boats.

rove, roove. A washer-like piece of metal, which is forced over the point of a nail before it is clenched.

scarf, scarph, scarve. A tapered or wedge shaped joint between pieces of similar section at the join.

scrive or scrieve board the floor of the mould loft or (more frequently) a board erected in the open air for the same purpose.

shake a crack or split forming in wood, usually during drying or seasoning. Examples are: cup shakes - curved clefts between the growth rings; heart shakes - splits radiating from the centre of a tree.

sheer, sheer line. The curve of the upper edge of the hull.

sheer strake the top strake of planking.

shell construction a method of boatbuilding in which the shell (i.e. the watertight envelope of stems, keel and planking) is built or partly built <u>before</u> the ribs and other internal strengthening members are fitted (see skeleton construction).

shrouds ropes leading from the masthead to the sides of the boat to support the mast athwartships.

skeleton construction a method of boatbuilding in which a framework of stems, keel <u>and ribs</u> is first erected, or partly erected. This skeleton is then clothed in a 'skin' of planking. Skeleton and shell construction merge into one another with the use of 'intermediate methods' of uncertain age.

spall a light batten used during building to brace parts of the boat from floor, rafters or other strong point.

Spanish windlass a simple rope and rod device for forcing two elements closer together, and holding them there.

spile to transfer a curved line on to a pattern which, when laid flat, will give the shape to cut the lower edge of a plank

station the horizontal position of the transverse sections on the lines. They are used as datum lines when building from drawings, and moulds are generally made of transverse sections at some or all of these stations.

stealer a tapered plank or plate worked into the entry or run of a vessel to preserve the general lie of the strakes without making the ends too thick or thin.

Stockholm tar a blackish semi-liquid prepared by the destructive distillation of various trees of the <u>Pinaceae</u> family.

stocks set-up. The temporary wooden support on which a boat is built.

strake a single plank or combination of planks which stretches from one end of a boat to the other.

stretcher an athwartships length of timber against which a rower braces his feet.

template a shaped pattern of an element or section of a boat, made of plywood or hard-board etc.

thole a pin projecting upwards at sheer level to provide a pivot for an oar. Strictly thole-pin, but frequently used as above.

thwart a transverse member used as a seat.

timber an element of a frame or rib. May also be used generally referring to any piece of wood used in boat-building. One piece ribs or frames, especially those steamed or bent into place, are frequently called timbers.

trenail, treenail, trunnel. Wooden peg or through fastening used to join two members. It may be secured at each or either end by the insertion of a wedge.

The additional terms below are taken from The North Ferriby Boats by E.V. Wright. (National Maritime Museum Monograph No. 23).

bilge the curved surface between the bottom and the sides of a boat.

cleat the integral perforated blocks and ridges of varied shape left standing on the surface of the planks when they were hewn out of the solid timber. These 'cleats' could be used to secure the transverse bars or, if the interpretation is correct, the girth lashing. The same definition is commonly used for the similar features on the planks of Norse ships.

girth-lashing a binding of rope or similar material around the girth of a boat to hold the planks together.

hogging truss a heavy lengthwise rope attached (generally by passing right round the hull) at bow and stern and tightened up by one or more props amidships (three in the case of the ancient Egyptian examples). Its purpose is to prevent the hull of a boat from fracturing (hogging) when the midships section is supported on a wave and the bow and stern are unsupported.

lath a narrow strip of wood, used in the North Ferriby boats for capping the seams and preventing the luting (moss) from being washed out.

seam the juncture of two members required to be watertight.

strake generally a single plank or combination of planks which stretches from one end of a boat to the other. I use the term here only for the planks of the sides, not the bottom of the North Ferriby boats.

British Archaeological Reports

122 Banbury Road, Oxford OX2 7BP, England

List of Titles

(Cheques & money orders should be made payable to 'British Archaeological Reports' and sent to the above address. Prices include postage.)

B.A.R. 1, 1974, "Cuddesdon and Dorchester-on-Thames, Oxfordshire: two early Saxon 'princely' sites in Wessex", by Tania M. Dickinson. ISBN 0 904531 00 7. 54 pp., 4 figs., 4 plates. Price £0.75 post free.

B.A.R. 3, 1974, "A Corpus of Early Bronze Age Dagger Pommels from Great Britain and Ireland", by Ron Hardaker. ISBN 0 904531 02 3. 65 pp., 7 figs., 2 plates. Price £0.80 post free.

B.A.R. 4, 1974, "Coins and the Archaeologist", ed. John Casey and Richard Reece. ISBN 0 904531 06 6. 271 pp., 44 figs., 5 plates. Price £3.50 post free.

B.A.R. 6, 1974, "Anglo-Saxon Settlement and Landscape", ed. Trevor Rowley. ISBN 0 904531 05 8. 138 pp., 15 figs., 7 plates. Price £2.00 post free.

B.A.R. 7, 1974, "A Corpus of Pagan Anglo-Saxon Spear-Types", by M. J. Swanton. ISBN 0 904531 04 X. 90 pp., 4 figs. Price £1.10 post free.

B.A.R. 9, 1974, "Grooved Ware Sites in Yorkshire and the North of England", by T. G. Manby. ISBN 0 904531 09 0. 133 pp., 43 figs., 2 plates. Price £2.00 post free.

B.A.R. 10, 1975, "Stamp and Roulette Decorated Pottery of the La Tène Period in Eastern England: a Study in Geometric Designs", by Sheila M. Elsdon. ISBN 0 904531 11 2. 115 pp., 19 figs., 5 plates. Price £2.30 post free.

B.A.R. 11, 1975, "Anglo-Saxon Garnet Inlaid Disc and Composite Brooches", by Richard Avent. ISBN 0 904531 21 X. Part i, Discussion, 126 pp., 30 figs., 6 maps, 8 tables, 4 colour plates; Part ii, Catalogue and Plates, 52 pp., 78 plates. Price (parts i and ii together) £5.80 post free.

B.A.R. 12, 1975, "Cirencester: the Development and Buildings of a Cotswold Town", by Richard Reece and Christopher Catling. ISBN 0 904531 12 0. 78 pp., frontispiece and 11 figs., 9 plates. Price £1.50 post free.

B.A.R. 13, 1975, "Settlement Types in Post-Roman Scotland", by Lloyd R. Laing. ISBN 0 904531 13 9. 46 pp., 25 figs. Price £1.00 post free.

B.A.R. 14, 1975, "Clay Pipes for the Archaeologist", by Adrian Oswald. ISBN 0 904531 14 7. 207 pp., 23 figs., 6 plates. Price £3.80 post free.

B.A.R. 15, 1975, "The 'Small Towns' of Roman Britain: Papers presented to a Conference, Oxford, 1975", ed. Warwick Rodwell and Trevor Rowley. ISBN 0 904531 17 1. 237 pp., 70 figs., 17 plates. Price £4.80 post free.

B.A.R. 16, 1975, "Bar Hill: A Roman Fort and its Finds", by Anne Robertson, Margaret Scott and Lawrence Keppie. ISBN 0 904531 18 X. 185 pp., 57 figs. Price £3.50 post free.

B.A.R. 17, 1975, "New Forest Roman Pottery: manufacture and distribution, with a corpus of the pottery types", by M. G. Fulford. ISBN 0 904531 19 8. 200 pp., 61 figs. Price £3.90 post free.

B.A.R. 18, 1975, "The Roman Milestones of Britain: their petrography and probable origins", by Jeffrey P. Sedgley. ISBN 0 904531 20 1. 56 pp., 5 figs. Price £1.00 post free.

B.A.R. 19, 1975, "Rams Hill: A Bronze Age Defended Enclosure and its Landscape", by Richard Bradley and Ann Ellison. ISBN 0 904531 22 8. 264 pp., 59 figs., 21 plates. Price £5.50 post free.

B.A.R. 20, 1975, "Later Prehistory from the Trent to the Tyne", by A. J. Challis and D. W. Harding. ISBN 0 904531 23 6. Part i, Discussion, 247 pp; Part ii, Catalogue and Illustrations. 178 pp., 100 figs., 10 plates. Price (parts i and ii together) £6.50 post free.

B.A.R. 21, 1975, "Roman Fort-Defences to A.D. 117, with special reference to Britain", by Michael J. Jones. ISBN 0 904531 24 4. 192 pp., 21 figs., 1 map, 6 plates. Price £3.90 post free.

B.A.R. 22, 1975, "Hanging-Bowls, Penannular Brooches and the Anglo-Saxon Connexion", by David Longley. ISBN 0 904531 25 2. 49 pp., 19 figs. Price £0.95 post free.

B.A.R. 23, 1976, "Bordesley Abbey, Redditch, Hereford-Worcestershire: First Report on Excavations 1969-1973", by Philip Rahtz and Susan Hirst. ISBN 0 904531 26 0. 295 pp., 41 figs. in text, 3 colour plates, 17 black and white plates; wallet with a further 20 figures. Price £5.90 post free.

B.A.R. 24, 1976, "A Corpus of Religious Material from the Civilian Areas of Roman Britain", by Miranda J. Green. ISBN 0 904531 27 9. 321 pp., 21 figs., frontispiece and 30 plates. Price £5.90 post free.

B.A.R. 25, 1976, " Neolithic Flint Axes from the Cotswold Hills", by Alan Tyler. ISBN 0 904531 28 7. 98 pp., 9 figs. Price £1.90 post free.

B.A.R. 26, 1976, "The Black Prince's Palace at Kennington, Surrey", by Graham J. Dawson. ISBN 0 904531 29 5. 213 pp., 20 figs. Price £3.90 post free.

B.A.R. 27, 1976, "Wawcott III: A Stratified Mesolithic Succession", by F. R. Froom. ISBN 0 904531 30 9. 209 pp., 85 figs. Price £3.90 post free.

B.A.R. 28, 1976, "Barrow Mead, Bath 1964: Excavation of a Mediaeval Peasant House", by Jayne Woodhouse. ISBN 0 905431 31 7. 73 pp., 15 figs., 4 plates. Price £1.30 post free.

B.A.R. 29, 1976, "Occupation Sites on a Chiltern Ridge: Excavations at Puddlehill and Sites near Dunstable, Bedfordshire. Part I: Neolithic, Bronze Age and Early Iron Age", by C. L. Matthews, Foreword and Summary by C. F. C. Hawkes. ISBN 0 904531 32 5. 209 pp., 123 figs., 9 plates. Price £3.90 post free.

B.A.R. 30, 1976, "Studies in the Archaeology and History of Cirencester: based on papers presented to a research seminar on the post-Roman development of Cirencester held at the Corinium Museum, November 1975", ed. Alan McWhirr. ISBN 0 904531 43 0. 200 pp., 39 figs., 6 plates. Price £3.90 post free.

B.A.R. 31, 1976, "The Production and Distribution of Metalwork in the Middle Bronze Age in Southern Britain", by M. J. Rowlands. ISBN 0 904531 44 9. Part i, Discussion, 218 pp., 28 figs. Part ii, Catalogue and Illustrations, 314 pp., 60 plates, 26 maps. Price (parts i and ii together) £7.50 post free.

B.A.R. 32, 1976, "Roman Military Stone-built Granaries in Britain", by Anne P. Gentry. ISBN 0 904531 45 7. 98 pp., 14 figs., 3 plates, 4 tables, 2 maps. Price £1.95 post free.

B.A.R. 33, 1976, "Settlement and Economy in the Third and Second Millennia B.C. Papers delivered at a Conference organised by the Department of Adult Education, University of Newcastle upon Tyne, January 1976", ed. Colin Burgess and Roger Miket. Contributions by John Barrett, C. Stephen Briggs, Colin Burgess, P. D. Catherall, D. V. Clarke, David Coombs, George Eogan, H. Stephen Green, Peter Hill, J. V. S. Megaw, Roger J. Mercer, Roger Miket, Francis Pryor, M. J. Rowlands, Stephen Shennan, Ian A. G. Shepherd, D. D. A. Simpson and J. B. Stevenson. ISBN 0 904531 52 X. 331 pp., 100 figs., 20 plates. Price £5.90 post free.

B.A.R. 34, 1977, "Bronze Age Spearheads from Berkshire, Buckinghamshire and Oxfordshire", by Margaret R. Ehrenberg. ISBN 0 904531 61 9. 95 pp., 1 plate, 28 figs. Price £1.90 post free.

B.A.R. 35, 1977, "The Medieval Moated Sites of South-Eastern Ireland: Counties Carlow, Kilkenny, Tipperary and Wexford", by Terence B. Barry. ISBN 0 904531 69 4. 247 pp., 10 plates, 33 figs., 24 tables. Price £4.70 post free.

B.A.R. 36, 1077, "Edwardian Monetary Affairs (1279-1344); A Symposium held in Oxford, August 1976", ed. N. J. Mayhew. Contributions by M. M. Archibald, J. F. Hadwin, T. H. Lloyd, M. Mate, N. J. Mayhew, D. M. Metcalf, N. J. Palmer, M. C. Prestwich, S. E. Rigold and D. R. Walker. ISBN 0 904531 64 3. 186 pp., 9 figs. Price £3.50 post free.

B.A.R. 37, 1977, "Studies in Celtic Survival", ed. Lloyd Laing. Contributions by L. A. S. Butler, Ms Margaret Faull, P. S. Gelling, Lloyd Laing, A. D. S. Macdonald, Alex Morrison, V. B. Proudfoot and C. A. R. Radford. ISBN 0 904531 65 1. 123 pp., 8 figs. Price £2.40 post free.

BAR Supplementary Series

B.A.R. -S1, 1975, "The Lewis Collection of Engraved Gemstones in Corpus Christi College, Cambridge", by Martin Henig. ISBN 0 904531 33 3. 153 pp., 29 plates. Price £2.90 post free.

B.A.R. -S2, 1975, "Defended Sites of the Late La Tène in Central and Western Europe", by John Collis. ISBN 0 904531 34 1. 267 pp., 100 figs. Price £5.50 post free.

B.A.R. -S3, 1976, "Le Périple d'Hannon/The Periplus of Hanno", by Jacques Ramin. ISBN 0 904531 35 X. 125 pp., 4 maps. Bilingual volume, full texts in French and English. Price £2.40 post free.

B.A.R. -S4, 1976, "Coins of the Ancient Thracians", by Yordanka Youroukova: translated from the Bulgarian by V. Athanassov. ISBN 0 904531 36 8. 153 pp., 28 plates. Price £2.80 post free.

B.A.R. -S5, 1976, "The Metrology of the Roman Silver Coinage, Part I: from Augustus to Domitian", by D. R. Walker. ISBN 0 904531 37 6. 159 pp., 17 figs. Price £2.50 post free.

B.A.R. -S6, 1976, "Portraits and other Heads on Roman Historical Relief up to the Age of Septimius Severus", by Anthony Bonanno. ISBN 0 904531 38 4. 304 pp., 76 pages of plates. Price £5.90 post free.

B.A.R. -S7, 1976, "Prolegomena to the Study of the Second Jewish Revolt (A.D. 132-135)", by Shimon Applebaum. ISBN 0 904531 39 2. 100 pp., 2 maps. Price £1.95 post free.

B.A.R. -S8, 1976, "Dacian Trade with the Hellenistic and Roman World", by Ioan Glodariu, revised by the author, translated from the Romanian by Nubar Hampartumian. ISBN 0 904531 40 6. 337 pp., 62 plates and figs., 4 tables. Price £6.80 post free.

B.A.R. -S9, 1976, "Greek and Roman Coins from Aphrodisias", by David J. MacDonald. ISBN 0 904531 41 4. 53 pp., 3 plates. Price £1.25 post free.

B.A.R. -S10, 1976, "The Roman Cemetery at Gerulata Rusovce, Czechoslovakia", by Ľudmila Kraskovská, revised by the author, translated from the Slovak by Hana Schuck. ISBN 0 904531 42 2. 114 pp., 85 figs., 6 plates. Price £2.50 post free.

B.A.R. -S11, 1976, "Oppida: the Beginnings of Urbanisation in Barbarian Europe", ed. Barry Cunliffe and Trevor Rowley. Contributions by J. Collis, C. C. Haselgrove, Éva F. Petres, J. Břeň, D. Nash, B. Cunliffe, J. May, W. Rodwell. ISBN 0 904531 46 5. 367 pp., 121 figs., 2 plates. Price £6.75 post free.

B.A.R. -S12, 1976, "The Dal Pozzo Copies of the Palestrina Mosaic", by Helen Whitehouse. ISBN 0 905431 48 1. 98 pp., 43 pp. of figures and plates. Price £1.90 post free.

B.A.R. -S13, 1976, "Prehistoric Art in Bulgaria from the Fifth to the Second Millennium B.C.", by Anna Raduncheva: translated from the Russian by Caroline M. Schuck. ISBN 0 904531 59 7. 113 pp., 30 pp. of plates. Price £2.70 post free.

B.A.R. -S14, 1976, "L'Art pariétal de la Grotte de Gargas/Palaeolithic Art in the Grotte de Gargas", by Cl. Barrière avec la collaboration du Ali Sahly et des élèves de l'Institut d'Art Préhistorique de Toulouse. Translated from the French by W. A. Drapkin. Mémoires de l'Institut d'Art Préhistorique de Toulouse no. III. ISBN 0 904531 47 3. Parts i and ii: 409 pp., 144 figs., 93 plates. Bilingual volume, full texts in French and English. Price (Parts i and ii together) £8.50 post free.

B.A.R. -S15, 1976, "Aspects of the Notitia Dignitatum: papers presented to a Conference in Oxford, December 13 to 15, 1974", ed. R. Goodburn and P. Bartholomew. Contributions by J. C. Mann, J. J. G. Alexander, J. P. Wild, M. M. Roxan, J. S. Johnson, M. W. C. Hassall, A. L. F. Rivet, R. M. Price, J. F. Matthews, R. S. O. Tomlin and C. E. Stevens. ISBN 0 904531 58 9. 224 pp., 28 plates, 1 colour plate. Price £4.20 post free.

B.A.R. -S16, 1976, "The Archaeology and History of the Carpi from the Second to the Fourth Century A.D.", by Gh. Bichir, revised by the author, translated from the Romanian by Nubar Hampartumian. ISBN 0 904531 55 4. Part i, Text, 211 pp., 23 figs. and plates. Part ii, Illustrations, 191 pp., 188 figs. and plates. Price (Parts i and ii together) £8.90 post free.

B.A.R. -S17, 1976, "The Thraco-Roman Villa Rustica near Chatalka, Stara Zagora, Bulgaria", by Dimitur Nikolov, translated from the Bulgarian by Michael Holman. ISBN 0 904531 60 0. 169 pp., 118 figs. and plates. Price £3.60 post free.

B.A.R.-S18, 1977, "Roman Lamps from Ulpia Traiana Sarmizegetusa", by Dorin Alicu and Emil Nemeș, Introduction by Hadrian Daicoviciu; translated from the Romanian by Nubar Hampartumian. ISBN 0 904531 62 7. 209 pp., 51 plates, 38 figs. Price £4.50 post free.

B.A.R.-S19, 1977, "Archaeology and Anthropology: Areas of Mutual Interest", ed. Matthew Spriggs. Contributions by R. W. Chapman, Roy F. Ellen, Roland Fletcher, John Gledhill, L. M. Groube, Colin Haselgrove, Ian Hodder, Sir Edmund Leach, M. J. Rowlands and Matthew Spriggs. ISBN 0 904531 63 5. 176 pp., 11 figs. Price £3.30 post free.

B.A.R.-S20, 1977, "Monte Irsi, Southern Italy: The Canadian Excavations in the Iron Age and Roman Sites, 1971-1972", ed. Alastair Small. Contributions by Graeme Barker, R. J. Buck, Gisèle Deschênes, H. Hjelmqvist, Marie-Odile Jentel, Jeremy Rossiter, Alastair Small and Edith Mary Wightman. ISBN 0 904531 66 X. 296 pp., 38 plates, 42 figs. Price £5.50 post free.

B.A.R.-S21, 1977, "Typologie et Préhistoire de L'Asturien du Portugal/ The Asturian in Portugal: Typology and Chronology", by Jean Maury: translated from the French by Elisabeth Weeks. ISBN 0 904531 67 8. 105 pp., 46 figs., 10 plates. Price £2.10 post free.

B.A.R.-S22, 1977, "The Metrology of the Roman Silver Coinage, Part II: from Nerva to Commodus", by D. R. Walker. ISBN 0 904531 68 6. 120 pp., 12 figs. Price £2.30 post free.

B.A.R.-S23, 1977, "The Aes Coinage of Emporion", by Leandre Villaronga; translated from the Spanish by Elisabeth Weeks. ISBN 0 904531 70 8. 99 pp., 15 plates. Price £2.20 post free.